# DIRECTIONALLY SOLIDIFIED MATERIALS FOR HIGH TEMPERATURE SERVICE

PREDICTIVE AND QUANTITATIVE METALLURGY SERIES

general editor

D. McLean

# DIRECTIONALLY SOLIDIFIED MATERIALS FOR HIGH TEMPERATURE SERVICE

by
M. McLean

Division of Materials Applications
National Physical Laboratory
Teddington, Middlesex

THE METALS SOCIETY
1983

Book 296
published in 1983 by
The Metals Society
1 Carlton House Terrace
London SW1Y 5DB



ISBN 0 904357 52 X


Printed and made in England by
J. W. Arrowsmith Ltd, Bristol

# Contents

# Preface

The continual drive for increased thrust and improved efficiency from gas turbine engines, particularly for aircraft propulsion, has led to a progressive increase in operating temperatures which has made severe demands on certain critical components such as the turbine blades. The temperature capabilities of blading alloys have been increased by a combination of alloy and processing development; directional solidification has been an important stage of this process and is now critical to the production of components for several aero-engines where either columnar-grained or single-crystal superalloys are required. This has made new demands on the founder to produce such components and on the engineer to design with highly anisotropic materials. Both have done so extremely successfully. However, the technology of directional solidification has been implemented so rapidly that it is arguable that the scientific implications are not yet fully appreciated. Since in the future directionally solidified components will undoubtedly be more widely used both in aero- and land-based applications it is important to understand the full potential and limitations of the process.

This book is an attempt to present the current state of knowledge of the processing and behaviour of directionally solidified materials. It deals both with existing superalloys that are in commercial use and, more speculatively, with developmental *in situ* composites that may or may not find eventual applications. While borrowing heavily from work carried out at the National Physical Laboratory over the past ten years I have made every attempt to present a balanced view of the subject.

The shortcomings of the book are attributable directly to the author. However, these would have been greater without the help of advice, information and photographs supplied by many friends and colleagues both at NPL and elsewhere. Particular thanks are due to B. F. Dyson, P. J. Henderson, K. Menzies and P. N. Quested for their collaboration over several years. My perception of the subject has especially benefited from discussions with J. Northwood of the National Gas Turbine Establishment, with G. W. Meetham and M. Goulette of Rolls Royce Ltd, and with A. Kelly of Surrey University. D. McLean prompted the writing of this book and commented constructively on the text as it was produced. Finally, my family has shown remarkable patience while the book was being written; my wife, Malinda, improved the presentation while typing the text and my sons, Andrew and Calum, helped with calculations and drawings of figures.

M. McLean
Teddington
October 1982

# 1
# Requirements and limitations of existing materials

## 1.1 THERMODYNAMIC CONSIDERATIONS

The efficiencies of all heat engines and of most chemical processes can be related directly to the upper temperatures of the various thermodynamic or reaction cycles. The limiting factor for most systems is the capability of critical components to withstand the combination of relatively high stresses and temperatures often in aggressive environments that can lead to oxidation, corrosion or erosion of the components. There is, therefore, a continuing requirement to increase the temperature capabilities of the materials from which such components can be manufactured in order to improve the efficiencies of various processes and systems. Many recent developments in high temperature materials have been direct responses to the needs of the aero-gas turbine designer being motivated by either military requirements for increased thrust or commercial considerations of improved fuel consumption. However, similar requirements for non-aero gas turbines and, indeed, for other applications have led to more general utilization of developments in high temperature materials.

The theoretical efficiencies of heat engines may be described by an idealized Carnot cycle as illustrated schematically in Fig. 1.1 for the case where the working fluid is an ideal gas (i.e. satisfies both Boyle's and Charles's laws.) Curves *AD* and *BC* represent, for a fixed mass of gas, the interdependence of pressure $p$ and volume $v$ at two constant temperatures $T_1$ and $T_2$ ($T_2 > T_1$). By carrying out a series of operations on the fixed mass of gas that cause its state to follow the circuit *ABCDA* to its original condition heat can be translated to mechanical energy. Starting at point *A*, the various idealized stages of the cycle can be described as follows:

(a) *AB* is the sudden adiabatic compression of the gas that is accompanied by a rise in temperature from $T_1$ to $T_2$

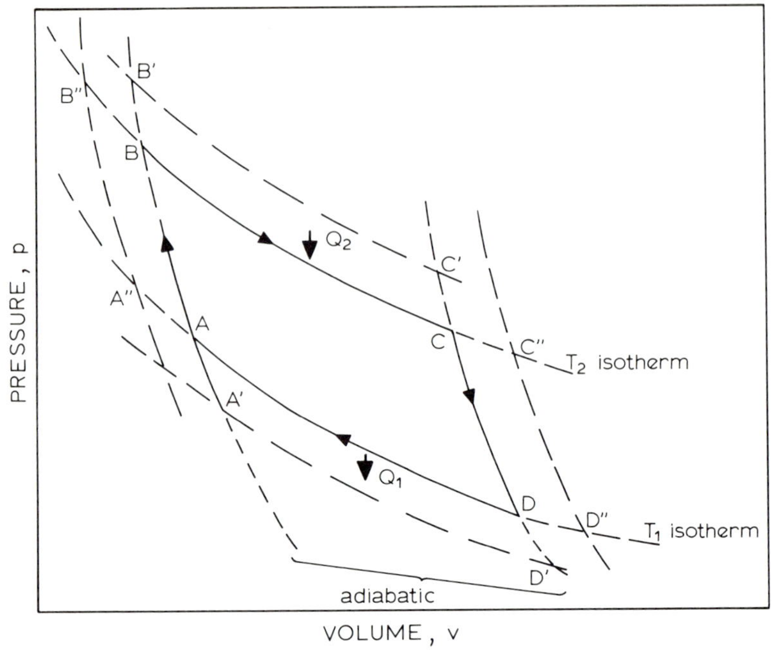


**Figure 1.1 Schematic variation in pressure, volume and temperature of a fixed mass of gas during an ideal Carnot cycle**

(b) $BC$ is the expansion of gas at constant temperature $T_2$ that is caused by the absorption of heat energy $Q_2$
(c) $CD$ is an adiabatic expansion of the gas with an associated temperature fall from $T_2$ to $T_1$
(d) $DA$ is an isothermal compression at temperature $T_1$ which releases a quantity of heat $Q_1$.

The mechanical work performed by the cycle is given by the area $ABCD$ and must equal $(Q_2 - Q_1)$ and the ideal efficiency $\varepsilon$ of the engine is the ratio of the work done to the heat taken from the energy source $Q_2$

$$\varepsilon = \frac{Q_2 - Q_1}{Q_2} \tag{1.1}$$

Clearly, if $T_2$ were increased ($B'C'$) or $T_1$ decreased ($A'D'$) the work extracted from the heat source, represented by the cycle, would be increased and the efficiency improved. In practice the upper temperature is the practicable variable. Alternatively (or additionally), improved efficiencies might result from increasing the pressure differences during the cycle ($A''B''C''D''$).

The working cycle of a gas turbine is continuous but the thermodynamic principles are similar to those of the idealized intermittent Carnot engine discussed above. Figure 1.2 relates the state of the gas to different regions of the gas turbine.[1] A continuous stream of ambient air (*A*) is taken into the engine and compressed (*BC*) using part of the mechanical energy generated by the engine; this is heated by combustion of a suitable fuel (*CD*) to a high temperature and the gas expands into the ambient air (*DA*) through a turbine which extracts sufficient energy to power the compression cycle, the remainder providing the mechanical work that can be used to propel an aircraft, pump oil etc. Point *D* of the gas turbine cycle in Fig. 1.2 probably represents the most severe combination of high temperature and pressure and this defines the operating environment of the high pressure gas turbine blades. It is

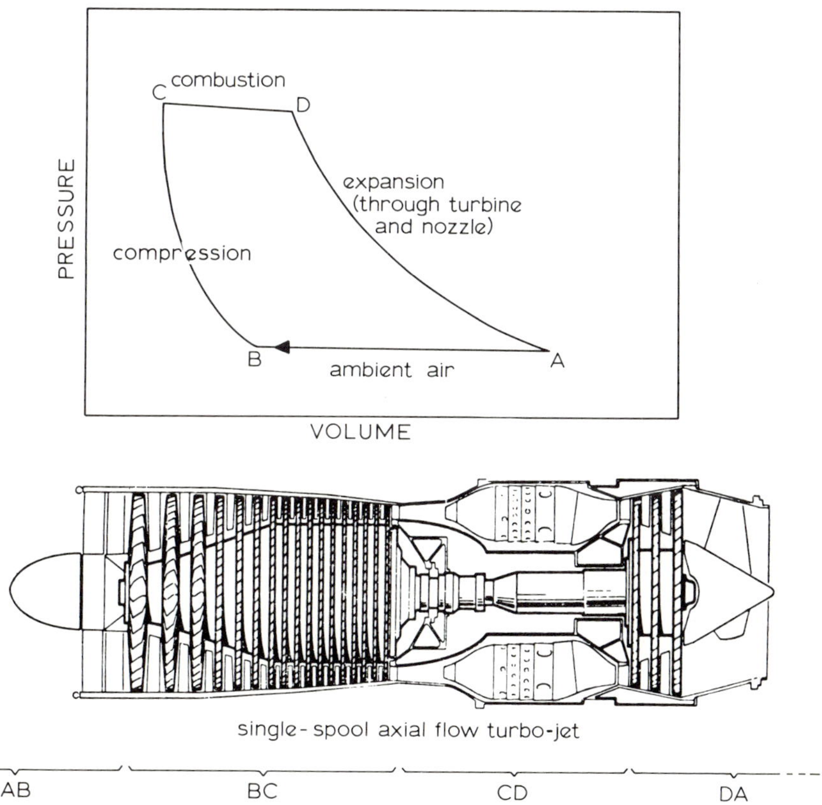


**Figure 1.2 Idealized working cycle of a gas turbine on a pressure/volume diagram and its relationship to various parts of the gas turbine structure [Ref. 1]**

arguable that progress in gas turbine development has been determined by the temperature capabilities of the rotating blades in the high temperature turbine and these depend on both the properties of the alloys available and design innovations, such as the introduction of air cooling around 1960.

The precise targets of materials developments for gas turbine blades depend very much on the specific application. Thus for military aero-gas turbines where component lives can be relatively short and the principal requirement is improved thrust the chief objective is for materials that retain their strengths to higher temperatures; engines powering commercial aircraft must be made more cost effective by enabling higher pay loads to be carried and/or reducing fuel consumption. The implications of changes in turbine inlet temperature and gas pressure ratio (the latter determining the turbine speed and therefore the centrifugal stress on blades) to the specific fuel consumption and engine thrust are shown in Fig. 1.3.[2] This indicates that improvements in fuel consumption may be best achieved by increased compression ratio at a constant turbine entry temperature whereas optimum thrust depends on operating at the maximum turbine entry temperature. However, in other applications where blade lives are oxidation/corrosion limited there may be a need to upgrade the high temperature properties of more corrosion

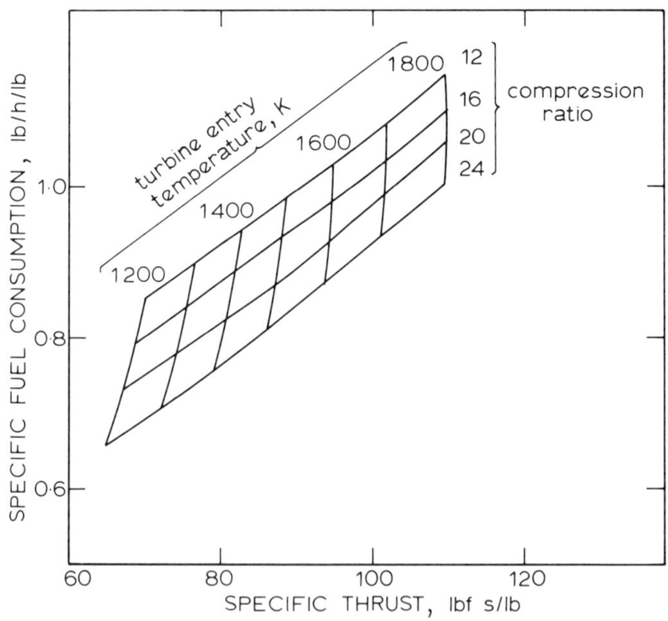


**Figure 1.3 Effect of turbine entry temperature and pressure ratio in an aero-turbine engine on thrust and fuel consumption [Ref. 2]**

resistant alloys rather than to optimize mechanical behaviour. Gas turbines operating in marine environments and industrial gas turbines, where long lives are essential, fall into this latter category. Emphasis on the retention or improvement of oxidation and corrosion resistance while improving mechanical properties of turbine blade alloys is likely to become more important with the increased use of lower grade and alternative fuels, such as those derived from coal.

Similar requirements for high temperature strength arise from a wide range of chemical processing plant where the efficiencies and rates of reactions are very sensitive to temperature. In general, the procedures used in the chemical industry have tended to be tailored to the materials available and limited effort has been directed to alloy development to allow critical components, such as reaction tubes, to operate at high temperatures. However, it may be necessary to re-examine this approach, particularly in petrochemical applications, where depleted resources will force the industry to refine crude oils of low yield and high impurity contents that would have previously been rejected. This would be achieved by increased reaction temperatures and therefore require improved containment tubes.

## 1.2 MATERIALS DEVELOPMENTS

Since the invention of the gas turbine in the 1930s, the turbine entry temperatures have been progressively raised by a combination of engineering and materials developments and this has been reflected in improved engine efficiency measured by several features; this is illustrated by the trend in the development of Rolls-Royce gas turbines[3] shown in Fig. 1.4. The materials contribution to this evolution, amounting to an average increase in temperature capability of gas turbine blading alloys of about 10 K per annum,[4] is shown in Fig. 1.5. The early materials improvements can be attributed directly to alloy selection and development through the introduction and optimization of solid solution and $\gamma'$ strengthened nickel-base superalloys. However, since about the mid-1950s process development has played an increasingly important role. Thus the introduction of vacuum melting enabled the development of investment casting of turbine blades and led to the identification of new alloys optimized for the cast rather than wrought condition.

The most important trend in the development of nickel-base superalloys has been the increase in the volume fraction of strengthening $\gamma'$ particles and the temperature at which they dissolve in the $\gamma$ matrix. Wrought superalloys can only be produced when the temperature difference between the $\gamma'$ solution and the incipient melting temperatures is sufficiently wide to allow practicable temperature control during thermomechanical processing. The ability to produce cast-to-shape components avoided this restriction and allowed the development of superalloys with high $\gamma'$ volume fractions

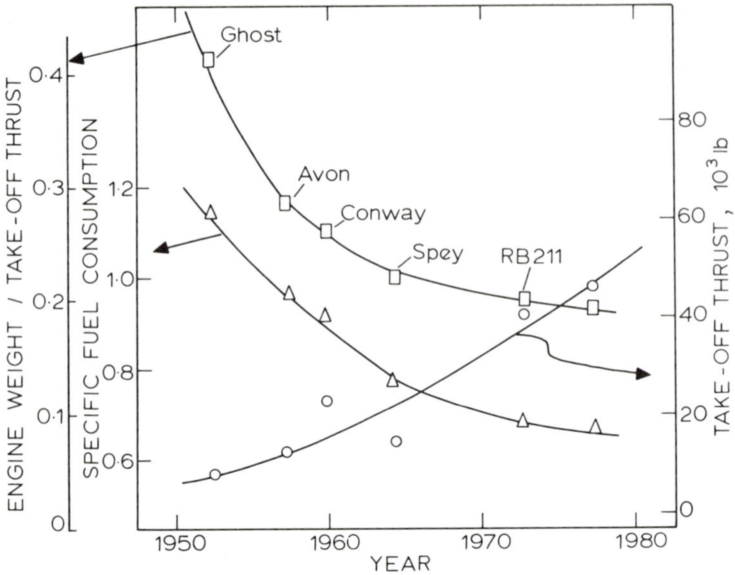


**Figure 1.4 Civil engine trends in thrust, specific fuel consumption and specific weight [Ref. 3]**

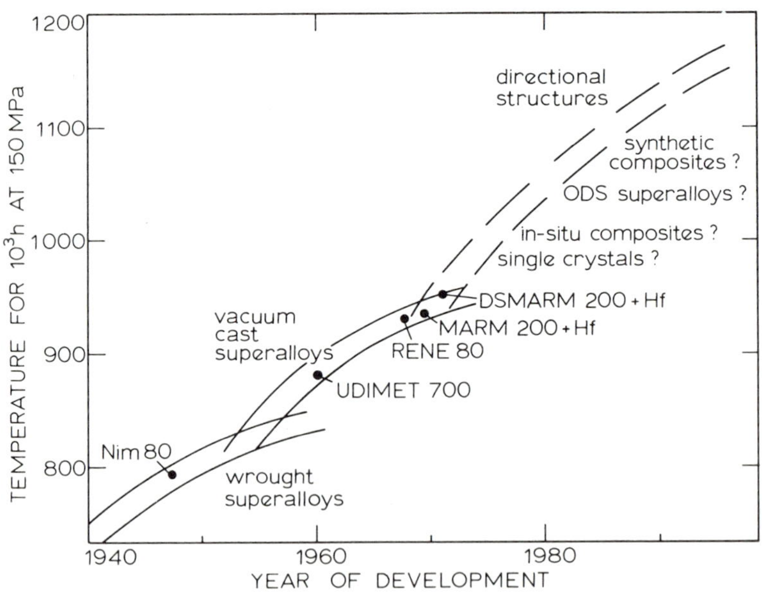


**Figure 1.5 Temperature capability of gas turbine blade materials for $10^3$ h life with a stress of 150 MPa displayed as a function of the year of introduction into service**

and high $\gamma'$ solution temperatures that had substantially better stress rupture and creep strengths than the earlier wrought materials. However, this was achieved at the expense of reduced ductility and, in some cases, thermal fatigue resistance. Although acceptable ductilities are obtained in many superalloys (e.g. IN100, IN738), this problem created a barrier to further developments.

Creep fracture and thermal fatigue failures in both wrought and cast alloys that have equiaxed grain structures are almost always associated with grain boundaries that are transverse to the applied stress. VerSnyder and co-workers[5] reasoned that by controlling the grain shape to reduce the density of transverse grain boundaries (i.e. by producing grains elongated parallel to the stress axis) grain boundary fracture might be inhibited and ductility increased. Northcott[6–8] had demonstrated in the 1930s that such elongated grains could be produced in a wide range of alloys by a fairly simple directional casting procedure. A typical microstructure in a Cu–Zn alloy is shown in Fig. 1.6. The approach was applied to nickel-base alloys by VerSnyder and Guard[9,10] who showed a correlation between the elongated structure and

**Figure 1.6 Columnar grain structure in directionally cast ingot of Cu–45%Zn [Ref. 7]**

improved high temperature mechanical properties. Since these early experiments directional solidification has evolved into a viable commercial process leading to very substantial advances in component performance and is now state-of-the-art in turbine blade production for aero engines. Subsequent related developments have included the casting of single crystal rather than columnar grained structures and, more speculatively, composite materials by directional solidification of near-eutectic melts. Figure 1.7 shows the macrostructures of experimental turbine blades of a nickel-base superalloy in the conventionally cast, directionally solidified and single crystal forms; the blades have been etched to indicate the different grain structures. The aim of this book is to describe the theory and practice of directional solidification in relation to gas turbine materials and to assess the properties of various directionally solidified materials.

**Figure 1.7 Experimental turbine blades, cast at the National Gas Turbine Establishment, of a nickel-base superalloy in the (from left to right) conventionally cast, directionally solidified and single crystal forms**
*(courtesy of J. E. Northwood, NGTE)*

## REFERENCES CHAPTER 1


1 Rolls Royce (1971) Ltd, 'The jet engine', Publication Ref. TSD1302, 3rd edition, July 1969, reprinted with revision in 1973, Derby, England

2 G. W. MEETHAM: *Metall. Mater. Technol.*, 1976, **8**, 589

3 G. L. WILDE: *J. Roy. Aero. Soc.*, 1978, **78**, 281

4 H. R. GRAY: 'The promise of eutectics for aircraft turbines', NASA TM-73714, NASA Lewis Research Center, Cleveland, USA, October 1977

5 F. L. VERSNYDER AND M. E. SHANK: *Mater. Sci. Eng.*, 1970, **6**, 213

6 L. NORTHCOTT: *J. Inst. Met.*, 1938, **62**, 101

7 L. NORTHCOTT: *J. Inst. Met.*, 1939, **65**, 173

8 L. NORTHCOTT AND D. E. THOMAS: *J. Inst. Met.*, 1939, **65**, 205

9 F. L. VERSNYDER in 'High temperature materials', edited by R. F. Hehemann and G. Mervin Ault, 1959, New York, Wiley-Interscience

10 F. L. VERSNYDER AND R. W. GUARD: *Trans. ASM*, 1960, **52**, 485

# 2
# Basics of directional solidification: superalloys

Although the principal motivation for the development of directional solidification was a desire to control the grain morphology of superalloys, the process also influences several other microstructural characteristics. Thus a sharp crystallographic texture is obtained due to the dominance of the fast growing ⟨100⟩ crystal orientation parallel to the direction of solidification and the dendrites are similarly aligned. Less obviously, but perhaps also of importance in relation to the mechanical behaviour of the material, there is lower porosity than in the conventionally cast material and there are significant differences in the nature and distribution of the precipitated phases. This chapter will consider the factors controlling these various aspects of the structure.

Superalloys are extremely complex, often consisting of ten or more atomic constituents which solidify to form several phases. For example, a modern nickel-base superalloy such as IN738LC contains a face-centred-cubic nickel matrix strengthened by ~40% by volume of the coherent intermetallic $\gamma'$-$Ni_3Al$ and small volumes ($\lesssim 1\%$) of various carbides, borides and carbosulphides. Additional phases such as topologically close packed phases (e.g. $\sigma$) may be deposited and reactions between the carbides may occur during service. However, the solidification process itself essentially involves the formation of a single phase ($\gamma$-nickel) constituting ~99% by volume. The $\gamma'$ largely precipitates from the solid state at a lower temperature and the minor phases have little effect on solidification. Cobalt- and iron-base superalloys also solidify to form a major $\gamma$ phase but do not have $\gamma'$ precipitates. Consequently, the general solidification behaviour of superalloys appears to be described reasonably well by the formalism that has been developed for single phase freezing.

The microstructures obtained on solidifying complex alloys depend in a complicated manner on how the heat extracted from the solidifying material is distributed between normal heat flow mechanisms, such as radiation or conduction, and the latent heat $L_f$ created on freezing. *Directional solidification* provides a particularly potent means of controlling the heat flow characteristics and the associated microstructures in order to optimize particular properties. The approach has developed from an experimental investigative procedure to a reliable industrial processing technique that is now used in the manufacture of certain gas turbine blades from advanced nickel- and cobalt-base superalloys.

## 2.1 HEAT-FLOW CONSIDERATIONS

In conventional casting, the alloy melt is normally poured into a mould of relatively low thermal conductivity $K_m$. In the case of a cylindrical mould of internal radius $a$ and of thickness $d$ where the only heat loss is through the mould walls the principal temperature gradient is within the ceramic since the thermal conductivities of the solid and liquid metals are generally very much greater than $K_m$. Flemings[1] considered the heat flow resulting from pouring a charge of liquid metal at its melting point $T_m$ (i.e. no superheat) into a mould at temperature $T_0$ and, for mathematical simplicity, assumed that $T_0$ was maintained at the outside surface. This configuration is shown schematically in Fig. 2.1*a*. The heat flow must satisfy the time-dependent

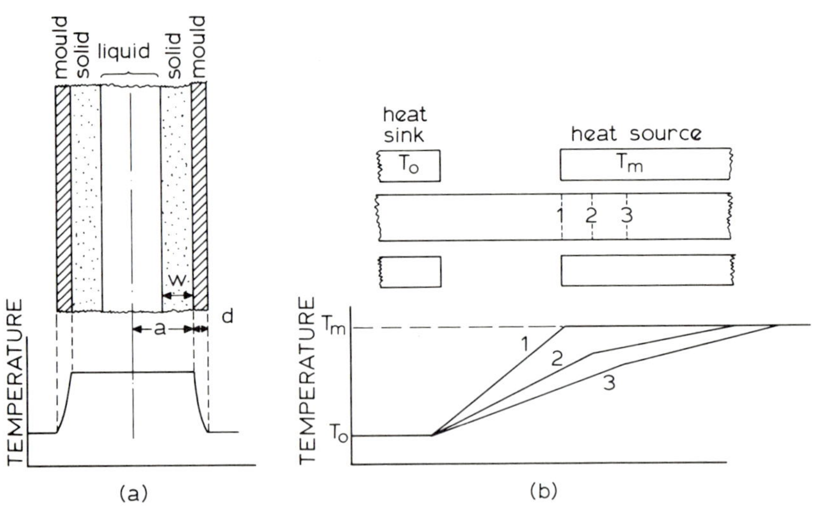


**Figure 2.1** **Schematic diagrams of the mould-furnace configurations and temperature distributions during (a) conventional casting and (b) directional solidification: *1* $V_I = V_I^{max}$; *2* and *3* indicate decreasing $V_I$**

Laplace equation

$$\frac{\partial^2 T}{\partial r^2}+\frac{1}{r}\frac{\partial T}{\partial r}=\frac{1}{\alpha_m}\frac{\partial T}{\partial t} \tag{2.1}$$

where $\alpha_m = K_m/\rho_m C_m$ is the heat diffusivity, $\rho_m$ = mould density, $C_m$ = specific heat of mould, $t$ is time and $r$ is the radial distance from the cylinder axis. Solving eqn (2.1) with appropriate boundary conditions gives an expression for the thickness of metal solidified, $w$

$$w=\left(\frac{T_m-T_0}{\rho_m L_f}\right)\left[\frac{2}{\sqrt{\pi}}(K_m/\rho_m C_m)^{1/2} t^{1/2}+\frac{K_m t}{2a}\right] \tag{2.2}$$

Thus the growth rate decreases as the solid/melt interface progresses towards the centre of the casting.

With directional solidification the casting is usually removed at a predetermined withdrawal rate $V_w$, normally constant, relative to the heating source, which maintains part of the charge molten, to some cooler zone below the alloy melting point. Consequently a temperature gradient is imposed along the solidifying material. Assuming that the solidifying interface is planar and advancing at a rate $V_I$, the heat balance there equates the heat of fusion to the differences in thermal conductivity of the solid and liquid alloy.

$$K_s G_s - K_L G_L = \rho_s L_f V_I \tag{2.3}$$

where the subscripts s and L refer to the solid and liquid alloy respectively and $K_i$ is the thermal conductivity, $G_i$ is the temperature gradient and $\rho_i$ is the density (i = s or L). This shows that $V_I$ depends on the difference between $K_s G_s$ and $K_L G_L$ and that $V_I$, $G_s$ and $G_L$ are not independent variables. We can consider various extremes in the solidification behaviour that will depend on the practical details of heat input and removal

(a) If $G_L$ and $G_s$ can be fixed, then $V_I$ is completely determined by eqn (2.3) and, consequently, is independent of the withdrawal rate $V_w$. However, this is physically unlikely.

(b) For a given macroscopic temperature gradient (determined by the furnace/casting arrangement) the maximum freezing rate $V_I$ attainable without undercooling of the melt, i.e. $V_I^{max}$, occurs as $G_L \to 0$ and $V_I^{max}$ increases with increasing $G_s$.

(c) For $V_I < V_I^{max}$, the values of $G_s$ and $G_L$ will adjust to satisfy eqn (2.3) and also will be compatible with the macroscopic temperature gradient imposed by the geometry of the furnace. Schematic temperature profiles for different solidification rates are shown in Fig. 2.1*b*. It is clear that low values of $V_I$ will give rise to relatively high values of $G_L$ which is shown later to be an important parameter influencing the microscopic solidification morphology.

In practice, conditions between these extremes will apply. Although it cannot be assumed that the rate of mould withdrawal $V_w$ is identical to the velocity of the alloy melt/interface, $V_I$, the major differences will be transient effects at the beginning and end of solidification; for most of the solidification length where the temperatures on either side of the solid/liquid interface attain steady state profiles $V_I = V_w$. However, the position of the solid/liquid interface will vary, as indicated in Fig. 2.1*b*, giving values of $G_L$ and $G_s$ that depend on $V_w$.

A more detailed discussion of heat flow analyses in relation to the practice of directional solidification is given in Chapter 4.

## 2.2 EQUILIBRIUM SOLIDIFICATION

Practical engineering materials are invariably multicomponent alloys which often form several phases during or after solidification and have quite wide melting ranges. Consequently freezing occurs over a range of temperatures and may give rise to microsegregation within the alloy. Although a typical superalloy may consist of about eight solute elements in a nickel, cobalt or iron solvent we shall first of all consider the solidification behaviour of a simple binary alloy.

Figure 2.2 shows portions of idealized phase diagrams where increasing solute concentration leads to either decreased (Fig. 2*a*) or increased (Fig. 2*b*) melting temperatures. In both cases the addition of solute gives rise to a finite melting range. The liquidus and solidus boundaries are represented by straight lines of slopes $m_L$ and $m_s$ respectively and it is convenient to relate the liquidus and solidus concentrations at a given temperature (i.e. $C_L$, $C_s$ respectively) by a partition coefficient $k$:

$$k = \frac{C_s}{C_L} = \frac{1 - m_s^{-1}(T' - T_L)}{1 - m_L^{-1}(T' - T_L)} \tag{2.4}$$

where $T'$ is the melting temperature of the pure solvent, and $T_s$, $T_L$ are the solidus and liquidus temperatures of an arbitrary alloy. In the following discussion particular reference will be made to the type of behaviour illustrated in Fig. 2.2*a* which constitutes the most common solvent/solute interaction. Most of the conclusions remain valid for the case shown in Fig. 2.2*b*; any differences will be discussed as they arise.

Consider a cylinder of binary liquid alloy of solute concentration $C_0$ solidifying under conditions where equilibrium is always maintained. The freezing sequence is illustrated in Fig. 2.3. At the liquidus temperature $T_L$ the first solid appears having a solute concentration $kC_0$ and rejecting excess solute, thus enriching the solvent marginally. As the temperature falls below $T_L$ and solidification proceeds, the composition of the solid portion progressively approaches and that of the remaining liquid increasingly deviates from

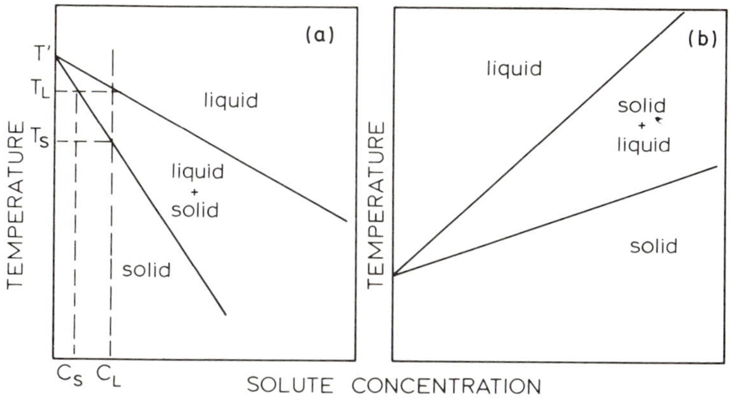


**Figure 2.2 Schematic diagrams of simple binary phase diagrams showing (a) decreasing and (b) increasing melting temperatures with increasing solute concentration**

$C_0$. Eventually the final liquid fraction will have a solute concentration $C_0/k$, which will solidify at $T_s$ to give a solid rod of uniform concentration $C_0$.

Although possible in principle such behaviour is unlikely to occur in practice since it requires that diffusion in the solid and liquid phases be sufficiently fast to allow the complete mixing that is necessary to maintain equilibrium. The conditions necessary to allow sufficient solid state diffusion, the most likely rate determining process, can be estimated readily. The length of solidified ingot will increase from an arbitrary $x$ to $(x+\Delta x)$ in a period $\Delta t$ where

$$\frac{dx}{dt} = V_I \tag{2.5}$$

and $V_I$, as before, is the velocity of the solid/liquid interface. Complete equilibration will only occur if the characteristic diffusion distance is greater than the total *new length* of solid:

$$x+\Delta x \leqslant (D_s \Delta t)^{1/2} = \left(\frac{D_s \Delta x}{V_I}\right)^{1/2} \tag{2.6}$$

where $D_s$ is the rate determining solid state solute diffusion coefficient. For a given $V_I$ the maximum length of ingot capable of equilibrium solidification is approximately given by the equation:

$$x^2 + 2x\, dx = \frac{D_s}{V_I} dx \tag{2.7}$$

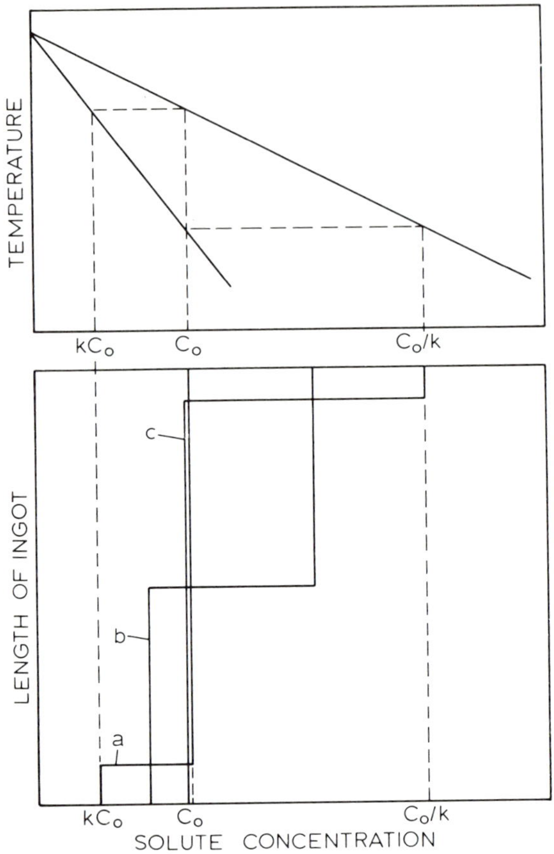


**Figure 2.3 Idealized solute concentrations during equilibrium solidification of an ingot with an initial composition $C_0$: (a) start of solidification; (b) during solidification; (c) end of solidification**

which has an upper limit solution

$$x = \frac{D_s}{2V_I} \tag{2.8}$$

For substitutional and interstitial solutes where $D_s \sim 10^{-12}$ and $\sim 10^{-9}$ m$^2$ s$^{-1}$ respectively, the maximum solidification velocities compatible with equilibrium solidification over ~10 mm are:

*substitutional* $V_I \sim 0{\cdot}5 \times 10^{-14}$ m s$^{-1}$ (or 1 mm in $8 \times 10^8$ years)
*interstitial* $V_I \sim 0{\cdot}5 \times 10^{-11}$ m s$^{-1}$ (or 1 mm in $8 \times 10^5$ years).

Such rates are only likely to be approached in geological conditions. In commercial and laboratory practice solute concentrations at various stages of solidification will be determined by the appropriate kinetic limitations. In practice, therefore, conventional solidification of alloys always results in some casting segregation.

## 2.3 DIFFUSION LIMITED PLANE FRONT SOLIDIFICATION

The conditions encountered in most cases of practical directional solidification are such that diffusion in the solid is very limited and that in the liquid is also too slow to allow equilibrium solute concentrations to be maintained in the melt; equilibrium concentrations, given by the phase diagram, are only maintained at the solid/liquid interface. As in the previous case the composition of the first solid will be $kC_0$. However, as solidification proceeds and the solute concentration in the solid increases from $kC_0$ towards $C_0$ a concentration gradient is frozen in the solid and the rejected solute forms a solute enriched zone in the liquid ahead of the solid/liquid interface. Tiller *et al.*[2] and Smith *et al.*[3] have shown that when diffusion and heat flow occur in the same single dimension and the melt/solid interface is planar then the solute concentration in the solid increases from $kC_0$ to $C_0$ over a characteristic distance $x_c$:

$$x_c = \frac{D_L}{kV_I} \tag{2.9}$$

(Taking $D_L = 5 \times 10^{-9}\ \mathrm{m^2\,s^{-1}}$, $k = 0{\cdot}1$, $V_I = 10^{-4}\ \mathrm{m\,s^{-1}}$ gives a typical initial solidification transient $x_c \sim 0{\cdot}5$ mm.)

After the initial transient when the solute concentration in the solid approaches $C_0$ a steady state configuration is achieved. The concentration profile in the liquid decays from the equilibrium value $C_0/k$ at the interface ($x = 0$) to $C_0$ at $x = \infty$ as indicated in Fig. 2.4*a*. In any further solidification the diffusion flux of solute in the melt will be exactly that required to form solid at a rate $V_I$ and consequently the system will retain the same solute profile.

The steady state solute distribution in the liquid must satisfy the Laplace diffusion equation

$$D_L \frac{d^2C_L}{dx^2} + V_I \frac{dC_L}{dx} = 0 \tag{2.10}$$

The first term represents the diffusion flux and the second the contribution from the moving boundary. Equation (2.10) must be solved with the boundary conditions

$$\begin{aligned} C_L &= \frac{C_0}{k} \quad \text{at } x = 0 \\ C_L &= C_0 \quad \text{at } x = \infty \end{aligned} \tag{2.11}$$

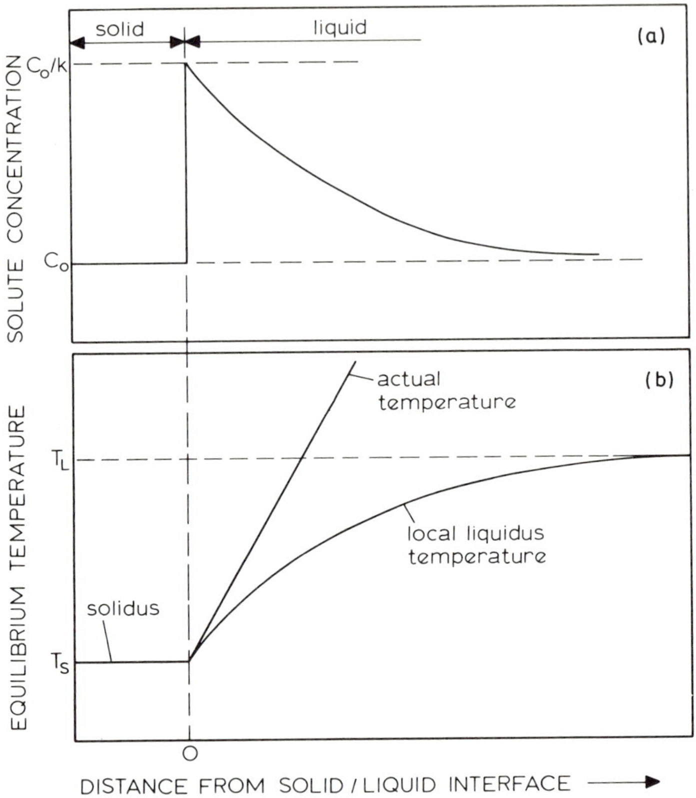


**Figure 2.4 (a) Steady state concentration profile in the melt ahead of the solidifying interface. (b) Equilibrium liquidus temperatures associated with the varying composition ahead of the solidifying interface**

and the flux continuity condition

$$-D_{\mathrm{L}}\frac{\mathrm{d}C_{\mathrm{L}}}{\mathrm{d}x}\bigg|_{x=0} = V_{\mathrm{I}}C_{0}(1-k) \tag{2.12}$$

Equation (2.12) equates the solute flux in the liquid at $x=0$ to the rate of solute rejection on solidification. The solution, first given by Tiller *et al.*,[2] is

$$C_{\mathrm{L}} = C_{0}\left[1+\frac{1-k}{k}\exp\left(-\frac{V_{\mathrm{I}}x}{D_{\mathrm{L}}}\right)\right] \tag{2.13}$$

As a consequence of this solute segregation in the liquid there is a local variation in the liquidus temperature ahead of the solidifying interface, shown schematically in Fig. 2.4*b* and defined by

$$m_L(C_L - C_0) = T_L - T_S = m_L \frac{C_0(1-k)}{k} \exp\left(-\frac{V_I x}{D_L}\right) \tag{2.14}$$

The gradient in liquidus temperature $dT_L/dx$ at the solid/liquid interface is given by the expression

$$\left.\frac{dT_L}{dx}\right|_{x=0} = -m_L \frac{C_0 V_I (1-k)}{D_L k} \tag{2.15}$$

$$= -\frac{V_I \Delta T}{D_L} \tag{2.16}$$

where $\Delta T$ is the melting range at the composition $C_0$. If the actual temperature gradient in the liquid $G_L$ is less than $dT_L/dx$ given by eqn (2.16) then the melt temperature locally falls below the freezing temperatures of some compositions and nucleation of solid can occur in the liquid ahead of the solidifying interface (i.e. constitutional supercooling can occur). This leads to the condition given first by Chalmers and co-workers[2,4] for the stability of a planar solidifying interface.

$$\frac{G_L}{V_I} \geqslant \frac{-m_L C_0 (1-k)}{k D_L} \tag{2.17}$$

$$= \frac{\Delta T}{D_L} \tag{2.18}$$

It has been demonstrated for a wide range of fairly simple single phase alloys that a critical value of $G_L/V_w$ (shown in § 2.2 to be equal to $G_L/V_I$ under steady state conditions) must be exceeded to achieve plane front solidification (see, for example, Flemings[1]). Particular evidence relating to the directional solidification of complex engineering materials such as superalloys is sparse. Tien and Gamble[5] probably achieved the first plane front solidification of a complex alloy in producing single crystals of the alloy MarM200 at a growth rate of $10^{-3}$ mm s$^{-1}$ in a temperature gradient of about 10 K mm$^{-1}$. They determined a melting range of 125 K (i.e. 1563–1688 K) for the alloy using differential thermal analysis and showed that their results were compatible with eqn (2.18). Subsequently plane front solidification has also been obtained by Bhambri *et al.*[6] on Inconel 713C and by Quested and McLean[7] on MarM246, MarM002, and IN738LC. The various experimental values of $G_L$, $V_w$ and $\Delta T$ from these studies are listed in Table 2.1 as are the values of $D_L$ calculated from these data using eqn (2.18). Bearing in mind the experimental inaccuracies, particularly in determining $G_L$, these results

**Table 2.1 Solidification parameters for plane front solidification of nickel-base superalloys**

| Alloy | $G_L$, $K\,mm^{-1}$ | $V_I$, $mm\,s^{-1}$ | $\left(\frac{G_L}{V_I}\right)_C$, $K\,s\,mm^{-2}$ | $\Delta T$, K | $D_L = \Delta T\left(\frac{G_L}{V_I}\right)_C^{-1}$, $mm^2\,s^{-1}$ |
|---|---|---|---|---|---|
| IN738LC | 20 | $1{\cdot}67\times10^{-3}$ | $1.20\times10^4$ | 85 | $7{\cdot}1\times10^{-3}$ |
| MarM246 | 20 | $1{\cdot}67\times10^{-3}$ | $1{\cdot}2\times10^4$ | ~100 | $\sim8{\cdot}3\times10^{-3}$ |
| MarM002 | 20 | $1{\cdot}67\times10^{-3}$ | $1{\cdot}2\times10^4$ | ~100 | $\sim8{\cdot}3\times10^{-3}$ |
| MarM200 | 10 | $10^{-3}$ | $10^4$ | 139 | $13{\cdot}9\times10^{-3}$ |
| IN713C | 11 | $0{\cdot}83\times10^{-3}$ | $1{\cdot}3\times10^4$ | 30 | $2{\cdot}31\times10^{-3}$ |

are remarkably close to the generally accepted values for liquid state diffusion at the melting point.

The variation in liquidus temperature resulting from segregation in the melt ahead of the solid/liquid interface, given by eqn (2.14), is a sensitive function of $V_I$ for a given temperature gradient $G_L$. Figure 2.5 shows that as $V_I$ is progressively increased beyond the critical value given by eqn (2.17), constitutional supercooling can occur over an increasing temperature range that approaches asymptotically the equilibrium melting range of the alloy. In addition, the range of compositions of the solid formed increases with increasing $V_I$ leading to local transverse segregation.

## 2.4 CELLULAR AND DENDRITIC SOLIDIFICATION

Plane front solidification occurs when the solute diffusional flux through the liquid normal to the solidifying interface is sufficiently large to supply enough latent heat of fusion to satisfy the heat flow condition (i.e. eqn (2.3)). When heat flow is more rapid than that supplied by normal diffusion, for example at high rates of solidification, then radial components of flux supplement the longitudinal contribution to give the additional solidification required to satisfy the heat conservation condition. This leads to the formation of *cells* and *dendrites*. The structures of the solid/liquid interfaces of the nickel-base superalloy MarM246 directionally solidified at two different rates in an approximately constant temperature gradient of $\sim13\,K\,mm^{-1}$ and frozen in by suddenly quenching the solidifying ingot are shown in Fig. 2.6. At the lower solidification rate the solidification front is almost planar but at the higher withdrawal rate long solid columns that are branched (called dendrites) project into the melt.

As the experimental value of $G_L/V_I$ increasingly deviates from the critical value of eqn (2.17), a progression of microstructural forms, shown schemati-

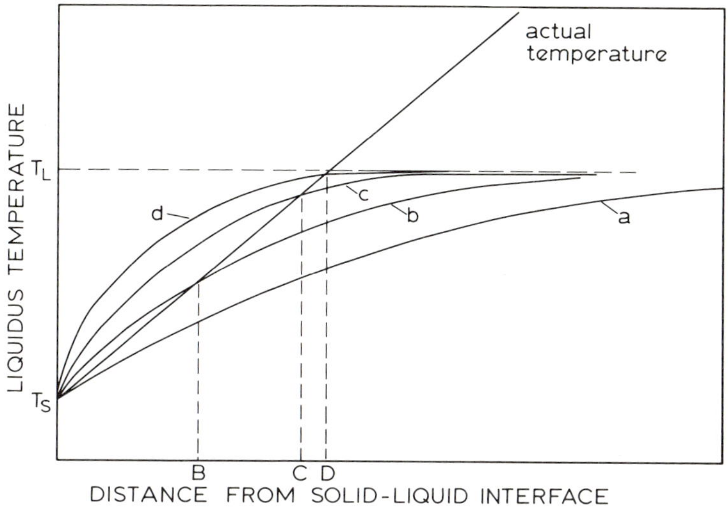


**Figure 2.5 Schematic diagram showing the change in liquidus temperature profile ahead of a solidifying interface for increasing values of $V_I$ and a constant $G_L$. $V_I$ increases from *a* to *d*. (a) Plane front; constitutional supercooling occurs to distances *B*, *C* and *D* ahead of the interface in cases (b), (c) and (d) respectively**

cally in Fig. 2.7, has been observed on many materials. The categorization of these morphologies is somewhat arbitrary and here we adopt the scheme described by Flemings:[1]

(a) *Plane front* which was the subject of § 2.3.
(b) *Cellular* with cylindrical cells parallel to the direction of heat flow.
(c) *Cellular* with near cylindrical cells parallel to the crystallographic direction of easy growth (⟨100⟩ for cubic materials). Normally the heat flow and the easy growth directions are only slightly misoriented.
(d) *Cellular/dendritic* with flanged cells, having sections shaped like maltese crosses, parallel to the crystallographic direction.
(e) *Dendritic* where the flanges of the cellular/dendritic structure have broken down further to form secondary, and higher order, arms.
(f) *Equiaxed* with no obvious directionality.

Examples of plane front, cellular and dendritic solidification microstructures are shown in Fig. 2.8 for directionally solidified IN738LC.

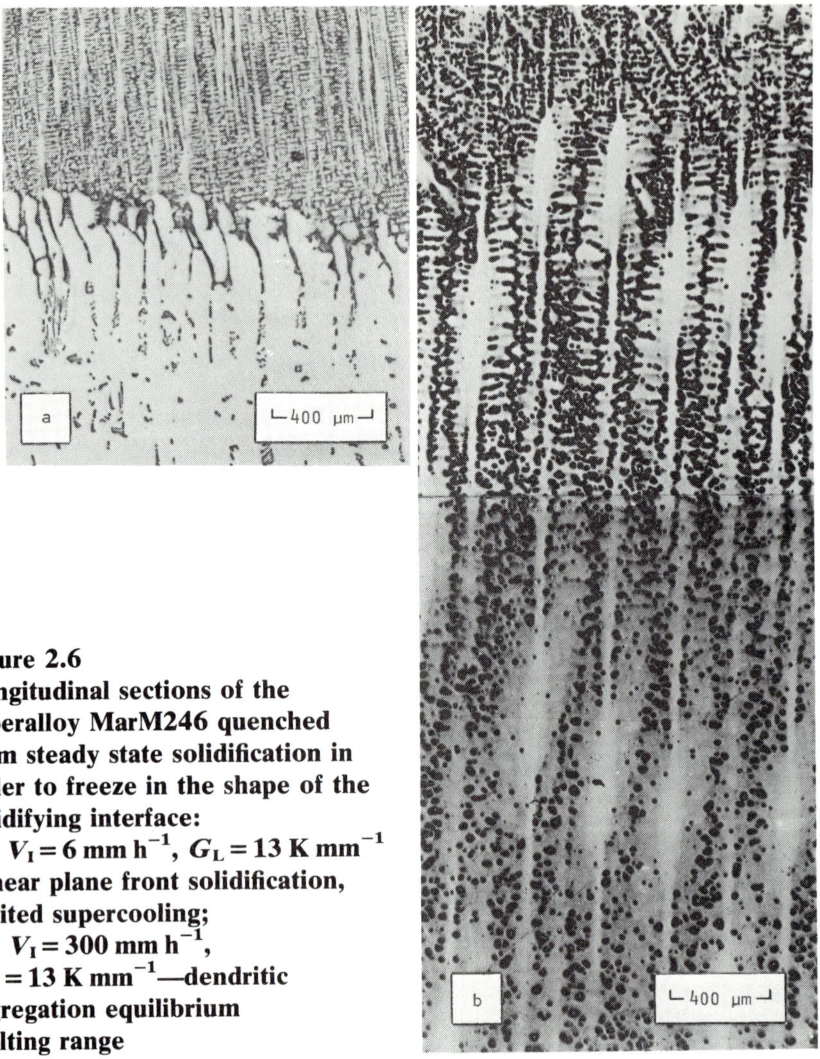


**Figure 2.6**
**Longitudinal sections of the superalloy MarM246 quenched from steady state solidification in order to freeze in the shape of the solidifying interface:**
**(a) $V_I = 6$ mm h$^{-1}$, $G_L = 13$ K mm$^{-1}$—near plane front solidification, limited supercooling;**
**(b) $V_I = 300$ mm h$^{-1}$, $G_L = 13$ K mm$^{-1}$—dendritic segregation equilibrium melting range**

Although as yet there are no rigorous theories describing the evolution of these solidification morphologies some qualitative observations can be made. There is clearly a progressive increase in the ratio of surface area to volume for the cells with increasing $V_I$ and $G_L$. This is necessary to generate by radial diffusion the greater latent heat of solidification required to balance the greater heat losses at high $V_I$ and $G_L$. The factors leading to the transitions between the various solidification morphologies are quite similar and it is probably possible to identify a different critical value of $G_L/V_I$ for each transition, although a theoretical expression is only available for the plane

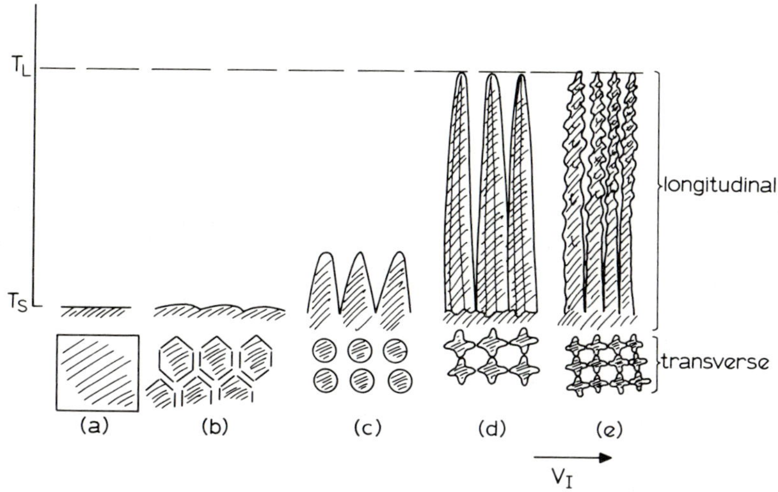


**Figure 2.7 Schematic illustration of the changing shape of the solidifying front with increasing $V_I$ for a constant $G_L$ as the range of constitutional supercooling extends to the equilibrium melting range of the alloy**

front to cellular transition (eqn (2.17)). Figure 2.9 indicates the types of microstructures obtained on six directionally solidified nickel-base superalloys in a range of conditions with various values of $G_L$ and $V_I$. The different types of solidification morphology fall into clear zones in the plot of log $G_L$ versus log $V_I$ that are consistent with transitions between the zones being given by $G_L/V_I$ bands that are similar for all nickel-base alloys.

Most attempts to model dendritic solidification have considered the growth of a single isothermal dendrite into an infinite liquid (e.g. Horvay and Cahn,[8] Nash and Glicksman[9]). These treatments describe most closely the dendrites which form in the absence of a temperature gradient at the solid/liquid interface during conventional casting. Of more interest in the present context is the steady state growth of an array of dendrites at a predetermined growth rate $V_I$ in the presence of a temperature gradient $G_L$. The following approximate treatment given by Burden and Hunt[10] and Hunt[11] clearly indicates the dependence of dendrite dimensions on $G_L$ and $V_I$.

It is assumed that the real branched dendrite shape can be approximated by a smooth steady state shape as indicated in Fig. 2.10 where $\lambda_p$ is the radius and $L_p$ the nearest neighbour separation of dendrites. Whereas in § 2.3 the equations describing solute diffusion during plane front solidification were one-dimensional, the analogous equations describing cellular solidification

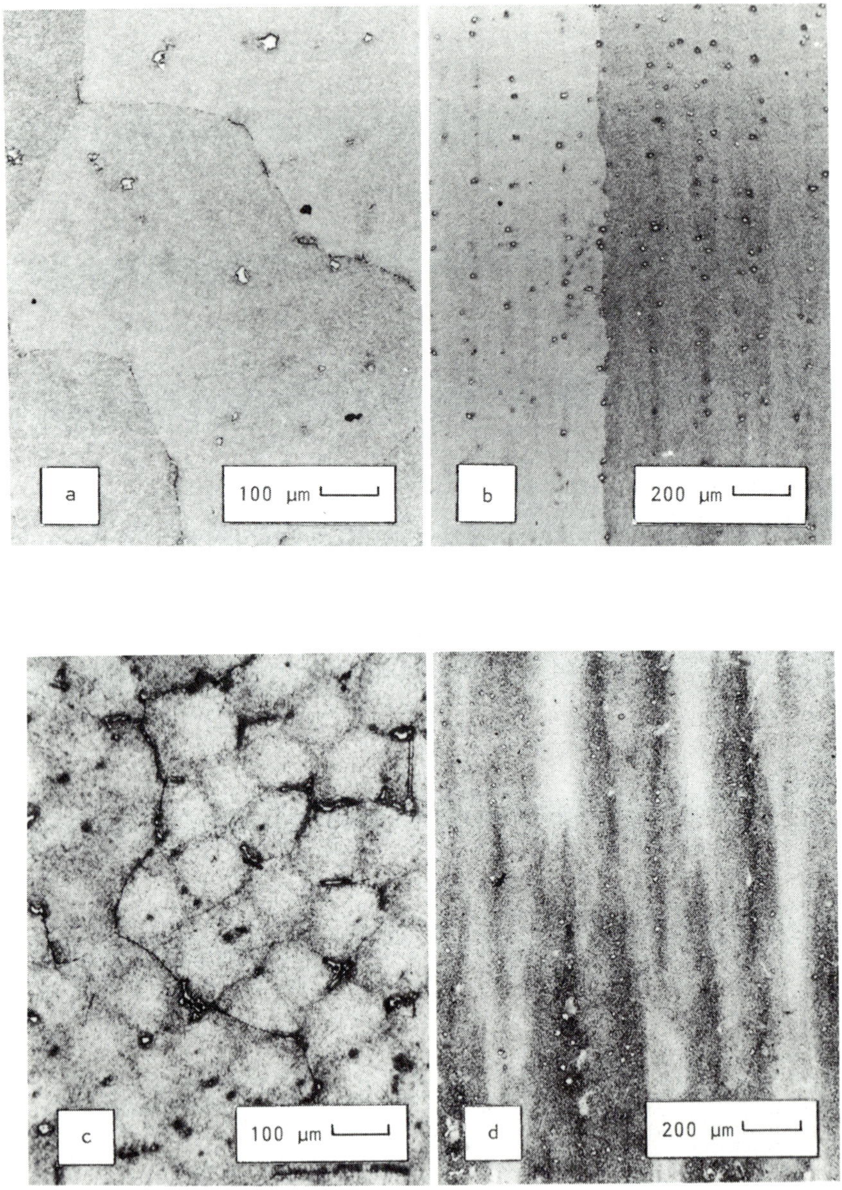
a
100 μm
b
200 μm
c
100 μm
d
200 μm

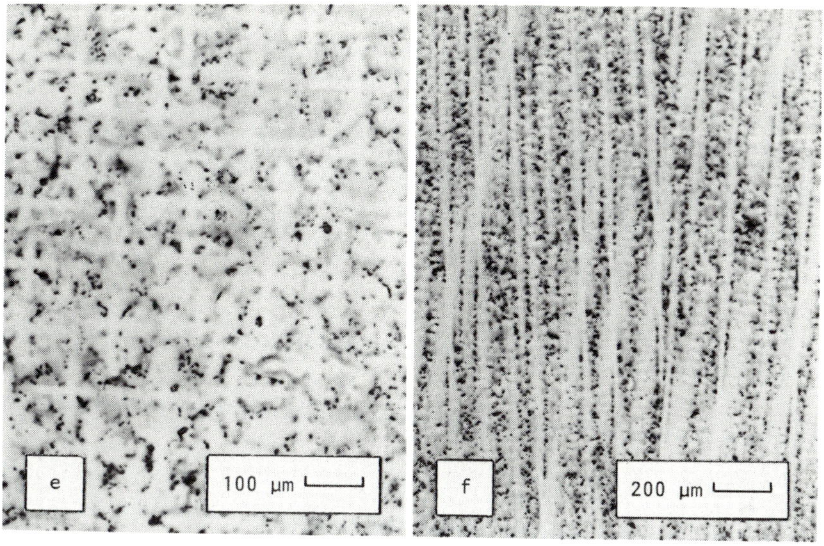


**Figure 2.8 Transverse and longitudinal sections of the superalloy IN738LC directionally solidified to give various solidification morphologies:**
**(a, b) plane front, $V_I = 9{\cdot}6\ \mathrm{mm\ h^{-1}}$, $G_L = 20\ \mathrm{K\ mm^{-1}}$**
**(c, d) cellular, $V_I = 60\ \mathrm{mm\ h^{-1}}$, $G_L = 13\ \mathrm{K\ mm^{-1}}$**
**(e, f) dendritic, $V_I = 600\ \mathrm{mm\ h^{-1}}$, $G_L = 13\ \mathrm{K\ mm^{-1}}$**

are 3-dimensional and cannot easily be solved rigorously. The requirements of flux conservation in the melt and of continuity of matter at the interface, respectively, lead to the following equations:

$$D_L \nabla^2 C_L + V_I \frac{dC_L}{dx} = 0 \tag{2.19}$$

$$-D_L \frac{\partial C_L}{\partial \mathbf{n}} = V_n C_L^I (1-k) \quad \text{at the interface.} \tag{2.20}$$

where **n** designates the surface normal to the dendrite/melt interface and the superscript I refers to the interface.

As in the case of plane front solidification, the solidifying interface is undercooled relative to the equilibrium liquidus temperature associated with the composition $C_0$. A compositional undercooling $\Delta T_D$ is present in both plane front and dendritic solidification, but in the latter case the curvature of the dendrite leads to an additional undercooling $\Delta T_\gamma$. (In the following

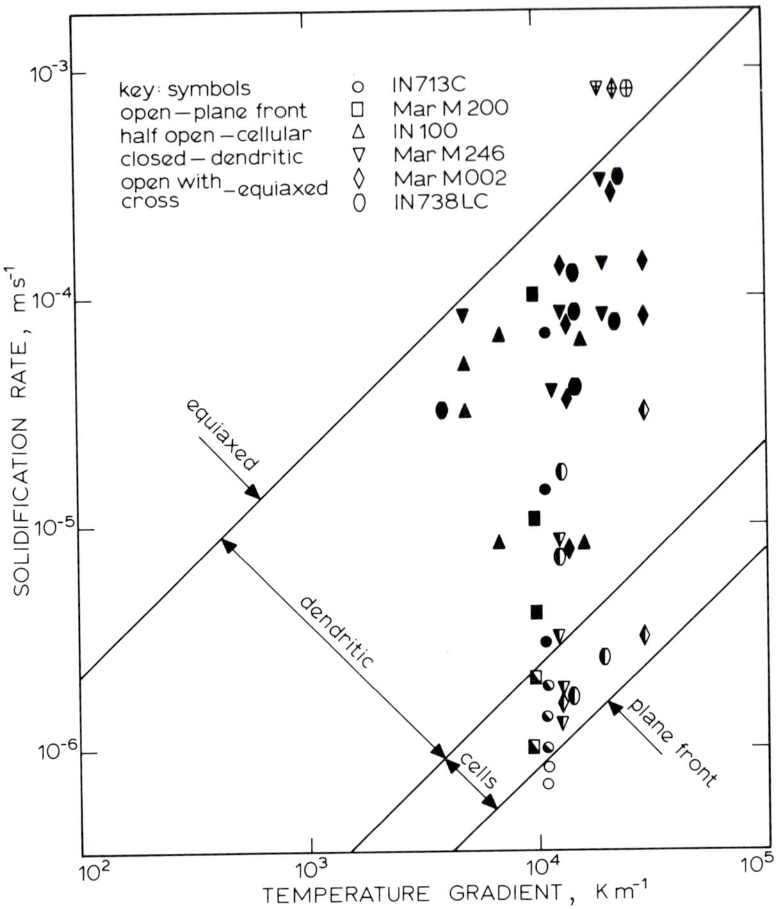


**Figure 2.9 Morphologies of directionally solidified nickel-base superalloys mapped on a plot of log $V_I$ versus log $G_L$. The lines drawn between the various zones are contours of constant $G_L/V_I$**

treatment kinetic undercooling is assumed to be negligible.) Then

$$\Delta T = \Delta T_D + \Delta T_\gamma \tag{2.21}$$

$$\Delta T_D = m_L(C_0 - C_L^I) \tag{2.22}$$

$$\Delta T_\gamma = \theta(R_1^{-1} + R_2^{-1}) \tag{2.23}$$

where $\theta$ is a constant proportional to the interfacial energy $\gamma$ and $R_1$, $R_2$ are the principal radii of curvature of the dendrite.

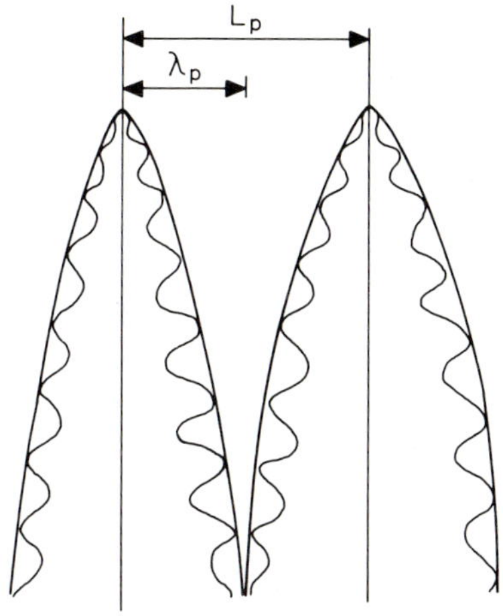


**Figure 2.10 Schematic diagram indicating the geometrical idealization of the dendrite shape and defining certain dimensional parameters [after Hunt[11]]**

It is convenient to consider separately the solutions to these equations (i) for the interdendritic liquid between the dendrite stalks and (ii) in the melt ahead of the tips.

**(i) Between dendrites**

The two principal assumptions in this regime are that the curvature undercooling $\Delta T_\gamma$ is negligible, since the curvatures of the dendrite shanks are much smaller than those of the dendrite tips, and that heat flow is rapid compared with diffusion. Consequently the temperature can be taken to vary only in the direction parallel to the dendrite axes. Moreover, since radial growth of the dendrite shank is likely to be small, it is reasonable to assume that equilibrium is attained between the liquid and the dendrite surface. The relationship between the average liquid composition $\bar{C}(x)$ and the liquid fraction of $g_L(x)$ is given by the Scheil equation:[1]

$$\bar{C}_L(x) = C_0(g_L(x))^{k-1} = C_I(x) \tag{2.24}$$

An essentially one dimensional equation relates the rate of change of solute concentration to the algebraic sums of the solute fluxes entering and leaving a toroidal element and solidifying on the dendrite shank (Fig. 2.11).

$$g_L(x)\frac{\partial \bar{C}_L(x)}{\partial t} = \frac{\partial}{\partial x}\left[D_L g_L(x)\frac{\partial \bar{C}_L(x)}{\partial x}\right] + \bar{C}_L(x)(k-1)\frac{\partial g_L}{\partial t} \tag{2.25}$$

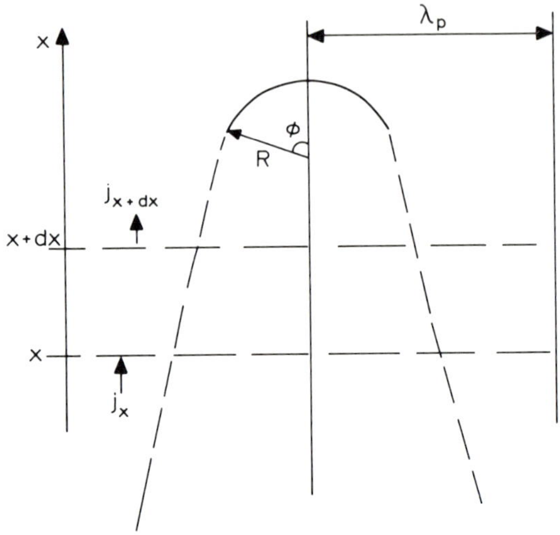


**Figure 2.11 Schematic view of a cellular (or dendritic) unit defining the parameters matching a spherical tip to the shank [after Hunt[11]]**

At steady state

$$\frac{\partial g_L(x)}{\partial x} = -\frac{1}{V_I}\frac{\partial g_L(x)}{\partial t} \quad \text{and} \quad G_L = m\frac{d\bar{C}_L}{dx} \quad \text{is constant.}$$

Thus eqn (2.25) may be written

$$g_L(x)\frac{\partial \bar{C}_L(x)}{\partial t} = \left\{-\frac{D_L G_L}{m_L V_I} + \bar{C}_L(k-1)\right\}\frac{\partial \bar{C}_L}{\partial t} \tag{2.26}$$

Equation (2.26) may be solved subject to the boundary condition $\bar{C}_L(x) = C_0$ when $g_L = 1$ (i.e. when solidification is complete), and $g_L$ may be expressed in terms of the dendrite geometry, $\pi r_I^2 = (1-g_L)\pi\lambda^2$, to give an expression describing the dendrite shape.

$$\left\{\frac{G_L}{m_L}x + \bar{C}_L(x=0)\right\}(1-k) + \frac{D_L G_L}{m_L V_I} = \left\{C_0(1-k) + \frac{D_L G_L}{m_L V_I}\right\}\left(1-\frac{r^2}{\lambda^2}\right)^{k-1} \tag{2.27}$$

**(ii) Beyond dendrite tips**

The solute concentration ahead of the dendrite tips is approximated by the sum of a unidirectional mean solute concentration $\bar{C}_L(x)$, similar to that occurring during plane front solidification, and a radial perturbation $\int C_L$ in the vicinity of the cell tips.

$$C_L = \bar{C}_L(x) + \int C_L \tag{2.28}$$

Then

$$\Delta T = m(C_0 - \bar{C}_T) - m_L \int C_T + \Delta T_\gamma \tag{2.29}$$

$$= \frac{m_L D_L}{V_I} \frac{d\bar{C}_L}{dx} - m_L \int C_T + \frac{2\theta}{R} \tag{2.30}$$

where we assume the cell tip to be a portion of a sphere of radius $R$ and the subscript $T$ refers to the tip.

Zener,[12] making a similar description of the growth of a sphere, has proposed that $\int C$ falls off over a distance about equal to the radius, i.e.

$$\left(\frac{\partial \int C}{\partial x}\right)_{x=x_T} = -\frac{\int C_T}{R} \tag{2.31}$$

Substitution of eqn (2.31), together with the matter continuity equation (eqn (2.22)) and $G_L = m_L\,(\partial\bar{C}_L/\partial x)$ into the differential of eqn (2.28) gives

$$\int C_T = \left\{\frac{V_I}{D_L}(1-k)\,C_T + \frac{G_L}{m_L}\right\}R \tag{2.32}$$

There is some arbitrariness in fitting a spherical cap to the dendrite shank, but Hunt[11] has shown that the range of possible values of $R$ vary by a factor of $\sqrt{2}$ only. For $\phi = 45°$ (see Fig. 2.11):

$$R = \frac{-G_L\lambda_p^2}{\sqrt{2}\left[m_L C_T(1-k) + \dfrac{D_L G_L}{V_I}\right]} \tag{2.33}$$

Combination of eqns (2.30), (2.32) and (2.33) gives a complete expression for dendrite undercooling in terms of the dendrite geometry

$$\Delta T = \frac{G_L D_L}{V_I} + \frac{G_L\lambda_p^2 V_I}{\sqrt{2}D_L} + \frac{2\theta(1-k)V_I}{D_L} - \frac{2\sqrt{2}\theta}{G_L\lambda_p^2}\left[m_L(1-k)C_0 + \frac{kG_L D_L}{V_I}\right] \tag{2.34}$$

In order to derive separate relationships between $\Delta T$, $V_I$ and $\lambda_p$ it is necessary to eliminate one of the variables and this requires another equation or criterion restricting eqn (2.34). This difficulty arises in many descriptions of growth phenomena. Various authors have argued that the system will adopt the growth conditions that will minimise $\Delta T$ or maximise $V_I$; this has become known as the extremum condition. Applying this to eqn (2.34) (i.e. $(\partial \Delta T/\partial \lambda_p)=0$) leads to the expression

$$V_I G_L^2 \lambda_p^4 = -4\theta D_L \left[ m_L(1-k)C_0 + \frac{kG_L D_L}{V_I} \right]$$

This equation is most conveniently expressed in terms of the dendrite spacing $L_p = 2\lambda_p$.

$$V_I G_L^2 L_p^4 = -64\theta D_L m_L(1-k)C_0\beta \tag{2.35}$$

where

$$\beta = 1 + \frac{kD_L}{m_L(1-k)C_0} \frac{G_L}{V_I} = 1 - \frac{D_L}{\Delta T} \frac{G_L}{V_I}$$

We note from eqn (2.17) that $\beta \to 0$ as the condition for plane front solidification is approached. For most experimental conditions $G_L/V_I \ll \Delta T/D_L$ so that $\beta \sim 1$.

Experimental data for three nickel-base superalloys are plotted in Fig. 2.12 showing the primary dendrite (or cell) spacings $L_p$ as functions of

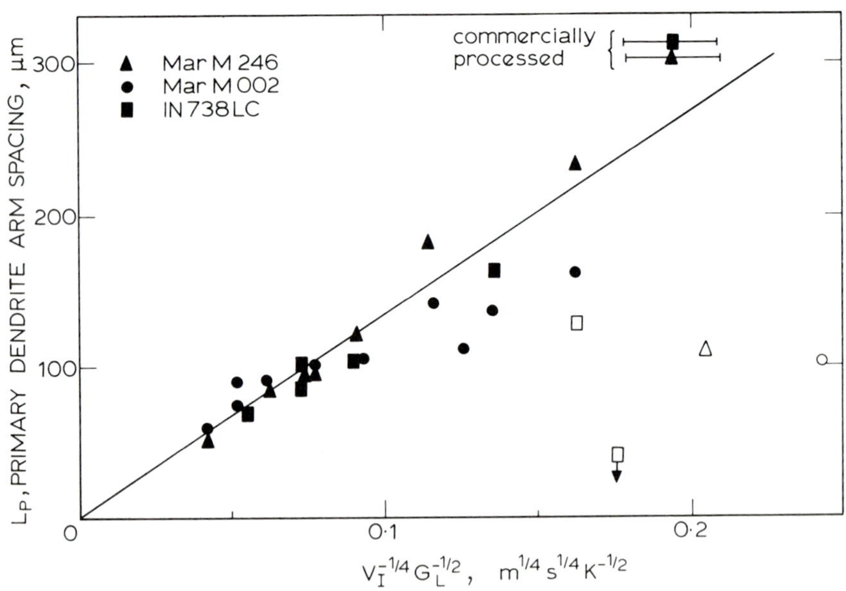


**Figure 2.12 The primary dendrite arm spacing $L_p$ plotted as a function of $G_L^{-1/2} V_I^{-1/4}$ for three directionally solidified nickel-base superalloys**

the composite term $V_I^{-1/4}G_L^{-1/2}$. All of the superalloys have been directionally solidified over a wide range of solidification rates (6 to 3000 mm h$^{-1}$) and in various temperature gradients (4 to 30 K mm$^{-1}$); the major uncertainty in the characterization arises from the difficulties in measuring $G_L$. The data for material having clearly established cellular or dendritic microstructures where $\beta \sim 1$ fall on a common, approximately linear curve (solid symbols); only the data corresponding to plane front, or near plane front, solidification (open symbols) deviate from the curve. The gradient of Fig. 2.12 ($1{\cdot}32\times 10^{-3}$ m$^{3/4}$ K$^{1/2}$ s$^{-1/4}$) may be identified with the right-hand side of eqn (2.35). Taking values of $D_L = 5\times10^{-9}$ m$^2$ s$^{-1}$, $m_L = 300$ K (atom fraction)$^{-1}$, $1-k = 0{\cdot}9$, $C_0 = 0{\cdot}2$ atom fraction and $\beta = 1$ gives an estimate of the capillarity constant $\theta = 1{\cdot}76\times10^{-7}$ K m. Burden and Hunt[10] show that for a dilute solution $\theta \sim \gamma/\Delta S$ where $\gamma$ is the solid/liquid interfacial energy and $\Delta S$ is the entropy of fusion. Typically $\Delta S \sim 10^6$ J m$^{-3}$ K$^{-1}$, giving a value of the solid/liquid interfacial energy of 176 mJ m$^{-2}$ which is compatible with previous measurements.[13]

Figure 2.13 indicates the range of primary dendrite dimensions that may be achieved using current directional solidification technology. The typical commercially processed alloy has a clear dendritic structure with $L_p \sim 300$ μm; use of the same solidification rate (60 mm h$^{-1}$) with a high $G_L$ obtained by liquid metal cooling leads to a cellular microstructure with a finer cell spacing. Further increases in both $G_L$ and $V_I$ cause a return to dendritic shapes but of very much finer dimensions. Control of the principal solidification parameters enables the periodicity of dendrites, and therefore of the associated segregation, to be determined over a wide range of dimensions, and current procedures can give aligned dendrites that are finer than those produced by current conventional casting procedures.

Figure 2.14 shows elemental concentration maps of different solutes on a transverse section of directionally solidified IN738LC; the solute is concentrated in the darker areas of the maps. The dendrite morphology is clearly profiled in each case but there are large differences in both the sign and magnitudes of segregation of various species. In general, the relevant binary phase diagrams provide a useful guide to the nature of segregation in multicomponent systems. Thus, where the Ni–*X* phase diagram has the form of Fig. 2.2*a* then solute *X* concentrates in the last liquid to solidify (i.e. to the interdendritic regions); examples of this type of behaviour are Ti, Ta and Mo in Ni. On the other hand, where the Ni–*X* diagram has the form of Fig. 2.2*b* as for *W* in Ni, then the solute segregates to the dendrites. The nature of segregation varies with solidification conditions. For example material prepared under conditions of plane front solidification is transversely homogeneous in composition. On the other hand, the fine dendrites arising from high solidification rates have an associated fine transverse periodicity of segregation. The implications of this to the homogenization of these materials will be discussed in Chapter 7.

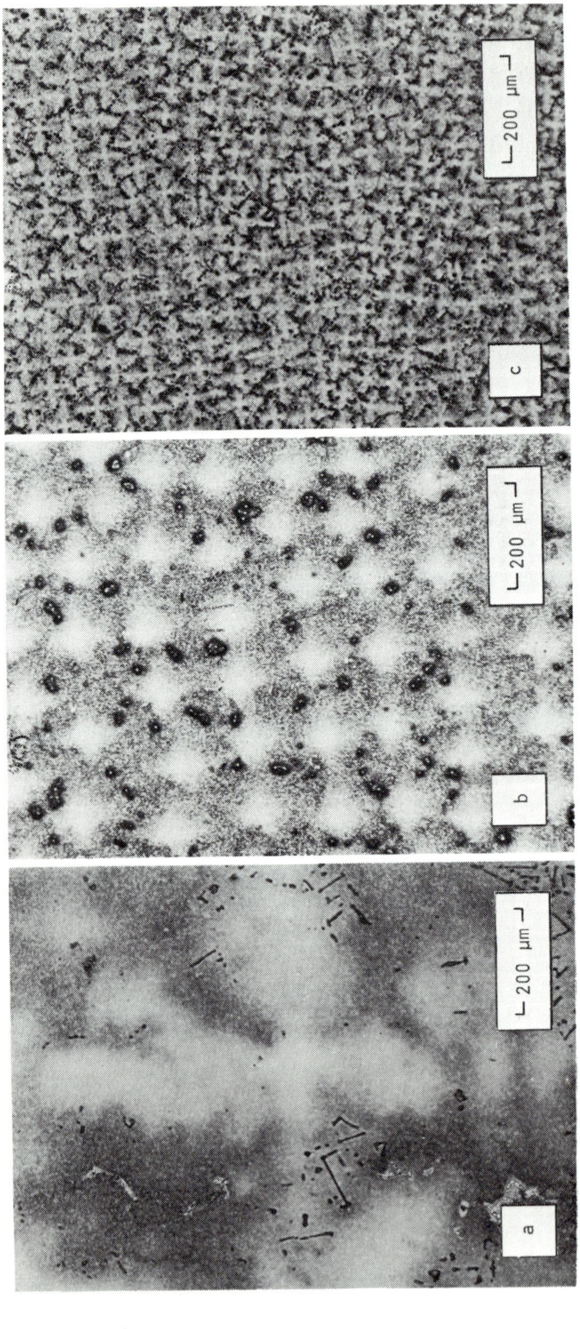


**Figure 2.13** **Transverse sections of the alloy IN738LC directionally solidified in various conditions to produce a wide range of cell or dendrite spacings:**
**(a) $V_I = 60$ mm h$^{-1}$, $G_L = 4$ K mm$^{-1}$ (commercial)**
**(b) $V_I = 60$ mm h$^{-1}$, $G_L = 13$ K mm$^{-1}$**
**(c) $V_I = 1200$ mm h$^{-1}$, $G_L = 20$ K mm$^{-1}$**

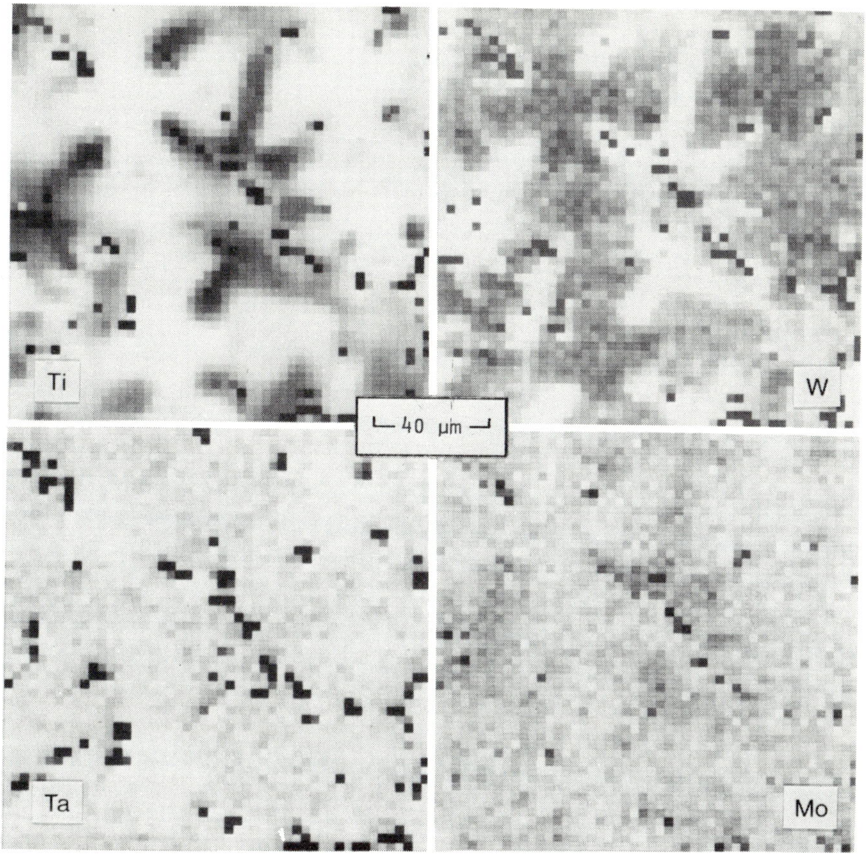


**Figure 2.14 Elemental concentration, determined by microprobe analysis, on a transverse section of the alloy IN738LC directionally solidified at 600 mm h$^{-1}$ with $G_L = 13$ K mm$^{-1}$ to produce a dendritic microstructure**

## 2.5 SECONDARY DENDRITES

The development of dendrites during directional solidification can be observed directly on material that has been suddenly quenched after steady state solidification has been established. Such experiments indicate that, whereas the initial primary dendrite arm spacings are retained throughout the freezing range of most alloys, the secondary dendrite arms which grow normal to the macroscopic solidification direction are finest and most dense at the dendrite tips and that they progressively coarsen with distance along the dendrite shank.

It is thought[14] that the higher curvatures of the secondary relative to the primary dendrites give rise to a high driving force for surface energy motivated mass transport by solute diffusion through the melt which leads to different types of morphological change that will be discussed more fully in Chapter 7 in relation to diffusional instabilities in the solid state. Three types of secondary dendrite instability, illustrated schematically in Fig. 2.15, have been proposed.

(a) The secondary dendrite arms may be subject to Ostwald ripening (Fig. 2.15*a*) in which smaller than average branches dissolve and larger than average branches grow. This phenomenon, which will be discussed in more detail in Chapter 7, follows a cubic growth law when, as here, volume diffusion is rate determining. The appropriate equation is normally framed to describe isothermal growth; here the problem is to account for arm growth as the material cools through the applied temperature gradient. Assuming that the secondary dendrites may be approximated by cylinders of radius $r$ then the increase in mean radius of the secondaries in an incremental distance $x$ to $x+\mathrm{d}x$ behind the dendrite tips may be expressed

$$r_{x+\mathrm{d}x}^3 = r_x^3 + \frac{4}{9}\frac{\bar{C}_{\mathrm{L}}(x) D_{\mathrm{L}} \gamma_{\mathrm{L}} \Omega^2}{R T_{\mathrm{L}}(x)} \frac{\mathrm{d}x}{V_{\mathrm{I}}} \tag{2.36}$$

where $\gamma_{\mathrm{L}}$ is the dendrite/melt interfacial energy, $R$ is the gas constant and $\Omega$ is the atomic volume. Expressing $D_{\mathrm{L}} = D_{\mathrm{L0}} \exp(-Q/RT)$ and

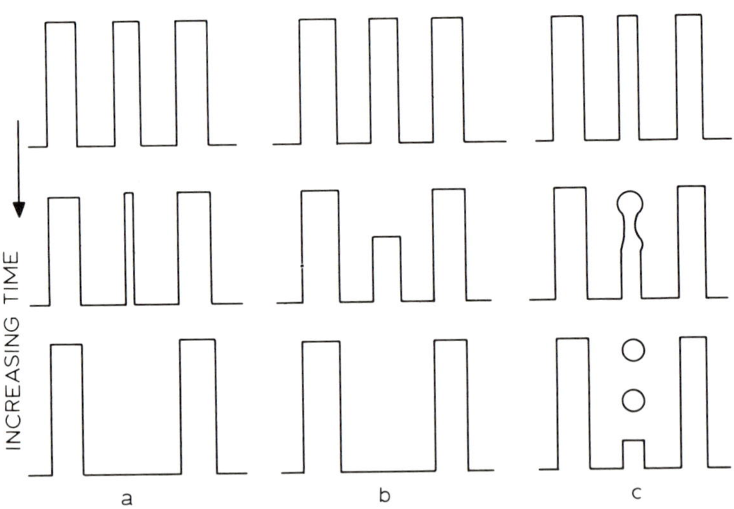


**Figure 2.15 Schematic diagrams showing three possible modes of diffusional aging of secondary dendrites in the semi-solid regime**

taking $T_0$ to be the temperature of the dendrite tips, eqn (2.36) can be expressed in the form

$$\frac{dr_x^3}{dx}=\frac{4}{9}\frac{K}{V_I}\frac{\exp[-Q/R(T_0-G_Lx)]}{(T_0-G_Lx)} \tag{2.37}$$

where $K=[\bar{C}_L(x)D_{L0}\gamma_L\Omega^2]/R$. This differential equation may be solved readily giving

$$r^3=r_0^3+\tfrac{4}{9}K\Psi(G_LV_I)^{-1} \tag{2.38}$$

where

$$\Psi=\int_{Q/RT_0}^{\infty}\frac{1}{u}e^{-u}\,du$$

which is a tabulated integral. Typically $\Psi\sim0{\cdot}1$ and $\sim10^{-6}$ for liquid and solid state diffusion respectively.

(b) The secondary arms may retract into the primary cell as shown in Fig. 2.15*b*. When mass transport occurs by solute diffusion in the interdendritic liquid it can readily be shown (McLean[16]) that the retraction of the secondary branch $h$ is given by an expression of the form

$$h=h_0-[0{\cdot}33\,K\Psi(G_LV_I)^{-1}]/r_0^2 \tag{2.39}$$

This mechanism would lead to a reduction in the density of the secondary dendrites, but not to a growth of those remaining.

(c) The secondary dendrites may pinch-off from the primary stalk, as illustrated in Fig. 2.15*c*, forming spherical dendritic islands adjacent to the longitudinal primary cells. Again following McLean[16] it can be shown for the case of liquid state solute diffusion that the maximum radius of secondary dendrites to have spheroidized is given by

$$r_{max}^3=0{\cdot}035K\Psi(G_LV_I)^{-1} \tag{2.40}$$

The bulk of experimental evidence supports mechanism (a); however, it is clear that the cooling rate $G_LV_I$ is an important parameter in each case. Figure 2.16 shows a plot of log $\lambda_s$ as a function of log $G_LV_I$ for superalloys prepared by a wide range of techniques. The linear form of the curve and the slope of ~3 are compatible with the concept of capillary motivated changes in secondary dendrite structure during freezing.

## 2.6 TEMPERATURE GRADIENT ZONE MELTING OF SECONDARY DENDRITES

Allen and Hunt[17] have pointed out that the presence of a temperature gradient can provide a driving force for solute diffusion in interdendritic liquid that may lead to some solute redistribution in the solid and migration of the secondary dendrites up the temperature gradient. Following Allen and Hunt

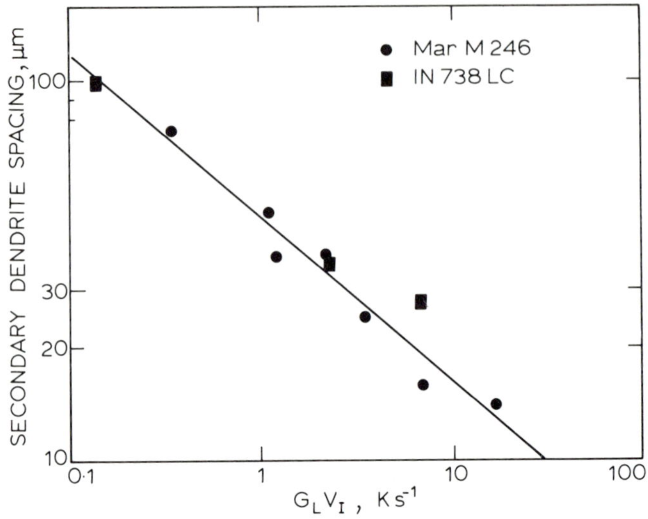


**Figure 2.16 Secondary dendrite arm spacings $\lambda_s$ of directionally solidified nickel-base superalloys as a function of the cooling rate $G_L V_I$**

we idealize the structure of secondary dendrites as shown in Fig. 2.17. The temperature gradient is taken to be perpendicular to the secondary and parallel to the primary dendrite arms. At steady state eqn (2.10) remains valid but the equilibrium solute concentrations $C_L$ given by the phase diagrams can be achieved only at the dendrite/liquid interfaces (i.e. $x = \pm L/2$ where $L$ is the secondary dendrite spacing). Integration of eqn (2.10) subject to this boundary condition leads to the expression

$$\frac{dC_L}{dx} = \frac{G_L}{m_L}\left(1 - \frac{V_I x}{D_L}\right) \tag{2.41}$$

Substitution into eqn (2.12) gives the expressions for the velocity of the cold and hot surfaces of the secondary arm.

$$v_c = -\frac{D_L}{C_0(1-k)}\frac{G_L}{m_L}\left(1 + \frac{V_I}{2D_L}L\right) \tag{2.42}$$

$$v_h = -\frac{D_L}{C_0(1-k)}\frac{G_L}{m_L}\left(1 - \frac{V_I}{2D_L}L\right) \tag{2.43}$$

When $V_I L/2D_L \gg 1$, the velocities of the hot and cold faces are of approximately the same size and are in opposite directions. This leads to no change

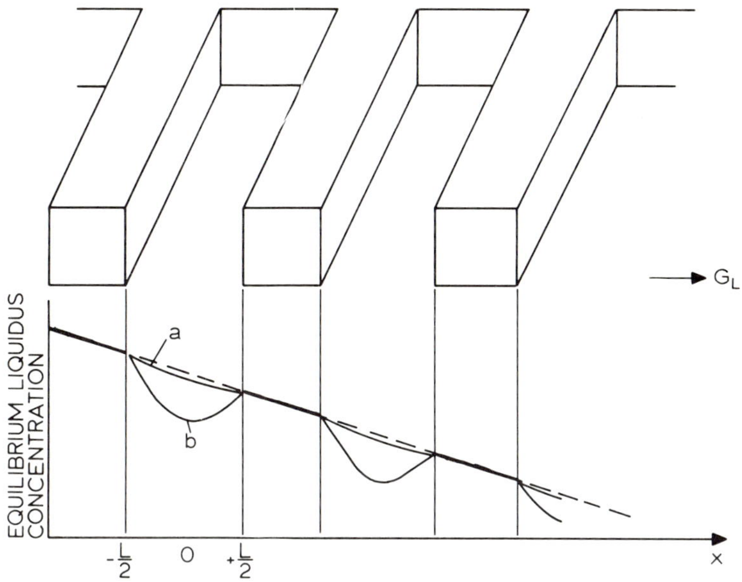


**Figure 2.17 Schematic diagram showing solute concentrations in the liquid between secondary dendrites: (a) $V_I L/2D_L \leqslant 1$; (b) $V_I L/2D_L \geqslant 1$ [after Allen and Hunt[17]]**

in the position but to an increase in thickness of the secondary dendrite by deposition of solute from the melt. One consequence of such freezing is a coring of, or change in composition through the thickness of, the secondary dendrite. However, when $V_I L/2D_L \ll 1$ the hot and cold surfaces move in the same direction, leading to a translation of secondary dendrites up the temperature gradient. This is achieved by continual dissolution of solute at the cold and redeposition at the hot ends of the dendrite. This has a homogenizing effect reducing the level of dendrite segregation. The real behaviour of dendrites will be intermediate between these two extremes. The total distance $d$ that a secondary dendrite moves during solidification is given approximately by

$$d = \int \bar{v} \frac{dt}{dC_L} dC_L \qquad (2.44)$$

where

$$\bar{v} = \frac{v_c + v_h}{2} \quad \text{and} \quad \frac{dC_L}{dt} = -\frac{G_L V_I}{m_L}$$

Then

$$d \sim \frac{D_L}{G_L V_I (1-k)} \ln\left(\frac{C_2}{C_1}\right) = \frac{D_L}{V_I(1-k)} \ln(1/k) \tag{2.45}$$

where $C_1 = C_0$ and $C_2 = C_0/k$ are the solute concentrations in the liquid at the beginning and end of solidification respectively. Thus the fraction of the dendrite that has been remelted, $f_m$, is given by

$$f_m \sim \frac{d}{\lambda_s} \tag{2.46}$$

Taking $D_L \sim 5 \times 10^{-9}$ m$^2$ s$^{-1}$ and $V_I = 10^{-4}$ m s$^{-1}$ which are typical of many processes for the directional solidification of superalloys, eqn (2.45) would predict a significant contribution of TGZM for $\lambda_s \leqslant 100$ μm. Typically $\lambda_s \sim 50$ μm for conventionally cast superalloys and $\lambda_s \sim 100$ μm for alloys directionally solidified using current commercial processes, which suggests that TGZM is likely to be significant in modifying the levels of segregation in most directionally solidified alloys.

## 2.7 CONVECTIVE MIXING IN THE LIQUID

Our discussion so far has assumed that the solute distribution is determined solely by diffusion. However, the presence of a temperature gradient and the variation in density of the melt with temperature will lead inevitably to a degree of mixing by convective flow. Burton, Prim and Slichter[18] considered the situation where there was complete mixing in the liquid except in a hydrodynamically stagnant layer of thickness $\delta$ ahead of the interface where only diffusion occurred. The solute distribution is therefore obtained by solving eqn (2.14) with new boundary conditions.

Unlike the convection-free case this has no steady state solution. It is convenient to define an effective distribution coefficient $k_e$ which is the ratio of the composition of the solid being deposited $C_S^I$ and the bulk liquid composition $C_0$.

$$k_e = \frac{C_S^I}{C_0} \tag{2.47}$$

Burton, Prim and Slichter[18] show that

$$k_e = \frac{k}{k + (1-k)\exp(-V_I \delta / D_L)} \tag{2.48}$$

and the solid and liquid compositions are given by analogues to the Scheil equation

$$C_s = k_e C_0 g_L^{(k_e - 1)} \tag{2.49}$$

$$C_L = C_0 (1 - g_L)^{k_e - 1} \tag{2.50}$$

where $g_L$ is the fraction liquid. Figure 2.18 shows schematically the solute variation obtained by (a) convection-free solidification ($\delta = \infty$, $k_e = 1$), (b) complete mixing in the liquid ($\delta = 0$ and $k_e = k$) and (c) partial mixing in the liquid ($k_e = 1 + \frac{1}{2}k$).

No experimental study of boundary layers and effective distribution coefficients appears to have been reported for superalloys. However, taking values of $\delta \geqslant 0{\cdot}1$ mm found for other metals, $D_L \sim 5 \times 10^{-9}\ m^2\ s^{-1}$ and $V_I \sim 10^{-4}\ m\ s^{-1}$ gives $k_e \sim 1$ which indicates that the solute profiles are essentially defined by diffusion and that, under normal circumstances, convection has relatively little effect on the general solidification characteristics.

Convection, however, is thought to be the cause of longitudinal chains of equiaxed and randomly oriented grains, called 'freckles', that can develop particularly at the outside of castings. By studying the solidification of a transparent analogue system Copley *et al.*[19] showed that the grains nucleated on debris from eroded dendrite tips which were damaged by upward flowing convective jets of liquid from the interdendritic regions. The cause of the jet is thought to be a density inversion in the liquid when segregation of certain solutes lowers the melt density locally below that of the bulk liquid. The problem can be eliminated by increasing the temperature gradient $G_L$ since this reduces the distance over which solid and liquid coexist and prevents the convective instability from developing.

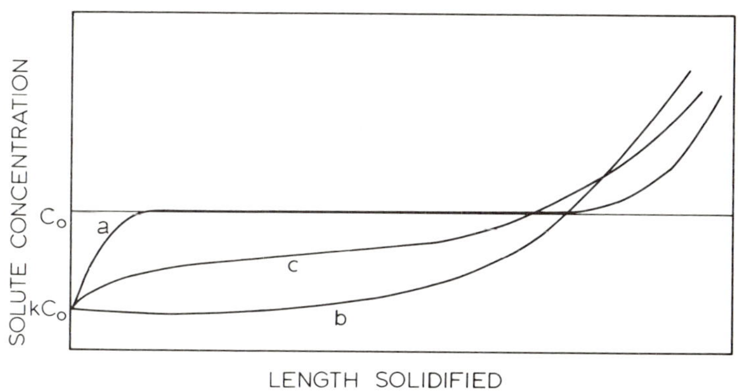


**Figure 2.18 Solute concentration profiles after directional solidification of an ingot of initial solute concentration $C_0$: (a) convection-free solidification ($k_e = 1$); (b) complete mixing in the liquid ($\delta = 0$, $k_e = k$); (c) partial mixing in the liquid ($k_e = 1 + \frac{1}{2}k$) [after Burton *et al.*[18]]**

## 2.8 MICROPOROSITY

The volume decrease on solidification can lead to the formation of porosity particularly in conventional castings where the final liquid fractions are likely to be completely surrounded by solid. There is likely to be much less porosity in directionally solidified material because a head of molten alloy is constantly feeding the solidifying surface. However, some pore formation may occur at the roots of the dendrites where flow of the liquid may be insufficient to supply the volume mismatch between solid and liquid. Since the channels at the dendrite roots through which the liquid must flow decrease both with reductions in the remaining fraction that is liquid and in the primary dendrite spacing, both are likely to be important parameters in the formation of porosity.

Flemings and co-workers[20,21] have adapted the theories of fluid flow through porous media to describe interdendritic liquid feeding. The net pressure at the dendrite root consists of the sum of the pressures due to the atmosphere and to the height of liquid $p_0$ less the pressure drop $\Delta p$ due to viscous drag at the dendrite surfaces; this must exceed the capillary pressure associated with a pore if porosity is to be inhibited:

$$\text{pore unstable if } p_0 - \Delta p < \frac{2\gamma_L}{r} \tag{2.51}$$

where $\gamma_L$ = liquid surface energy, $r$ = radius of pore. Flemings[1] argues that $\Delta p$ should be inversely proportional to the total dendrite/liquid interfacial area

$$\Delta p = \alpha L_p^{-2} g_L^{-2} \tag{2.52}$$

where $\alpha$ is a constant of proportionality containing the viscosity of the melt. This indicates that porosity is most likely to form in the last fraction to solidify and that it is most likely to occur in material having the finest microstructure.

The most extensive experimental study of microporosity in directionally solidified superalloys, that of Julien *et al.*,[22] showed that the levels of microporosity in alloy variants of IN738 and C263 were sensitive to minor compositional changes. The extent of porosity decreased with increased content of elements such as titanium, zirconium and especially hafnium in association with carbon which leads to increased volume fractions of the $\gamma$–$\gamma'$ eutectic that constitutes the last liquid. This behaviour can be understood qualitatively in terms of the implications of $\gamma$–$\gamma'$ eutectic formation on the configuration at the roots of dendrites. As shown schematically in Fig. 2.19 the presence of $\gamma$–$\gamma'$ eutectic causes a flattening of the dendrite roots because all of the remaining liquid must solidify at the same temperature and therefore leads to a reduction in the viscous drag term $\Delta p$ due to the smaller dendrite surface area. Thus pores are less likely to form in the presence than in the absence of $\gamma$–$\gamma'$ eutectic.

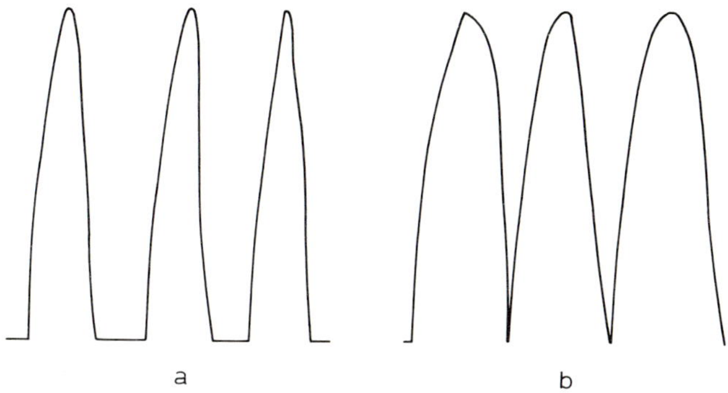


**Figure 2.19** **Schematic diagram showing the wider liquid channels at the base of dendrites (a) when the final liquid fraction forms eutectic, than (b) when no eutectic liquid is present**

## 2.9 PRECIPITATE DISTRIBUTION

The high strengths of superalloys derive from various precipitates, principally the ordered $\gamma'$ and various carbides, that act as barriers to deformation. The nature and distribution of these particles can be influenced significantly by the solidification conditions.

### 2.9.1 Carbides

A wide range of carbides has been identified in various superalloys and it is clear that the existence of a specific carbide depends in a complicated manner on detailed alloy chemistry, processing conditions and thermal history (i.e. service conditions). It is not enough to examine the relative stabilities of the various carbides, displayed as free energies of formation in Table 2.2, which would always favour the formation of MC-type carbides (M = Ti, Zr, Ta, Nb etc.); rather, interactions between the many constituents of the superalloys can lead to a progressive degeneration of the apparently 'stable' MC carbides into those of lower intrinsic stability (e.g. $M_{23}C_6$) when a total energy reduction can be achieved and when kinetic rather than thermodynamic considerations dominate.

Several experimental studies[6,23,24] have attempted to characterize the sequence of formation and the morphological evolution of carbides during directional solidification. As in studying dendrite formation, the examination of the microstructure of an ingot which has been suddenly quenched after steady state directional solidification has been established has proved to be a potent experimental tool. There is a general consistency in the results of such experiments on various alloys directionally solidificd under conditions giving dendritic solidification.

**Table 2.2 Free energies of formation of various carbides at 1375°C**

| Carbide | $-\Delta G$, kJ mol$^{-1}$ |
|---|---|
| $Cr_3C_2$ | 590 |
| $Cr_7C_3$ | 710 |
| $Cr_{23}C_6$ | 786 |
| VC | 744 |
| NbC | 1360 |
| TaC | 1505 |
| TiC | 1595 |
| ZrC | 1647 |
| HfC | 2111 |

(i) In the melt ahead of the dendrite tips (i.e. $T > T_L$) coarse faceted MC carbides have been observed in MarM200,[5] IN713C[6] and IN100.[24] It is likely that these carbides never melted during processing but continually coarsened by liquid-phase diffusion. This was confirmed by Fernandez *et al.*[24] who showed that the nature of those carbides was insensitive to the cooling rate during solidification. Bhambri *et al.*[6] argue that the carbides, being less dense than the melt, can rise ahead of the solidifying interface due to both their natural buoyancy and convection effects. Applying concepts of Uhlmann *et al.*[25] and of Jaffrey and Chadwick,[26] they show that the absorption efficiency of carbides into the solid will increase with increasing solidification velocity and with increasing carbide size. Thus at slow growth rates near the conditions for plane front solidification there would be a rejection of this carbide leading to a progressive enrichment of the melt; this was noted by Tien and Gamble[5] who detected a significant accumulation of Nb along an ingot of MarM200 which had been directionally solidified at $10^{-6}$ m s$^{-1}$.

(ii) Behind the dendrite tips ($T \sim T_L - 30$ K) additional MC carbides are deposited interdendritically from the melt and these may differ in crystallography and composition from the 'unmelted' type referred to above. In IN100, the unmelted carbides are predominantly titanium carbides, whereas those that grow from the melt behind the dendrite tips are mixed (Ti, Mo, V) C.[24] On the other hand, Bhambri *et al.*[6] report both types to be mixed (Nb, Mo, Ti) monocarbides in IN713C. The morphologies of MC carbides that grow from the melt depend on the detailed solidification conditions. In both IN713C and IN100 they are blocky and aligned parallel to the solidification direction for high values of $G_L/V_I$ like those in Fig. 2.13*b* and have a 'Chinese script' appearance on polished sections (e.g. Fig. 2.13*a*) for material produced at low values of $G_L/V_I$. In the latter case the carbides have a complex three-dimensional shape, shown by the extracted carbide in Fig. 2.20, due to rapid growth along $\langle 111 \rangle$ directions. Data from Fernandez *et al.*,[24] reproduced in Fig. 2.21*a*, show the changing proportions of a constant 1·3%MC between blocky and Chinese script as a function of $G_L/V_I$. The sizes of each type decrease with increasing cooling rate $G_L V_I$ as shown in Fig. 2.21*b*.

(iii) At the end of the solidification range the final liquid fraction often forms a $\gamma$–$\gamma'$ eutectic mixture that solidifies over a temperature range of a few degrees. At this point there may be additional MC deposition in the form of a $\gamma$–$\gamma'$–MC eutectic.[6]

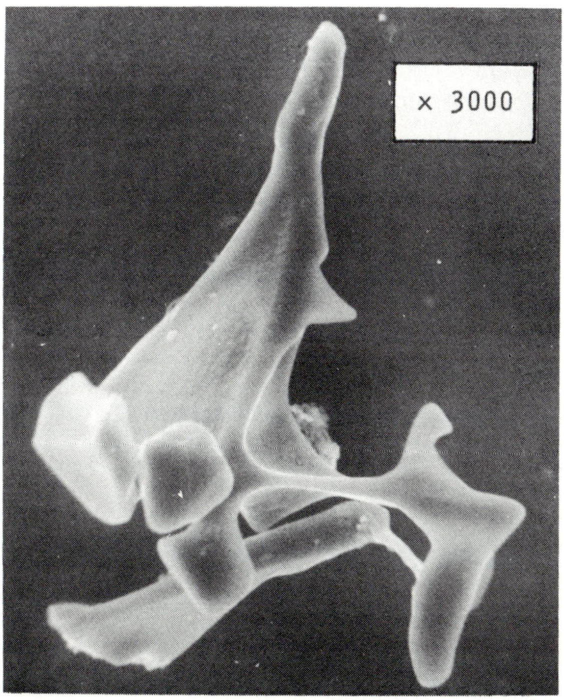


**Figure 2.20 Extracted 'Chinese script' carbide from IN738LC directionally solidified at 600 mm h$^{-1}$ in a temperature gradient of 13 K mm$^{-1}$ [Ref. 27]**

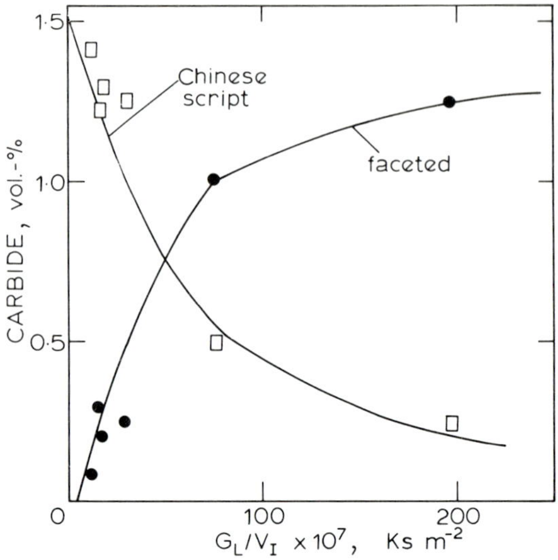


**Figure 2.21a Volume fractions of blocky and 'Chinese script' MC carbides as a function of $G_L/V_I$ for directionally solidified IN100 [Ref. 24]**

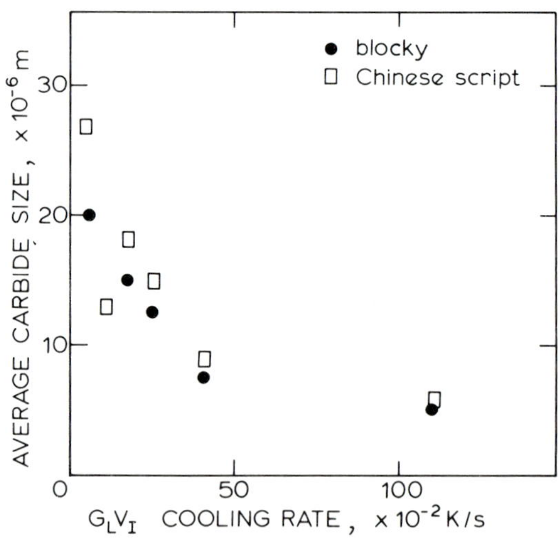


**Figure 2.21b Average sizes of blocky and 'Chinese script' MC carbides in directionally solidified IN100 as a function of cooling rate [Ref. 24]**

Some alloys, particularly those such as IN738LC containing relatively high chromium contents, show evidence of post-solidification degradation of MC to $M_{23}C_6$. This is thought to occur by the reaction

$$\gamma + MC \rightarrow M_{23}C_6 + \gamma'$$

and involves the rejection of Cr from the $\gamma$ matrix to form the predominantly chromium-base $M_{23}C_6$. This transformation can continue through heat treatments and in service, the $M_{23}C_6$ eventually coating the longitudinal grain boundaries to the detriment of certain mechanical properties. Figure 2.22 compares the microstructures in IN738LC of as-directionally solidified material that is virtually devoid of $M_{23}C_6$ and of the same alloy after creep testing that shows profuse intergranular $M_{23}C_6$.

It is interesting that the carbide morphology of virgin and revert (i.e. remelted) superalloys often exhibits a 'Chinese script'-to-blocky transition that can be accompanied by high porosity and reduced mechanical performance. The reasons are not yet clear but one suggestion is that stable TiN particles form on remelting which act as nuclei for profuse carbide formation and, consequently, influence the carbide growth characteristics.

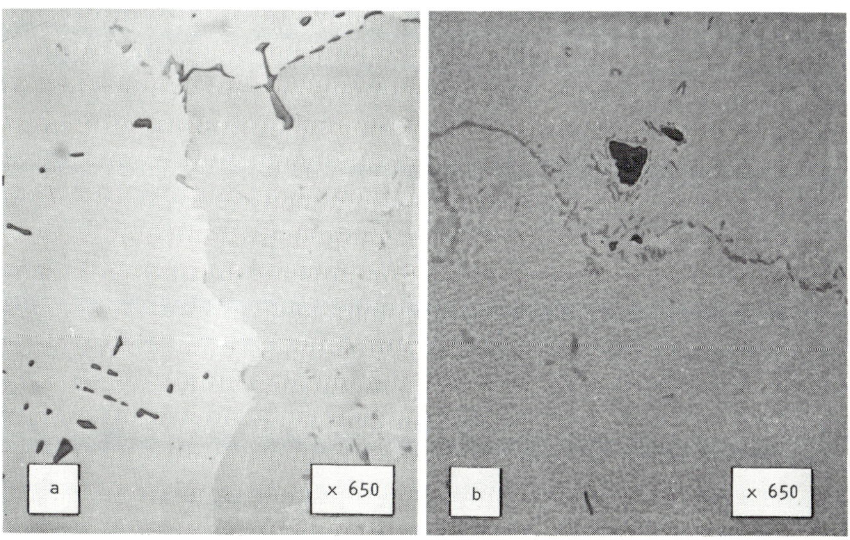


**Figure 2.22 Longitudinal sections of directionally solidified IN738LC: (a) as directionally solidified—no $M_{23}C_6$ apparent; (b) directionally solidified and aged for 10 000 h at 850°C—$M_{23}C_6$ coats the grain boundaries**

### 2.9.2 Gamma-prime precipitation

In nickel-base superalloys most of the coherent intermetallic precipitate which takes the form $Ni_3(Al, Ti)$, with an fcc structure that closely matches that of the $\gamma$-nickel matrix, deposits homogeneously from the solid state. The solution temperature and the detailed orientation relationships of the $\gamma'$ in the matrix are both sensitive to the specific alloy compositions. In cast alloys in general, and directionally solidified superalloys in particular, the segregation levels occurring during dendritic solidification can lead to significant variations in the crystallography, shape, size and distribution of $\gamma'$ on the scale of the dendrite structure. Since the mechanical behaviour of such materials depends on the form of these precipitate particles, heterogeneities in the distribution are likely to affect their performance.

When near plane-front solidification giving low levels of segregation is achieved, no spatial variation in the nature of the $\gamma'$ particles is noted in MarM200;[5] however, the same material directionally solidified at faster rates to produce dendrites exhibits relatively fine cubic particles in the dendrite cores and much coarser irregular $\gamma'$ particles in the interdendritic regions. Similar trends have been noted in other alloys such as IN713C[6] and IN738LC.[7] Figure 2.23 shows representative $\gamma'$ distributions in the core and interdendritic

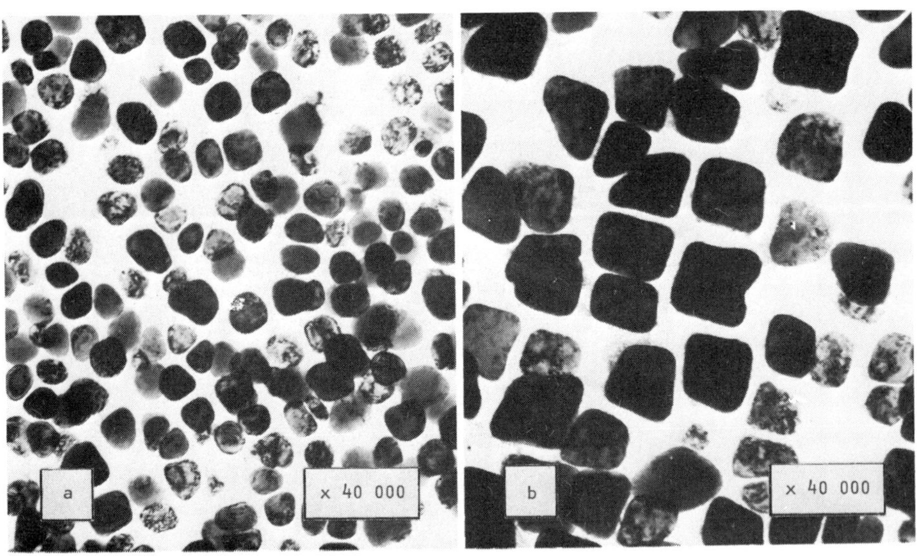


**Figure 2.23 Representative $\gamma'$ distribution (a) in and (b) between dendrites in IN738LC directionally solidified at 300 mm $h^{-1}$ in a temperature gradient of 13 K $mm^{-1}$ [Rosenthal and West[27]]**

regions of IN738LC directionally solidified at 300 mm $h^{-1}$ in a relatively high temperature gradient of ~13 K $mm^{-1}$. Quenching experiments indicate that the $\gamma'$ deposits at a significantly higher temperature in the interdendritic regions in spite of the local liquidus temperature being lower there. Indeed, in IN738LC the difference between the $\gamma'$ solvus and the liquidus temperatures in the interdendritic zone is very small (~10 K) and the solutioning stage of the manufacturer's recommended heat treatment,[28] designed to avoid incipient melting, dissolves the dendritic but coarsens the interdendritic $\gamma'$. Figure 2.24 shows the consequences of the full heat treatment on directionally solidified IN738LC (i.e. 2 h at 1120°C and 16 h at 845°C followed by air

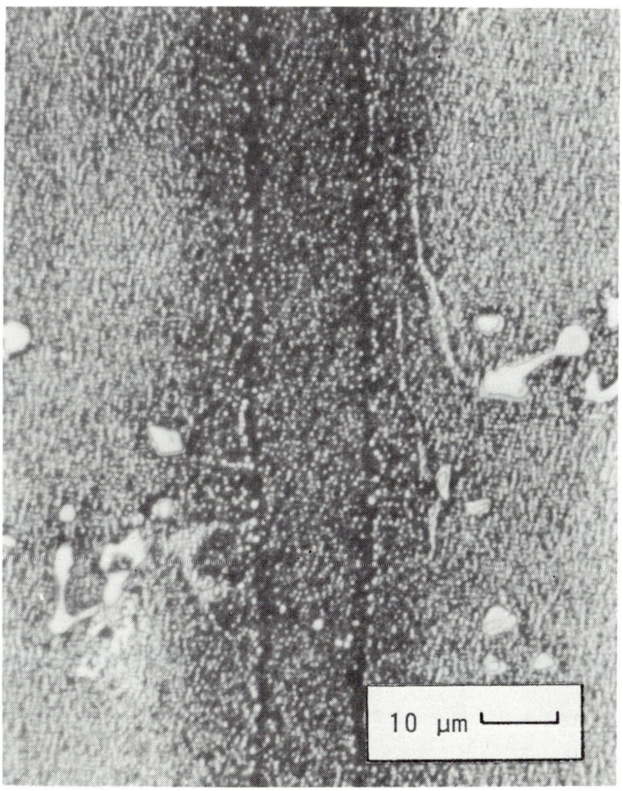


**Figure 2.24 Longitudinal section of IN738LC after directional solidification at 300 mm $h^{-1}$ in a temperature gradient of 13 K $mm^{-1}$ and a two-stage heat treatment of 2 h at 1120°C and 16 h at 845°C and 16 h at 845°C**

cooling at each stage). Two distinct types of $\gamma'$ distribution are obtained: uniform and fine within the dendrites, coexisting coarse and fine populations between the dendrites.

The dimensions of the $\gamma'$ particles also depend on the solidification conditions. Since it is thought that the growth of $\gamma'$ particles occurs by volume diffusion controlled Ostwald ripening[29] the final size should be determined by integrating the particle growth occurring as the material passes through the temperature gradient $G_s$ at a rate $V_I$. This is the same phenomenon invoked to describe the development of secondary dendrite arms; the mean $\gamma'$ size $\bar{a}$ (i.e. radius or cube edge) should be given by a modification of eqn (2.38) with a different numerical shape constant ($\frac{8}{9}$ instead of $\frac{4}{9}$) and containing solid- (rather than liquid-) state diffusion parameters.

$$\bar{a}^3 = \bar{a}_0^3 + \tfrac{8}{9}K'\Psi(G_s V_I)^{-1}$$

where $K' = CD_0\gamma\Omega^2/R$.

$$\Psi = \int_{Q/RT_0}^{\infty} \frac{1}{u} e^{-u}\, du$$

The work of Bhambri *et al.*[6] on IN713C confirms the expected increase in $\gamma'$ particle size with decreasing cooling rate for high values of $G_s V_I$. Their data shown in Fig. 2.25 as a plot of $\log(\bar{a}/2)$ versus $\log G_s V_I$ has the expected slope of $\sim 3$. However, at the lowest cooling rates the $\gamma'$ particle size appears to be independent of $G_s V_I$; Piearcey *et al.*[23] noted a similar constancy of $\gamma'$ dimensions with varying cooling rate in MarM200. No convincing explanation of this effect has been proposed.

Although most of the $\gamma'$ deposits homogeneously, varying fractions can nucleate in a heterogeneous manner depending on the solidification conditions. Two mechanisms for localized $\gamma'$ formation are possible:

(i) $\gamma$–$\gamma'$ eutectic solidifying from the melt
(ii) discontinuous growth of $\gamma'$ at grain boundaries.

Since most microstructural characterizations are based on appearance rather than the mechanism of phase formation, there can be considerable confusion over the differentiation of eutectic and discontinuous $\gamma'$.

Examination of IN738LC directionally solidified in a wide range of conditions[7] indicates that significant amounts of heterogeneously precipitated $\gamma'$ form when dendritic segregation occurs in combination with relatively slow cooling rates. The typical examples shown in Fig. 2.26 are from material prepared commercially ($V_I = 120$ mm h$^{-1}$, $G \sim 4$ K mm$^{-1}$). No discontinuous precipitate is observed at slow growth rates where plane front solidification is approached or at high cooling rates (e.g. $V_I = 300$ mm h$^{-1}$, $G \sim 13$ K mm$^{-1}$). Quenching experiments suggest that growth of these $\gamma - \gamma'$ islands can either

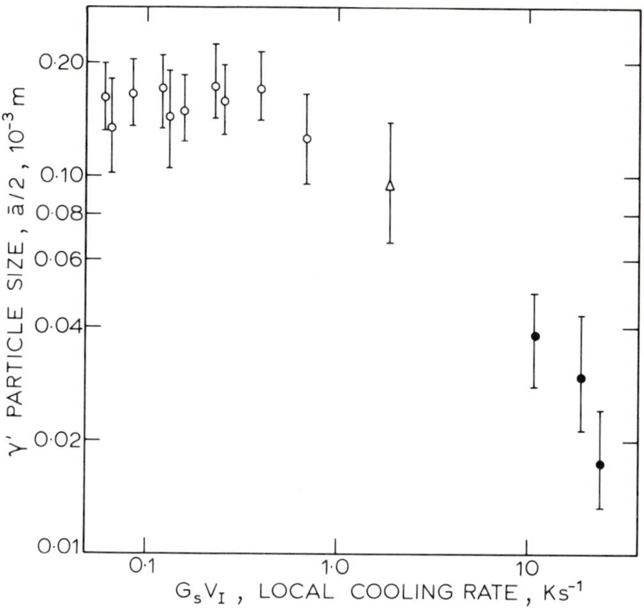


**Figure 2.25 Variation of $\gamma'$ particle size with cooling rate $G_s V_I$ for directionally solidified IN713C [Ref. 6]**

nucleate spontaneously at grain boundaries or can grow on small $\gamma$–$\gamma'$ eutectic nuclei. It is quite clear that a relatively small volume fraction of the so-called 'eutectic' $\gamma$–$\gamma'$ actually forms from the melt; rather the bulk of it deposits from the solid on small true eutectic pools.

The driving force for eutectic formation and presumably also for the nucleation of discontinuous $\gamma'$ depends on the extent of segregation during solidification and therefore, as discussed in § 2.4, is a function of $G_L^{-1/2} V_I^{-1/4}$. Thus at low solidification rates both types of precipitation are likely to be inhibited as is observed. At higher solidification rates where the dendrites extend over the entire equilibrium melting range the driving force reaches a maximum value and restrictions in the extent of discontinuous precipitation must be based on the kinetics of growth.

Gust[30] has recently reviewed the theoretical and experimental evidence relating to discontinuous precipitation in metallic alloys. Rapid diffusion along a grain boundary leads to faster decomposition of a supersaturated phase into a duplex lamellar structure of lower free energy (Fig. 2.27); in isothermal conditions the rate of advance of the precipitation front $\mathrm{d}r/\mathrm{d}t$ is of the form

$$\frac{\mathrm{d}r}{\mathrm{d}t} = \frac{\beta \delta D_I}{\lambda^2}$$

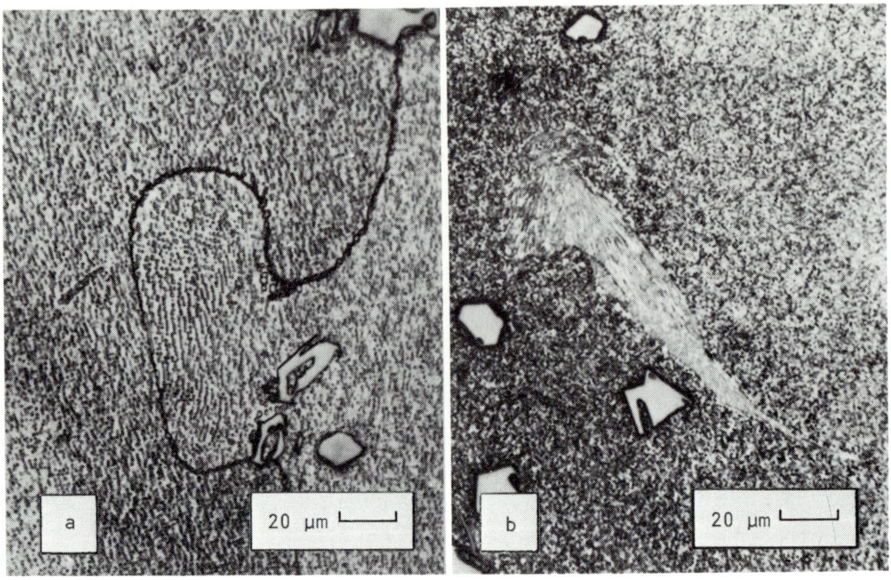


**Figure 2.26** **Discontinuous precipitation in IN738LC directionally solidified (a) in commercial conditions ($V_I = 120$ mm h$^{-1}$, $G_L \sim 4$ K mm$^{-1}$) and (b) in the laboratory ($V_I = 60$ mm h$^{-1}$, $G_L \sim 13$ K mm$^{-1}$)**

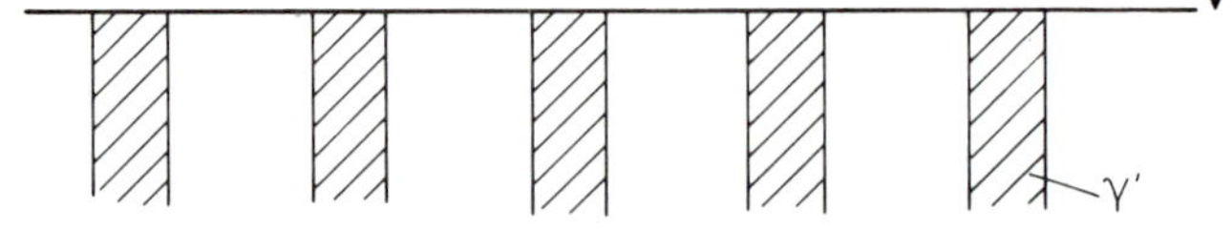


**Figure 2.27** **Schematic diagram showing rapid discontinuous precipitate growth by short circuit diffusion along a grain boundary of thickness $\delta$**

where $\beta$ is a measure of the driving force for the reaction, $\delta$ is the effective grain boundary thickness, $D_I$ is the interface diffusivity and $\lambda$ is the width of the lamellae formed.

Expressing $D_I = D_I^0 \exp(-Q/RT)$, the mean radius of a discontinuous zone of precipitation on passing through a temperature gradient $G_s$ at a rate $V_I$ can be estimated:

$$\bar{r} = \frac{\beta\delta}{\lambda^2} D_I^0 \int_0^x \exp\left(-\frac{Q}{RT}\right) \frac{dx}{V_I}$$

where $T = T_L - G_s x$ and $x$ is distance in the solidification direction. Integration in the limit when $x \to \infty$ gives

$$\bar{r} = \frac{\beta\delta D_I^L}{\lambda^2} \frac{RT_L^2}{Q} (G_s V_I)^{-1}$$

Again we see that the extent of discontinuous precipitation is inversely proportional to the cooling rate. This is consistent with the observed decrease in the occurrence of discontinuous precipitate at high growth rates.

## 2.10 GRAIN BOUNDARIES AND CRYSTAL ORIENTATION

Although one of the principal advantages claimed for directionally solidified superalloys is the virtual elimination of grain boundaries transverse to the solidification direction and the establishment of a columnar grain texture there is remarkably little information on the factors influencing the density and morphology of grains.

Directional freezing clearly encourages the production of long grains and transverse grain boundaries do not occur, in general, because the elongated grains grow freely into the melt and do not impinge on other grains growing in some other direction. The development of a preferred growth texture, ⟨100⟩ in the case of nickel-base superalloys, occurs because crystals with this orientation grow at a faster rate than others and spread transversely to eliminate the slowly growing orientations. Piearcey *et al.*[23] clearly show that, during directional solidification of MarM200, the density of longitudinal grains decreases as solidification proceeds (Fig. 2.28), competitive growth in an initially random population of crystals eliminating unfavourable grains and leaving the fast growing ⟨100⟩ crystals. The final product therefore consists of a series of long crystals oriented with ⟨100⟩ directions parallel to the solidification axis but misoriented with respect to each other by various rotations. Quested and McLean[7] have shown that the density of grains in IN738LC depends on the conditions of directional solidification (Fig. 2.29) decreasing with increasing temperature gradient $G_L$ and rate of solidification $V_I$. Although there is no theoretical basis, this suggests that the anisotropy

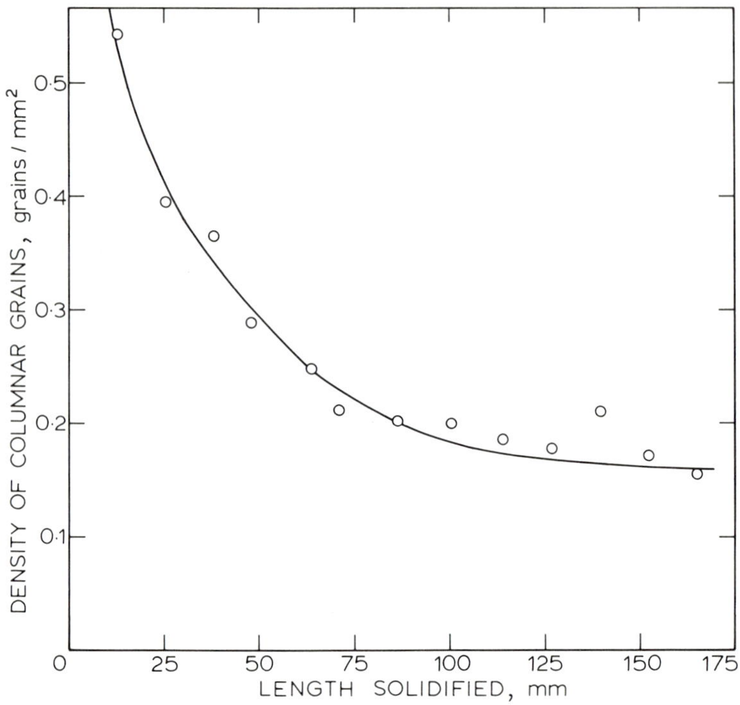


**Figure 2.28 Dependence of columnar grain size on the length of ingot directionally solidified for MarM200 [Ref. 23]**

of crystal growth is increased with increasing $G_L$ and $V_I$ allowing a more rapid establishment of the preferred texture.

Single crystals of superalloys are now produced routinely by simple extensions of the directional solidification process following the procedures established for crystal growth. Crystals with the naturally occurring ⟨100⟩ growth direction can be obtained by the introduction of a constriction to select only one of the ⟨100⟩ nuclei by preventing the others from growing. For ⟨100⟩ oriented single crystals the dendrite structures, and the processing parameters influencing them, are just the same as those described for the simpler columnar grained directionally solidified materials. Other single crystal orientations can be produced by partially melting back a suitable seed crystal before proceeding with the directional solidification stage through a grain selector; here, however, it is only the seed orientation that is allowed to grow. The dendrite forms and the values of the process parameters controlling them vary with orientation reflecting both the crystal symmetry (or lack of it) in the growth direction and lower rate of solidification.

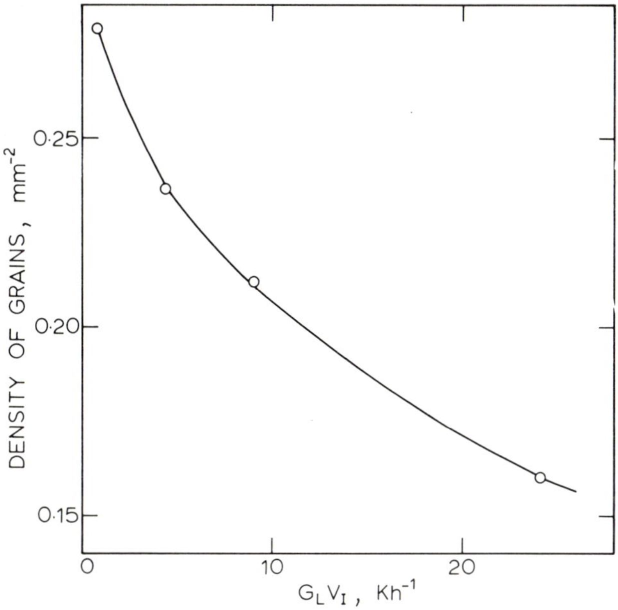


**Figure 2.29 Variation in the densities of grains, measured 100 mm from the start of directional solidification of IN738LC as a function of $G_L/V_I$**

## REFERENCES CHAPTER 2


1 M. C. FLEMINGS: 'Solidification processing', 1974, New York, McGraw-Hill

2 W. A. TILLER, K. A. JACKSON, J. W. RUTTER AND B. CHALMERS: *Acta Metall.*, 1953, **1**, 428

3 V. G. SMITH, W. A. TILLER AND J. W. RUTTER: *Can. J. Phys.*, 1955, **33**, 723

4 J. W. RUTTER AND B. CHALMERS: *Can. J. Phys.*, 1953, **31**, 15

5 J. K. TIEN AND R. P. GAMBLE: *Mater. Sci. Eng.*, 1971, **8**, 152

6 A. K. BHAMBRI, T. Z. KATTAMIS AND J. E. MORRAL: *Metall. Trans.*, 1975, **6B**, 523

7 P. N. QUESTED AND M. McLEAN: unpublished work

8 G. HORVAY AND J. W. CAHN: *Acta Metall.*, 1961, **9**, 695

9 G. E. NASH AND M. E. GLICKSMAN: *Acta Metall.*, 1974, **22**, 1283

10 M. H. BURDEN AND J. D. HUNT: *J. Cryst. Growth*, 1974, **22**, 99 and 109

11 J. D. HUNT: 'Solidification and casting of metals', 3, 1979, London, Metals Society

12 C. ZENER: *Trans. AIME*, 1946, **167**, 550

13 D. R. H. JONES AND G. A. CHADWICK: *Philos. Mag.*, 1970, **22**, 291

14 T. Z. KATTAMIS, J. COUGHLIN AND M. C. FLEMINGS: *Trans. Metall. Soc. AIME*, 1967, **239**, 1504

15 E. BULLOCK, M. McLEAN AND D. E. MILES: *Acta Metall.*, 1977, **25**, 333

16 M. McLEAN AND M. S. LOVEDAY: *J. Mater. Sci.*, 1974, **9**, 1104

17 D. J. ALLEN AND J. D. HUNT: 'Solidification and casting of metals' (Ref. 11), p. 39

18 J. A. BURTON, R. C. PRIM AND W. P. SLICHTER: *J. Chem. Phys.*, 1953, **21**, 1987

19 S. M. COPELY, A. F. GIAMEI, S. M. JOHNSON AND M. F. HORNBECKER: *Metall. Trans.*, 1970, **1**, 2193

20 T. S. PIWONKA AND M. C. FLEMINGS: *Trans. AIME*, 1966, **236**, 1157

21 R. MEHRABIAN, M. KEANE AND M. C. FLEMINGS: *Metall. Trans.*, 1970, **1**, 1209

22 P. JULIEN, J. C. LECOMTE, F. LAVAUD, G. LESOULT, F. WETS AND E. BACHELET: Ecole des Mines and SNECMA, Paris, final report on COST 50 programme F5/1, 1978

23 B. J. PIEARCEY, B. H. KEAR AND R. W. SMASHEY: *Trans. Q. ASM*, 1967, **60**, 634

24 R. FERNANDEZ, J. C. LECOMTE AND T. E. KATTAMIS: *Metall. Trans.*, 1978, **9A**, 1381

25 D. R. UHLMANN, B. CHALMERS AND K. A. JACKSON: *J. Appl. Phys.*, 1964, **35**, 2986

26 D. JAFFERY AND G. A. CHADWICK: *Philos. Mag.*, 1968, **18**, 573

27 R. ROSENTHAL AND D. WEST: private communication

28 H. WIGGIN AND CO. LTD: 'Nimocast alloys', Publ. No. 3610, 1973, Hereford, UK

29 A. J. ARDELL AND R. B. NICHOLSON: *Acta Metall.*, 1966, **14**, 1295

30 W. GUST: 'Phase transformations', Institution of Metallurgists conference proceedings, Series 3, Number 11, Vol. 1, 1979

# 3
# Basics of directional solidification: *in situ* composites

The consequences of directional solidification to the microstructures of *existing* superalloys were described in detail in Chapter 2. Kraft,[1] in 1960, proposed that the same processing technique, if applied to alloys that solidified to form two or more solid phases (for example by a eutectic reaction) could be used to produce an entirely new class of material—the so-called *in situ* composite. At the time of this suggestion there had already been considerable interest in developing composite materials for use at high temperatures by more conventional methods with the expectation of increasing the strength at high temperatures of existing alloys by incorporating strong, but brittle, refractory fibres within them. Indeed, the current status of this latter approach, which has continued in parallel with the development of *in situ* composites, is summarized in Chapter 8. The potential advantages of producing composite materials by directional solidification are manifest; it involves the extension of a well established technology, allows components to be produced from the melt in a single stage process and leads to a composite of phases that are (more or less) thermodynamically stable. On the debit side, the alloy compositions amenable to the production of *in situ* composites are restricted by the occurrence of appropriate phase equilibria, which are generally unknown in the multicomponent alloys designed to combine the various properties required in a modern high temperature alloy, and there is little scope for varying the relative volume fractions of the constituent phases. During the twenty years since Kraft's original proposal there has been an explosion of interest both in the fundamental understanding of the solidification and properties of polyphase alloys and in the technological development of *in situ* composites for particular applications. Some of this work concerned with the physical properties of composite materials (e.g. electrical, magnetic, optical) has led to interesting commercial applications; however, this is outside the scope of the present book and reference should be made to reviews by Albers[2] and Weiss[3] on the subject. By far the greatest effort has been devoted to developing replacement materials for superalloys in order to extend the capability of gas turbine engines and, to this end, has involved very substantial research investment by the major engine manufacturers and defence agencies.

Much of our understanding of solidification in polyphase alloys is based on comparisons of the predictions of quite idealized theoretical models with experimental data obtained on very simple binary alloys. On the other hand, very complicated multicomponent alloys have been evolved in attempts to satisfy the many service requirements of modern gas turbine materials. In this chapter, theory will be illustrated wherever possible by results obtained on complex alloys of potential practical importance. Although in principle composites can be produced by other reactions, most work on *in situ* composites has been carried out on alloys of eutectic or near-eutectic composition. Consequently, the emphasis here will also be on eutectic solidification.

## 3.1 CLASSIFICATION OF EUTECTIC MORPHOLOGIES

There have been many attempts to classify the various types of microstructure that can be obtained by solidifying alloys of eutectic composition. The early schemes were based simply on the appearance of the microstructures. A more useful approach by Hunt and Jackson[4] relates the eutectic structure to the nature of growth of the individual phases of the eutectic. Earlier work by Jackson[5] on single phase materials distinguished between 'non-faceted' phases that solidified with complete crystallographic isotropy via an atomically rough solid/liquid interface, and 'faceted' phases that had preferred crystallographic growth directions associated with atomically smooth solid/liquid surface facets. There is a correlation between solidification morphology and the entropy of fusion $S_f$, non-faceted solidification being preferred for metallic phases having a low $S_f$ and faceted solidification occurring for compounds with high $S_f$. Hunt and Jackson proposed three general categories of eutectic solidification associated with various combinations of faceted (f) and non-faceted (nf) phases although there are important subdivisions within each group.

(a) *non-faceted/non-faceted* (nf/nf) Each phase grows with an atomically rough interface and follows the local direction of diffusive mass-transport across the solid/liquid interface. Either rod or lamellar configurations can be obtained depending on the relative volume fractions of the phases. Most theoretical work refers to this type of alloy where the microstructures in the solid are completely determined by diffusion in the liquid. This is illustrated in Fig. 3.1*a* and *b* which shows schematically the regular aligned phases associated with a planar solid/liquid interface and the misaligned phases due to local growth normal to a curved solidification front. This

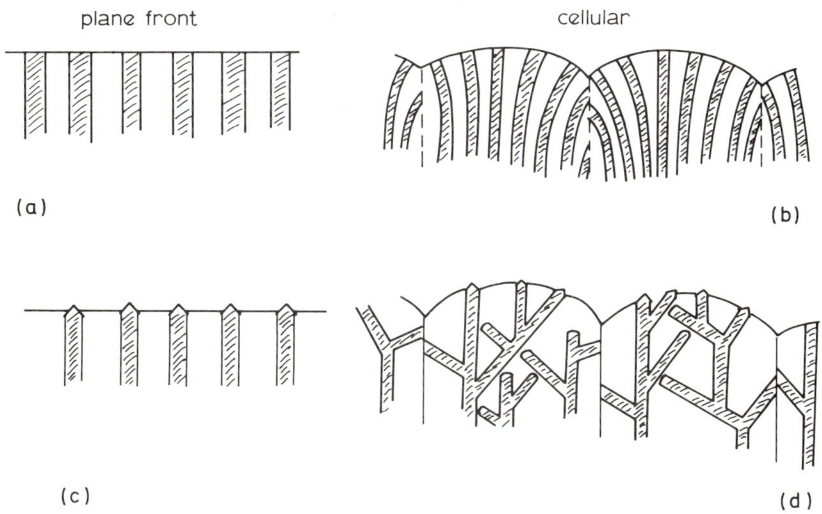


**Figure 3.1 Schematic illustrations of the phase alignments obtained by directionally solidifying different types of eutectics with planar and curved solidification fronts**
**(a) nf/nf, planar interface**
**(b) nf/nf, curved interface**
**(c) f/nf, planar interface**
**(d) f/nf, curved interface**

behaviour strictly applies to relatively few alloys that consist of two metallic phases (e.g. Ag–Cu, Cu–W); however, many more complex alloys exhibit the solidification morphologies associated with nf/nf systems.

(b) *faceted/non-faceted* (f/nf) Most *in situ* composites of potential industrial importance fall into this group. They generally consist of a relatively high volume fraction of an nf phase that constitutes the matrix of the composite (e.g. Ni, Co or Fe) and a smaller volume of an f phase (e.g. a carbide or intermetallic phase) in the form of either fibres or lamellae. At low rates of solidification where a planar solidification front can be maintained, the microstructure can be very similar to that of the nf/nf case as is illustrated in Fig. 3.1*c* and, indeed, the same theoretical description may remain valid. However, the conditions for maintaining an isothermal, flat solidification front are very stringent due to the different growth characteristics of the two phases and when the solid/liquid interface deviates from planarity, the pronounced growth anisotropy of the f phase can lead to a dramatic deterioration of the aligned structure (Fig. 3.1*d*).

Other differences between the morphologies of directionally solidified nf/nf and f/nf eutectics that are associated with the mechanisms of growth will be discussed in following sections.

(c) *faceted/faceted* (f/f) In this group the two faceted phases appear to grow independently and consequently are unlikely to produce the regular structure required for composite materials.

## 3.2 SOLIDIFICATION OF SIMPLE BINARY EUTECTICS

In the following discussion, for clarity, we shall first consider the solidification of a simple binary eutectic mixture of two types of atoms ($A$ and $B$) that solidify to form two phases $\alpha$ and $\beta$. The solidification behaviours of complex multicomponent eutectics that typify the materials of engineering interest will be described as modifications of this ideal response. Figure 3.2 shows a schematic eutectic phase diagram in which there is complete miscibility of $A$ and $B$ in the liquid-, but only partial miscibility in the solid state. The solid phases $\alpha$ and $\beta$ and the liquid only coexist in equilibrium at point E ($C_E$, $T_E$). The alloy of composition $C_E$ completely solidifies at the unique temperature $T_E$ to form an intimate mixture of $\alpha$ and $\beta$; all other alloys in the binary field exhibit a finite melting range.

When considering plane front solidification of a single-phase alloy in Chapter 2, limited diffusion in the liquid led to the development of a steady state concentration profile in the melt ahead of the solidifying interface and the solute distribution in the liquid could be described by eqn (2.10) which varies only in the direction of solidification. Eutectic solidification, however,

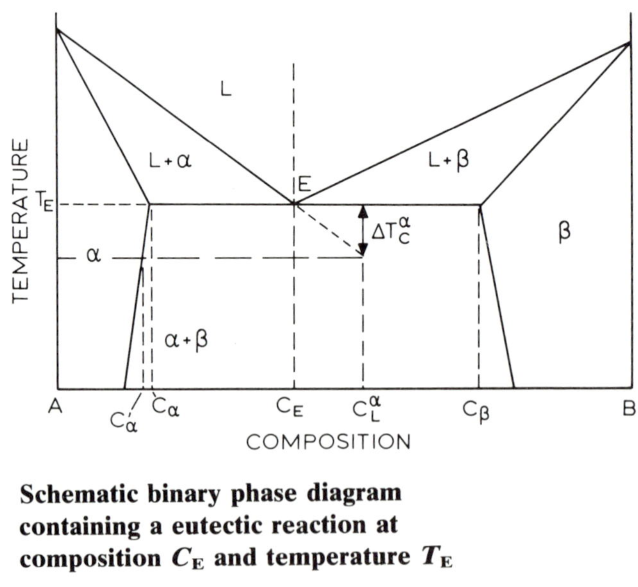


**Figure 3.2 Schematic binary phase diagram containing a eutectic reaction at composition $C_E$ and temperature $T_E$**

involves the partitioning of the constituent atoms ($A$ and $B$) between the product phases and this leads to an additional lateral diffusional flux in the melt. Consequently, diffusion in the melt must be described by the more general equation

$$D_{\mathrm{L}}\nabla^2 C_{\mathrm{L}} + V_I \frac{\mathrm{d}C_{\mathrm{L}}}{\mathrm{d}x} = 0 \tag{3.1}$$

which will take the two-dimensional form for lamellar eutectics and the three-dimensional for rod eutectics. We shall follow the arguments of Jackson and Hunt[6] for the lamellar case. Figure 3.3 illustrates schematically the formation of such a lamellar eutectic and defines the principal geometrical parameters.

As solidification proceeds, the $\alpha$ plates grow by the attachment of $A$ and $B$ atoms in the proportion required to form the solid of composition $C_\alpha$ and the excess $B$ atoms are rejected, accumulating in the melt ahead of the $\alpha$ lamellae. Similarly there will be an accumulation of $A$ atoms ahead of the $\beta$ lamellae. This means that there is a quasi-equilibrium between the solid $\alpha$ phase formed and the $B$-rich liquid, $C_{\mathrm{L}}^{\alpha}$, described by extrapolation of the $\alpha$ liquidus beyond the eutectic point as shown in Fig. 3.2. Thus the solid $\alpha$ phase formed will have a slightly different composition $C'_\alpha$ from the true equilibrium value and the melt ahead of the $\alpha$ lamellae will be undercooled by an amount $\Delta T_{\mathrm{C}}^{\alpha}$ due to this solute segregation

$$\Delta T_{\mathrm{C}}^{\alpha} = m_\alpha (C_{\mathrm{E}} - C_{\mathrm{L}}^{\alpha}) \tag{3.2}$$

where $m_\alpha$ is the slope of the $L/L+\alpha$ boundary. Similarly the constitutional undercooling ahead of $\beta$ lamellae is given by

$$\Delta T_{\mathrm{C}}^{\beta} = m_\beta (C_{\mathrm{E}} - C_{\mathrm{L}}^{\beta}) \tag{3.3}$$

We note that $\Delta T_{\mathrm{C}}^{\alpha}$ and $\Delta T_{\mathrm{C}}^{\beta}$ are not necessarily equal.

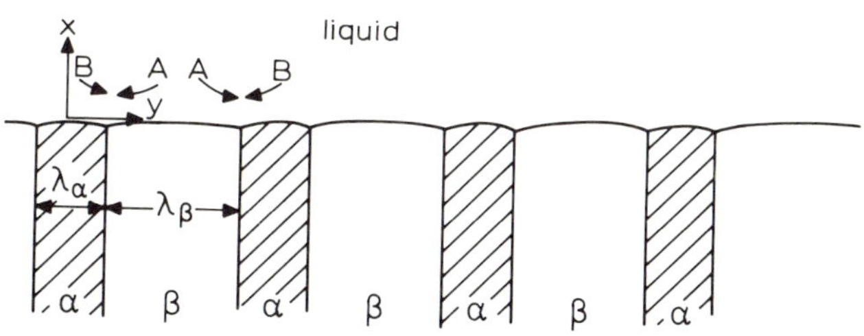


**Figure 3.3 Schematic illustration of a section through a planar solidification front for a lamellar nf/nf eutectic defining phase dimensions and coordinates used in the analysis. The lateral atomic diffusion is indicated**

There will not be a discontinuous change in solute concentration in the melt ahead of the boundary between the $\alpha$ and $\beta$ phases as shown in Fig. 3.4*a*; rather, lateral diffusion in the melt will lead to a gradation in the concentration of these solute-enriched layers with maximum and minimum values above the centres of the lamellae as illustrated in Fig. 3.4*b*. The associated undercooling of the melt in contact with the solid/liquid interface which is determined by the deviation of the melt concentration from $C_E$ will also vary locally as indicated in Fig. 3.4*c*.

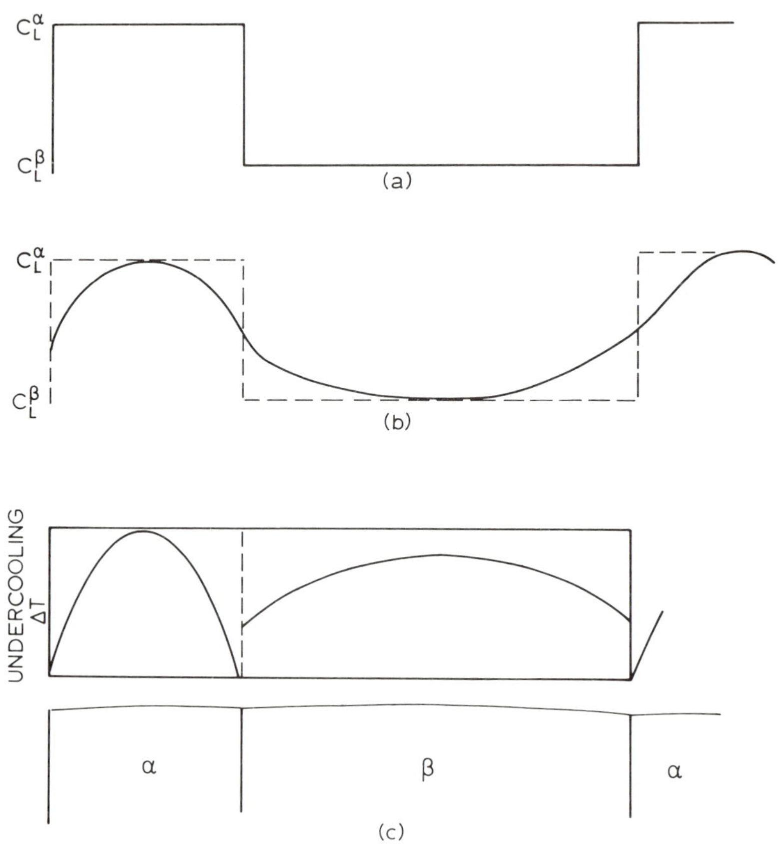


**Figure 3.4 Concentration in the melt at the solid/liquid interface ahead of $\alpha$ and $\beta$ lamallae: (a) assuming homogeneity ahead of each phase; (b) assuming equilibration at $\alpha\beta$ interface and lateral diffusion; (c) spatial variation in interface undercooling associated with (b)**

In plane front solidification the solid/liquid interface must be isothermal; if it were not, one of the phases would solidify ahead of the other and destroy the planarity. However, the lateral variation in constitutional undercooling would lead to local differences in the interface temperature. As we discussed for dendritic solidification, surface curvature can lead to an additional undercooling component $\Delta T_\gamma$. (Again, as for dendrites we shall assume that the solid/liquid interface is an efficient sink for impinging solute atoms so that there is no kinetic undercooling. This is certainly true for non-faceting phases. Kinetic undercooling will be larger for faceting phases, but even in this case is normally very much less than either $\Delta T_c$ or $\Delta T_\gamma$.) For lamellar geometry, Jackson and Hunt[6] show that

$$\Delta T^{i}_{\gamma} = \frac{\gamma_{iL} T_E}{L_f r \rho_{is}} \tag{3.4}$$

where $\gamma_{iL}$ is the local solid/liquid interfacial energy, $\rho_{is}$ is the density of solid formed, $i$ denotes $\alpha$ or $\beta$, $L_f$ is the latent heat of fusion of the alloy and $r$ is the local radius of curvature. The solidifying interface will adopt a shape with local curvatures that give a lateral variation in $\Delta T_\gamma$ which combines with $\Delta T_C$

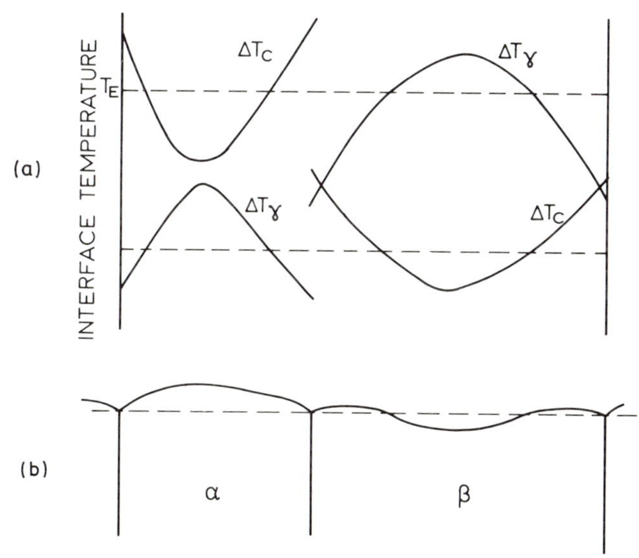


**Figure 3.5** **(a) Undercooling due to surface curvature $\Delta T_\gamma$ required to combine with the constitutional undercooling $\Delta T_c$ to give an isothermal solid/liquid interface. (b) Shape of solid/liquid interface to give $\Delta T_\gamma$ required in (a)**

to give a constant undercooling

$$\Delta T = \Delta T_{\gamma} + \Delta T_{C} = \text{constant} \tag{3.5}$$

This is shown schematically in Fig. 3.5 which indicates that negative curvatures of the lamellae are possible in certain circumstances particularly where there is a large difference in the volume fractions of the two phases.

The requirement that local thermodynamic equilibrium be maintained at the intersection of the solid $\alpha\beta$ interfaces with the solid/liquid surface introduces an additional constraint on the shape of the solidification front. If the solid/liquid interfacial energies $\gamma_{\alpha L}$ and $\gamma_{\beta L}$ are assumed to be isotropic, a simple triangle of forces relationship applies

$$\gamma_{\alpha\beta} = \gamma_{\alpha L} \cos \theta_1 + \gamma_{\beta L} \cos \theta_2 \tag{3.6}$$

where $\gamma_{\alpha\beta}$ is the energy of the solid $\alpha\beta$ interface and the dihedral angles $\theta_1$, $\theta_2$ are defined in Fig. 3.6. If the crystallographic anisotropy of $\gamma_{\alpha L}$ and $\gamma_{\beta L}$ are taken into account additional 'torque' terms are introduced into eqn (3.6). The magnitudes of these torque terms are not known and, to date, no analysis of solidification has been able to take them into account. Consequently in the following discussion we shall only consider the case of isotropic interfacial energies.

Jackson and Hunt[6] have used the concepts outlined above to provide an approximate analytical description of lamellar eutectic solidification. The configuration of the phases in the model is that shown schematically in Fig. 3.3. The first step is to solve the two-dimensional form of eqn (3.1) subject to some appropriate boundary conditions to determine the solute concentration in the melt $C_L$ as a function of position. Jackson and Hunt[6] assumed the following physically realistic conditions:

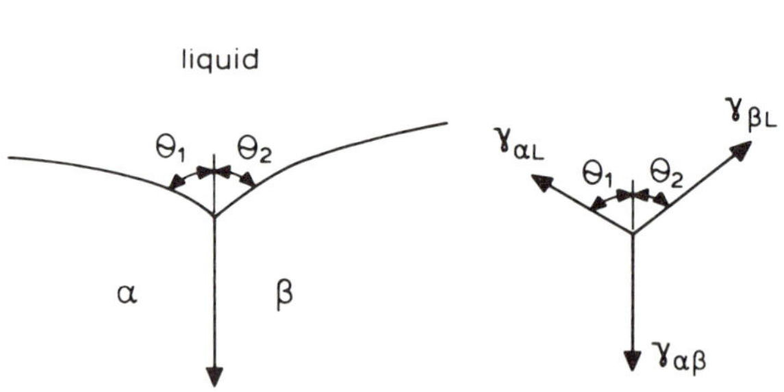


**Figure 3.6 Definition of the equilibrium configuration at the intersection of the solid $\alpha\beta$ interface with the solidification front assuming isotropic interfacial energies**

(i) $C_L \to C_E$ as $x \to \infty$ (the melt remains of eutectic composition far from the solid/liquid interface).

(ii) $dC_L/dy = 0$ above the centres of the lamellae (i.e. at $y = 0$ and $y = \frac{1}{2}(\lambda_\alpha + \lambda_\beta)$). This is a direct consequence of symmetry and is consistent with the argument leading to the construction of Fig. 3.4*b*.

(iii) The solid/liquid interface is completely planar. This is clearly untrue since we have shown that surface curvature is required to give an isothermal interface and that interface equilibrium leads to a groove at the edges of the lamellae. However, the assumption is required to simplify the mathematics. The effect of the shape of the interface is introduced later as a perturbation. Moreover, it turns out that the dimensional variations in interface shape are small compared with the scale of solute variation in the liquid, so that the assumption introduces little error.

(iv) At steady state the diffusive flux of $A$ and $B$ atoms through the liquid onto the interface leads to the formation of $\alpha$ and $\beta$ at the same rate $V_I$. It is further assumed that the impinging solute atoms are immediately incorporated into one of the solid phases. This is equivalent to assuming that both phases are non-faceting, i.e.

$$\left(\frac{dC_L}{dx}\right)_{x=0} = -\frac{V_I C_\alpha}{D_L} \quad \text{for} \quad -\tfrac{1}{2}\lambda_\alpha \leq y \leq \tfrac{1}{2}\lambda_\alpha \tag{3.7}$$

$$\left(\frac{dC_L}{dx}\right)_{x=0} = -\frac{V_I C_\beta}{D_L} \quad \text{for} \quad \tfrac{1}{2}\lambda_\alpha \leq y \leq \tfrac{1}{2}\lambda_\alpha + \lambda_\beta \tag{3.8}$$

Subject to conditions (i) to (iv), the solution to eqn (3.1) is found by Jackson and Hunt to be

$$C_L = C_E + \sum_{n=1}^{\infty} B_n \cos\left(\frac{2n\pi}{\lambda_\alpha + \lambda_\beta} y\right) \exp\left(\frac{-2n\pi}{\lambda_\alpha + \lambda_\beta} x\right) \tag{3.9}$$

where

$$B_n = \frac{(\lambda_\alpha + \lambda_\beta)}{(n\pi)^2} \cdot \frac{V_I}{D_L}(C_\beta - C_\alpha) \sin\left(\frac{n\pi\lambda_\alpha}{\lambda_\alpha + \lambda_\beta}\right) \tag{3.10}$$

By integrating eqn (3.9) with respect to $y$ separately over the widths of the $\alpha$ and $\beta$ lamellae, relatively simple expressions are obtained for the *average* solute concentrations in the melt in contact with the $\alpha$ and $\beta$ phases at the solid/liquid interface. The average solute concentration in contact with the $\alpha$ lamellae $\overline{C_L^\alpha}$ is given by the expression

$$\overline{C_L^\alpha} = \int_{-\lambda_\alpha/2}^{\lambda_\alpha/2} C_L \, dy \Big/ \int_{-\lambda_\alpha/2}^{\lambda_\alpha/2} dy \quad \text{evaluated at } x = 0$$

$$= C_E + \frac{1}{2} \frac{(\lambda_\alpha + \lambda_\beta)^2}{\lambda_\alpha} \frac{V_I}{D_L}(C_\beta - C_\alpha) P \tag{3.12}$$

where

$$P=\sum_{n=1}^{\infty}\left(\frac{1}{n\pi}\right)^{3}\sin^{2}\left(\frac{n\pi\lambda_{\alpha}}{\lambda_{\alpha}+\lambda_{\beta}}\right) \tag{3.13}$$

Similarly the average solute concentration in the liquid in contact with $\beta$, $\overline{C_{\mathrm{L}}^{\beta}}$ is

$$\overline{C_{\mathrm{L}}^{\beta}}=C_{\mathrm{E}}+\frac{1}{2}\frac{(\lambda_{\alpha}+\lambda_{\beta})^{2}}{\lambda_{\beta}}\frac{V_{\mathrm{I}}}{D_{\mathrm{L}}}(C_{\beta}-C_{\alpha})P \tag{3.14}$$

We can use eqns (3.12) and (3.14) to express the undercooling in terms of $V_{\mathrm{I}}$ and the lamellar dimensions. From eqns (3.2)–(3.5) the average undercooling at the $\alpha$ lamellae $\Delta T_{\alpha}$ is given by

$$\Delta T_{\alpha}=m_{\alpha}(C_{\mathrm{E}}+\overline{C_{\mathrm{L}}^{\alpha}})+\frac{2\gamma_{\alpha\mathrm{L}}T_{\mathrm{E}}\cos\theta_{\alpha}}{L_{\mathrm{f}}\lambda_{\alpha}} \tag{3.15}$$

where we have taken the radius of curvature $r$ of the $\alpha$ portion of the solid/liquid interface to be related to the dihedral angles $\theta_{\alpha}$:

$$\frac{1}{r}=\frac{2\cos\theta_{\alpha}}{\lambda_{\alpha}} \tag{3.16}$$

Then

$$\Delta T_{\alpha}\doteq\tfrac{1}{2}m_{\alpha}\frac{(\lambda_{\alpha}+\lambda_{\beta})^{2}}{\lambda_{\alpha}}\frac{V_{\mathrm{I}}}{D_{\mathrm{L}}}(C_{\beta}-C_{\alpha})P+\frac{2\gamma_{\alpha\mathrm{L}}T_{\mathrm{E}}\cos\theta_{\alpha}}{L_{\mathrm{f}}\lambda_{\alpha}} \tag{3.17}$$

Similarly

$$\Delta T_{\beta}=\tfrac{1}{2}m_{\beta}\frac{(\lambda_{\alpha}+\lambda_{\beta})^{2}}{\lambda_{\beta}}\frac{V_{\mathrm{I}}}{\mathrm{D}_{\mathrm{L}}}(C_{\beta}-C_{\alpha})P+\frac{2\gamma_{\beta\mathrm{L}}T_{\mathrm{E}}\cos\theta_{\beta}}{L_{\mathrm{f}}\lambda_{\alpha}} \tag{3.18}$$

Since the solid/liquid interface is isothermal, we can set $\Delta T_{\alpha}=\Delta T_{\beta}=\Delta T$ say, and manipulation of eqns (3.17) and (3.18) leads to the expression

$$\frac{\Delta T}{m}=V_{\mathrm{I}}A\bar{\lambda}+B/\bar{\lambda} \tag{3.19}$$

where

$$\frac{1}{m}=\frac{1}{m_{\alpha}}+\frac{1}{m_{\beta}},\qquad \bar{\lambda}=\tfrac{1}{2}(\lambda_{\alpha}+\lambda_{\beta})$$

$$A=\frac{P(C_{\beta}-C_{\alpha})}{D_{L}}\left(1+\frac{\lambda_{\beta}}{\lambda_{\alpha}}\right)^{2}\Big/\frac{\lambda_{\beta}}{\lambda_{\alpha}}$$

and

$$B=2\left(1+\frac{\lambda_\alpha}{\lambda_\beta}\right)\left[\frac{2T_E}{L_F}\left(\frac{\gamma_{\alpha L}\cos\theta_\alpha}{m_\alpha}+\frac{\gamma_{\beta L}\cos\theta_\beta}{m_\beta}\right)\right]$$

The terms $m$, $A$ and $B$ will be constants for a eutectic alloy, so eqn (3.19) expresses the relationship between $\Delta T$, $V_I$ and $\bar{\lambda}$.

Although eqn (3.19) allows a range of undercoolings for various combinations of $V_I$ and $\bar{\lambda}$, a great deal of experimental evidence points to the existence of a unique combination of $V_I$ and $\bar{\lambda}$. Clearly there must be an additional constraint on eqns (3.17) and (3.18) to do so. It has been postulated by Zener,[7] Tiller,[8] Hillert[9] and Jackson and Hunt[6] that the eutectic should solidify with the microstructure that minimizes $\Delta T$, i.e.

$$\frac{d}{d\bar{\lambda}}(\Delta T)=0 \tag{3.20}$$

This condition, which was also invoked in Chapter 2 to describe dendritic solidification, leads to a series of relationships by differentiating eqn (3.19).

$$\bar{\lambda}^2 V_I=\frac{B}{A} \tag{3.21}$$

$$\frac{\Delta T^2}{V_I}=4m^2AB \tag{3.22}$$

$$\bar{\lambda}\Delta T=2mB \tag{3.23}$$

There is no clear justification for the application of eqn (3.20), which is known as the extremum condition, except that it appears to work. It is thought to be equivalent to the assumption that the rate of entropy production always adopts a minimum value. However, that condition has been shown to be invalid when a system is far from equilibrium[10] which seems to be the case during solidification.

More recently Nash[11] has solved the relevant equations describing diffusion, heat conduction and solid/liquid interface motion numerically in a self-consistent manner and, in so doing, has avoided (i) the need to postulate minimum undercooling, and (ii) the contrary assumptions of a planar interface to solve the diffusion equations and a curved interface during heat flow. His results largely confirm the earlier predictions of Tiller[8] and of Jackson and Hunt[6] for the situation where the solidification front is almost planar, although there are significant differences when the solid/liquid interface has a large curvature. Figure 3.7 shows the solute concentrations in the melt calculated by Nash for surfaces with different curvatures and these are compared with results of the Jackson–Hunt analysis (i.e. for a flat interface). The two approaches give virtually identical results in the limit of small surface curvature

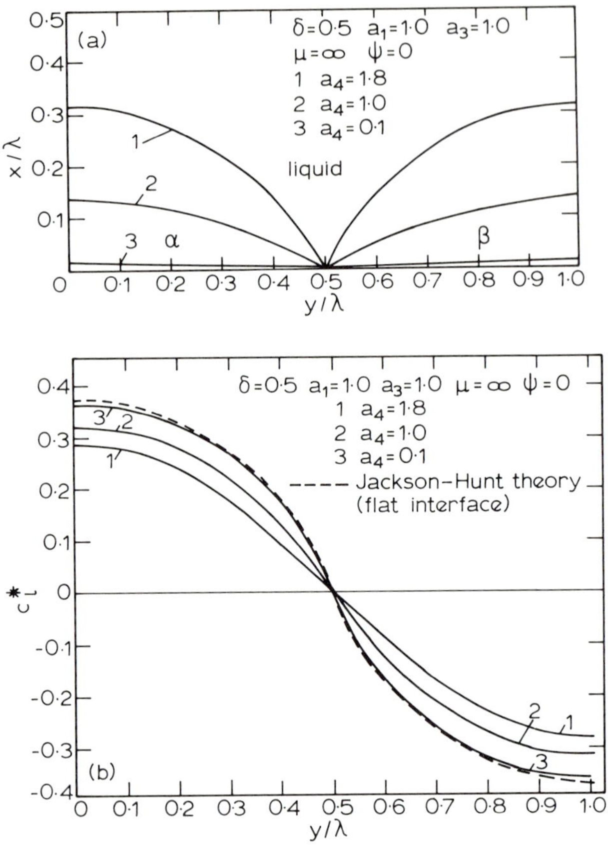


**Figure 3.7 (a) Interface shapes and (b) associated solute concentrations calculated by Nash.[11] The Jackson–Hunt result for a planar interface is shown in (b)**

and the deviations in solute profiles that occur for highly curved interfaces have little influence on the lamellar spacings.

Jackson and Hunt[6] have carried out a similar analysis for eutectics in which the minor phase adopts a rod or fibrous rather than a lamellar morphology. The general forms of the relationships between $V_I$, fibre radius $r_\alpha$ and undercooling $\Delta T$ closely parallel those derived for the lamellar geometry; the major difference is the need for a series of Bessel, rather than cosine, functions $J$ to represent the solution to the appropriate diffusion equation:

$$C_L = C_E + \sum_{n=1}^{\infty} A_n J_0\left(\frac{\gamma_n r}{r_\alpha + r_\beta}\right) \exp\left(-\frac{\gamma_n x}{r_\alpha + r_\beta}\right) \tag{3.24}$$

where

$$A_n = \frac{2V_I r_\alpha}{D_L}(C_\beta - C_\alpha) J_1\left(\frac{\gamma_n r_\alpha}{r_\alpha + r_\beta}\right)[\gamma_n J_0(\gamma_n)]^{-2} \tag{3.25}$$

Here $\gamma_n$ = root of the equation $J_1(x) = 0 \sim n\pi$; while $r_\alpha + r_\beta$ = radius of cylinder of $\beta$ that is associated with the rod of radius $r_\alpha$. The average compositions in front of the two phases are

$$\bar{C}_\alpha = C_E + \frac{4V_I}{D_L}(r_\alpha + r_\beta)(C_\beta - C_\alpha)M \tag{3.26}$$

$$\bar{C}_\beta = C_E - \frac{4V_I}{D_L}\frac{(r_\alpha + r_\beta)}{(r_\alpha + r_\beta)^2 - r_\alpha^2}(C_\beta - C_\alpha)M \tag{3.27}$$

where

$$M = \sum_{n=1}^{\infty} \gamma_n^{-3} \frac{J_1^2[(r_\alpha \gamma_n/(r_\alpha + r_\beta)]}{J_0^2(\gamma_n)} \tag{3.28}$$

Substitution of eqns (3.26)–(3.28) into (3.2)–(3.5) leads to the following expressions for average undercoolings in the melt adjacent to the $\alpha$ fibres and $\beta$ matrix:

$$\Delta T_\alpha = \frac{4m_\alpha V_I}{D_L}(C_\beta - C_\alpha)(r_\alpha + r_\beta)M + \frac{2\gamma_{\alpha L} T_E \cos\theta_\alpha}{L_f r_\alpha} \tag{3.29}$$

$$\Delta T_\beta = \frac{4m_\beta V_I}{D_L}(C_\beta - C_\alpha)\frac{(r_\alpha + r_\beta)}{(r_\alpha + r_\beta)^2 - r_\alpha^2}M + \frac{2\gamma_{\beta L} T_F \cos\theta_\beta}{L_f\{[(r_\alpha + r_\beta)^2 - r_\alpha^2]/r_\alpha\}} \tag{3.30}$$

Again, since the interface is isothermal $\Delta T_\alpha = \Delta T_\beta = \Delta T$. By equating (3.29) and (3.30) and algebraic manipulation it can readily be shown that

$$\frac{\Delta T}{m} = V_I A'(r_\alpha + r_\beta) + \frac{B'}{(r_\alpha + r_\beta)} \tag{3.31}$$

where

$$\frac{1}{m} = \frac{1}{m_\alpha} + \frac{1}{m_\beta}$$

$$B' = \frac{2T_E\sqrt{1+\xi}}{L_f}\left(\frac{\gamma_{\alpha L}\cos\theta_\alpha}{m_\alpha} + \frac{\gamma_{\beta L}\cos\theta_\beta}{m_\beta}\right) \tag{3.32}$$

$$\xi = \frac{V_f}{1 - V_f}; \qquad V_f = \text{volume fraction of fibres}$$

$$A' = \frac{4(1+\xi)}{D_L \xi}(C_\beta - C_\alpha)M \tag{3.33}$$

Extremum growth, defined by

$$\frac{\partial}{\partial(r_\alpha + r_\beta)}(\Delta T) = 0$$

occurs when

$$V_I(r_\alpha + r_\beta)^2 = \frac{B'}{A'} \tag{3.34}$$

or

$$\frac{\Delta T^2}{V_I} = 4m^2 A'B' \tag{3.35}$$

Equations (3.21) and (3.34) and eqns (3.22) and (3.35) are clearly of exactly the same form, indicating that the interrelationships between solidification velocity $V_I$, undercooling $\Delta T$ and microstructural dimensions $\bar{\lambda}$ (or $r_\alpha + r_\beta$) are common to both lamellar and fibrous eutectics and are completely defined, albeit by very complicated expressions, using a combination of the relevant equilibrium thermodynamic constants, kinetic parameters and geometrical factors.

The temperature gradient $G_L$, which was an important parameter in describing dendritic solidification, has not been invoked in this treatment of eutectic solidification and clearly this implies that both $\bar{\lambda}$ (or $r_\alpha + r_\beta$) and $\Delta T$ are completely defined by $V_I$ and are independent of $G_L$. When we extend our discussion to eutectic reactions in multicomponent alloys and to the solidification of off-eutectic compositions, $G_L$ will be reintroduced as an important variable.

There have been many attempts to verify the predictions of the Jackson–Hunt and related theories of eutectic solidification. Unfortunately, many of the important parameters in these theories have not been measured reliably. For example, the solid/liquid interfacial energies are often barely known to within an order of magnitude. It is not possible therefore to make an absolute comparison of theoretical and experimental findings. Rather, the trends in $\bar{\lambda}$ and $\Delta T$ caused by varying $V_I$ have been examined. Cline and Livingston[12] and Livingston *et al.*[13] have examined the microstructures of several binary eutectic alloys, both lamellar and fibrous, that had been directionally solidified over the very wide range of solidification rates from $\sim 8 \times 10^{-4}$ to 4 mm s$^{-1}$ (i.e. $\sim 2{\cdot}5$ to $\sim 1{\cdot}4 \times 10^4$ mm h$^{-1}$). In each case the results were consistent with a $\bar{\lambda} \propto V_I^{-1/2}$ relationship in accord with theory. Figure 3.8*a* shows the

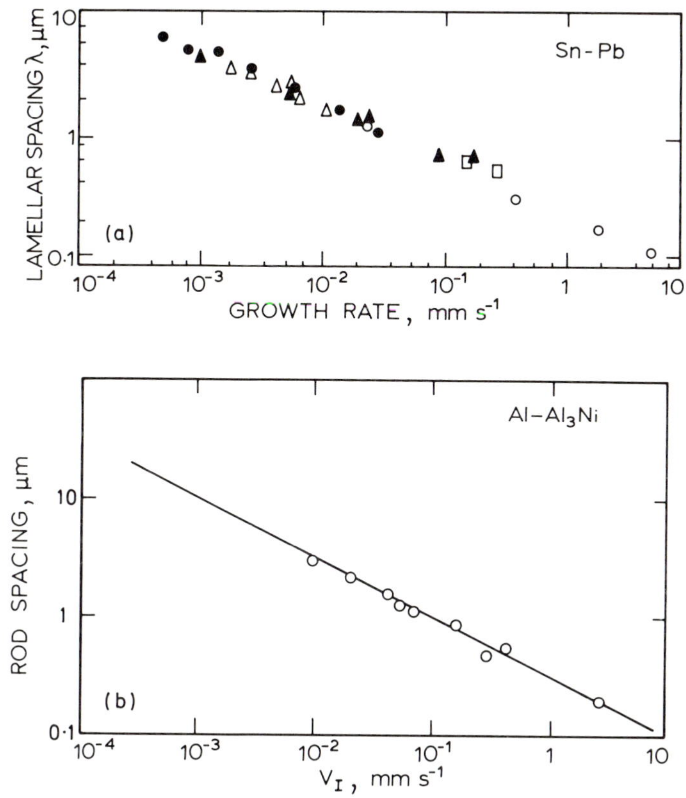


**Figure 3.8 Phase spacings as functions of solidification rate: (a) lamellar Sn–Pb eutectic,[12] (b) fibrous Al–$Al_3Ni$ eutectic[13]**

results of Cline and Livingston[12] for the Pb–Sn lamellar eutectic plotted as $\log \bar{\lambda}$ versus $\log V_I$; the slope of $-\frac{1}{2}$ is in complete agreement with eqn (3.21). A similar display of data for the fibrous Al–$Al_3Ni$ eutectic alloy shown in Fig. 3.8*b* is also consistent with the theoretical expectation. Courtney and Garmong,[14] however, have noted a deviation from the predicted behaviour for the Al–$Al_3Ni$ eutectic in the form of a variation in microstructural scale with temperature gradient $G_L$. We shall discuss possible reasons for this observation in § 3.6.

It is not possible to directionally solidify the multicomponent alloys of potential engineering importance over the wide range of solidification rates that have been applied to simple binary eutectics. Most available data for the more complex materials appear to be consistent with a $\bar{\lambda}^2 V_I = \text{constant}$ relationship; a typical set of results for the $\gamma$–$\gamma'$–$\delta$ alloy is shown in Fig. 3.9.

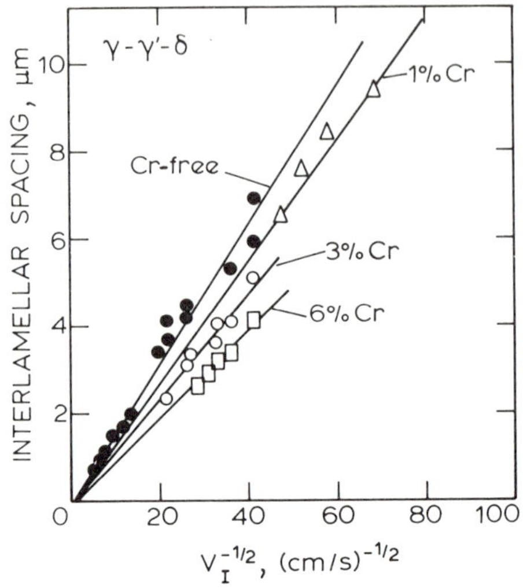


**Figure 3.9 Lamellar spacings of the modifications of the $\gamma$–$\gamma'$–$\delta$ eutectic as a function of solidification rate [Ref. 15]**

However, at least in some alloys the microstructural scale has been shown to be dependent on temperature gradient $G_L$ as well as solidification rate; an example of this type of behaviour for the $\gamma$–$\gamma'$–$Cr_3C_2$ eutectic alloy is shown in Fig. 3.31. This will be discussed further in § 3.6.

Hunt and co-workers[16–18] have measured undercoolings during eutectic solidification for several alloys. The particularly extensive set of results for the Al–$Al_2Cu$ alloy shown in Fig. 3.10 from Tassa and Hunt[18] confirm the $\Delta T/V_I^{1/2}$ = constant relationship expected from eqns (3.22) and (3.35).

## 3.3 CONDITION FOR LAMELLAR OR FIBROUS EUTECTIC GROWTH

The discussion in § 3.2 presupposed the existence of either lamellar or fibrous eutectic microstructures. It should be possible to establish a criterion for the occurrence of either type of growth form. Figure 3.11*a* shows schematically the surface defining $\Delta T$ as a function of $\bar{\lambda}$ and $V_I$ for lamellar and fibrous geometries. A section through such a schematic representation for an arbitrary $V_I$ is shown in Fig. 3.11*b*. A natural extension of the extremum growth criterion is to assume that the growth morphology for which the minimum $\Delta T$ is *least* will occur (i.e. lamellar in Fig. 3.11*b*). However, there is no guarantee that the same growth morphology will occur at all growth rates. Comparison of eqns (3.22) and (3.35) can provide a criterion for the occurrence of rods or lamellae. When rod undercooling is less than lamellar undercooling

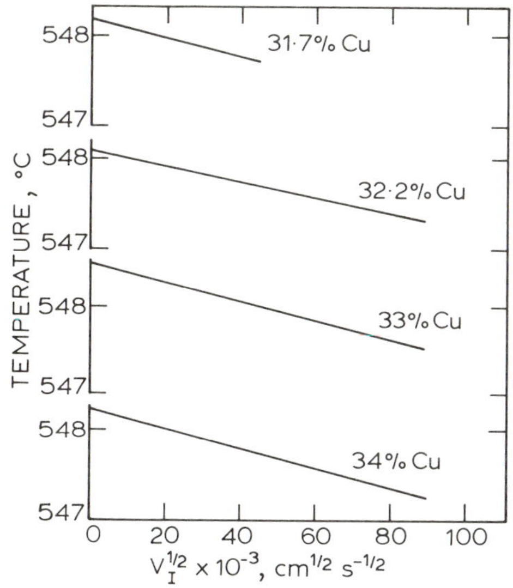


**Figure 3.10 Experimental plot of temperature at the solidifying interface against $V_I^{1/2}$ for different alloy compositions of Al–CuAl$_2$ [Ref. 18]**

then rods rather than lamellae should grow and this occurs when

$$4m^2A'B' < 4m^2AB$$

Algebraic manipulation leads to the condition for rods:

$$\frac{4M}{P}(1+\xi)^{-3/2} < 1 \tag{3.36}$$

This assumes that the interfacial and thermodynamic parameters are identical for lamellar and rod geometries. The composite term on the left of Inequality (3.36) is plotted in Fig. 3.12*a* as a function of the volume fraction of the minor phase; this shows that rods are preferred when the volume fraction of the minor phase is less than $1/\pi$ ($=0{\cdot}318$), otherwise lamellae should form.

Inspection of Fig. 3.12*a* shows that the curve is almost horizontal when the ordinate exceeds 1. Consequently a small change in $(4M/P)(1+\xi)^{-3/2}$ could make a big difference to the volume fraction at which lamellar solidification is preferred. The interfacial energies associated with lamellae, which can adopt orientations where mismatch between the phases is low, can often be significantly lower than for rods where a range of orientations are likely to occur. This may be taken into account by expressing the condition

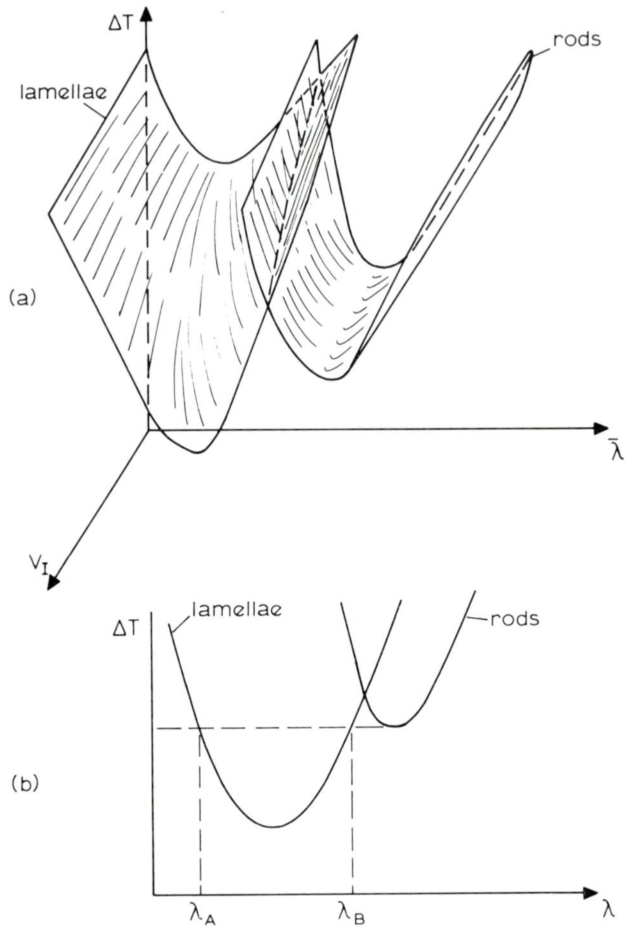


**Figure 3.11 (a) Schematic variations in undercooling of the solid/liquid interfaces of lamellar and rod eutectics with $V_I$ and $\lambda$. (b) Section of (a) for constant $V_I$**

for the formation of rods as

$$\frac{4M}{P}(1+\xi)^{-3/2} \leqslant \frac{(\gamma_{\alpha\beta})_{\text{lamellae}}}{(\gamma_{\alpha\beta})_{\text{rods}}} \tag{3.37}$$

Clearly a fairly small difference between $(\gamma_{\alpha\beta})_{\text{lamellae}}$ and $(\gamma_{\alpha\beta})_{\text{rods}}$ can lead to the formation of lamellar microstructures at quite low volume fractions of the minor phase.

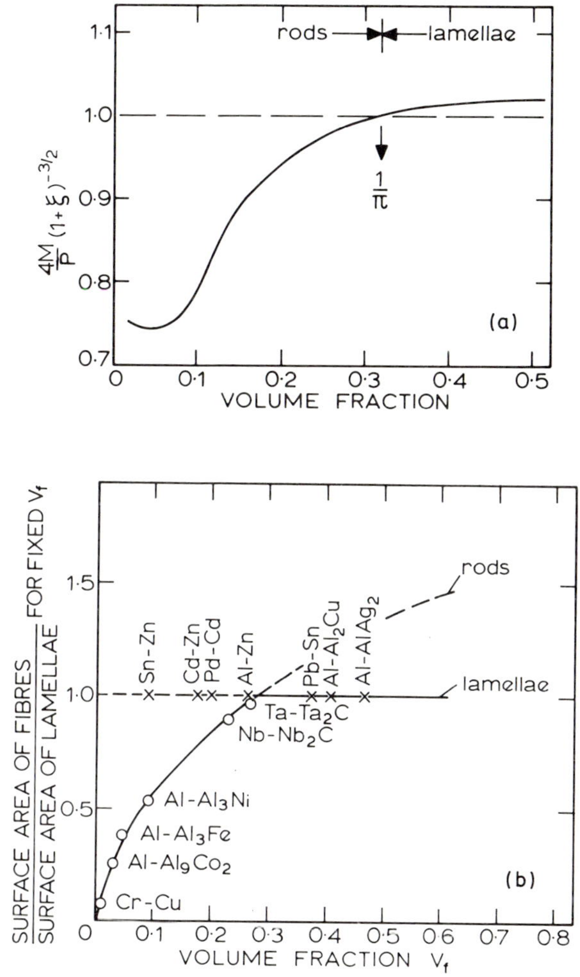


**Figure 3.12 (a) Plot of $(4M/P)(1+\xi)^{-3/2}$ as a function of volume fraction of the minor phase $\xi$; (b) relative surface areas of fibre and lamellar arrangements [Ref. 6]**

Although the predicted correlation between phase morphology and volume fraction is generally observed there are numerous exceptions. Where small volume fractions adopt a lamellar shape (e.g. Sn–Zn) there is usually a unique misorientation and growth texture of the phases which tends to lead to anisotropy of solid/solid interfacial energy; if only one low energy habit plane exists then lamellae are preferred whereas if several low energy planes are possible then faceted rods will form. In a few cases rods of a high volume fraction can occur (e.g. Al–$AlAg_2$); this is normally associated with large variations in volume fraction with temperature where the rods satisfy the volume fraction condition at the eutectic temperature but subsequently grow on cooling. Figure 3.12*b* shows a compilation of eutectic morphologies by Hellawell[39] displayed as the relative interfacial areas of lamellar and fibrous arrangements for various volume fractions.

There have been several observations of transitions between rod and lamellar morphologies as the solidification conditions are changed, but the direction of the transition appears to vary from system to system. Lamellar to rod transitions have been caused in the Al–$Al_3Ni$ and Al–$Al_2Cu$ eutectics by decreasing the solidification rate or by increasing the temperature gradient. Similar effects have been reported through the addition of small concentrations of impurities or by forcing the alloy to grow out of its preferred solidification direction.

## 3.4 MECHANISMS CONTROLLING PHASE SPACINGS

The theory of lamellar and fibrous eutectic growth outlined above assumes that extremum growth is achieved without enquiring as to how it might be brought about. In order to achieve the extremum growth condition there must be some mechanism whereby the lamellar (or fibre) spacings can readily adjust to satisfy eqn (3.21) or (3.34) for a particular growth rate. If the lamellae or fibres were perfectly aligned and fault-free the only way of modifying the phase spacings would be to either nucleate or destroy a certain proportion of the growing phases in the manner shown in Fig. 3.13 and there is a considerable energy barrier in doing so. It is generally agreed that adjustment of phase dimensions requires the presence of growth faults although the precise mechanism remains a subject of debate.

Jackson and Hunt[6] have proposed that the migration of lamellar terminations, illustrated schematically in Fig. 3.14, controls the change in spacing of

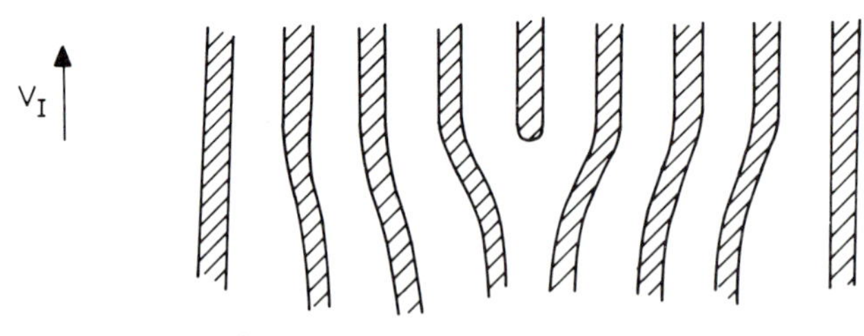


**Figure 3.13 Schematic illustration of the change in lamellar spacing by the homogeneous nucleation (or destruction) of a lamella**

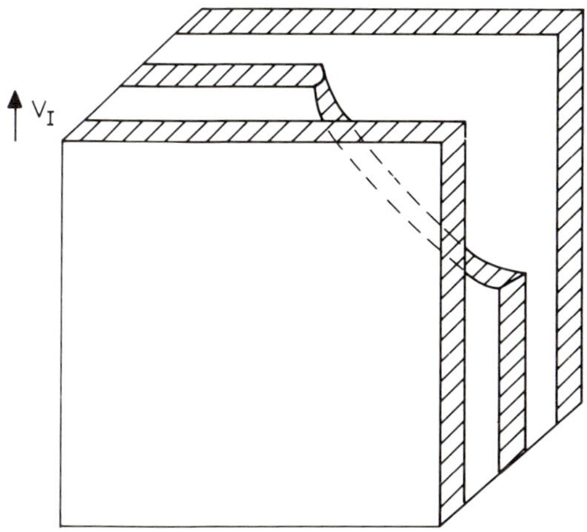


**Figure 3.14 Illustration of a lamellar fault; the lamellar spacing can be changed by the migration of the fault**

the eutectic microstructure; the spacing decreases if the termination moves to the left and vice versa. Consequently the condition for stable lamellar growth occurs when the fault neither advances nor recedes. Jackson and Hunt argue that since the lamellar termination can be regarded as part of a rod (which can adopt any effective radius) attached to a lamella then it probably adopts the minimum *rod* undercooling and consequently this rather than the minimum *lamellar* undercooling defines the condition for stable lamellar growth in such a faulted structure. This clearly leads to an ambiguity since examination of Fig. 3.11*b* shows the possibility of two lamellar spacings $\lambda_A$ and $\lambda_B$ corresponding to the minimum rod undercooling $\Delta T^R$. However, the smaller spacing $\lambda_A$ is inherently unstable. Figure 3.15 shows schematically the sequence of events following small reductions in spacing involving structures with spacings $\lambda_A$ and $\lambda_B$. In Fig. 3.15*a* a reduction in spacing below $\lambda_A$ leads to an *increased* undercooling component associated with the development of a concave growth front; since in nf/nf eutectics the phases grow normal to the solid/liquid interface this tends to lead to a further reduction in the interlamellar spacing and eventual elimination of such a structure. On the other hand, reducing the spacing of a structure with lamellar dimensions $\lambda_B$, shown in Fig. 3.15*b*, reduces the undercooling and this is compensated by a *convex* growth front which leads to an *increase* in lamellar dimensions back to the original value $\lambda_B$. Thus the $\lambda_B$ structure is stable to small reductions in spacing while the $\lambda_A$ structure is unstable. By similar arguments it can be

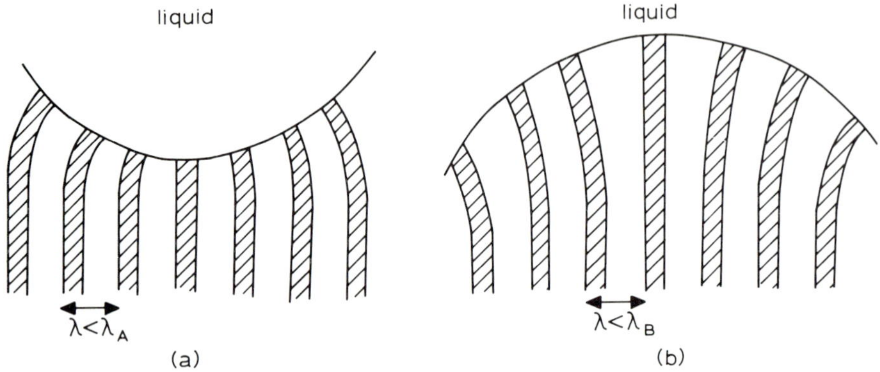


**Figure 3.15 Sequence of events following a reduction in lamellar spacing from (a) $\lambda_A$, a spacing smaller than the extremum, and (b) $\lambda_B$, a spacing greater than the extremum**

demonstrated that the $\lambda_A$ and $\lambda_B$ spacings are also respectively unstable and stable with respect to small increases in spacings.

Double[19] has advocated an alternative mechanism for maintaining extremum growth based on the motion of mismatch surfaces in the eutectic structure that are akin to subgrain boundaries in single phases. The mean lamellar spacing can readily be increased or decreased by motion of the mismatch surface in the direction where the local lamellar density is low or high respectively. Figure 3.16*b* shows an initially straight mismatch interface separating exactly the same number of lamellae on each side. The average lamellar spacing can decrease or increase by local migration of the interface into regions with higher (or lower) than average lamellar spacings. This causes the mismatch surface to become S-shaped as is illustrated schematically in Figs. 3.16*c* and *d* and eventually orient parallel to the alignment to create (or destroy) a new lamella (Fig. 3.15*e*). By performing careful metallography on specimens of the Al–$CuAl_2$ eutectic directionally solidified with progressively increasing growth rates, Double was able to demonstrate the existence of such S-shaped mismatch surfaces. Moreover, he argued that the necessary changes in interlamellar spacing with increasing growth rate could not be achieved by the few isolated lamellar imperfections that were observed, but that there were sufficient mismatch surfaces to do so.

There have been few systematic experimental studies into the mechanisms controlling phase dimensions. Undoubtedly both types of imperfection occur to differing extents in both simple binary and more complex eutectic alloys. Figure 3.17*a* and *b* shows a profusion of mismatch surfaces in Al–$Al_2Cu$ and

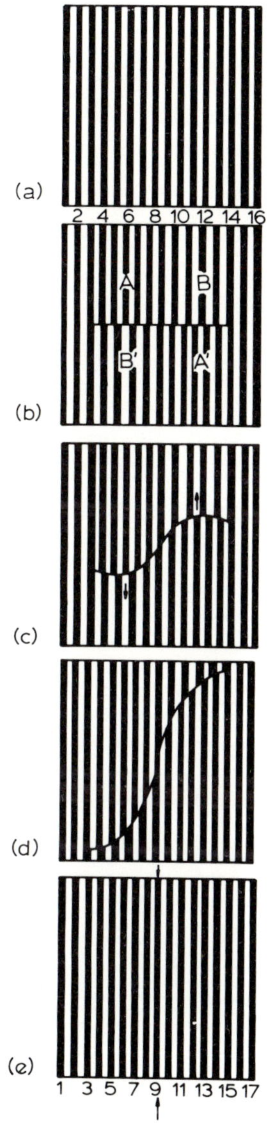


**Figure 3.16 Schematic illustration of the change in lamellar spacing by rotation of a mismatch surface [Ref. 19]**

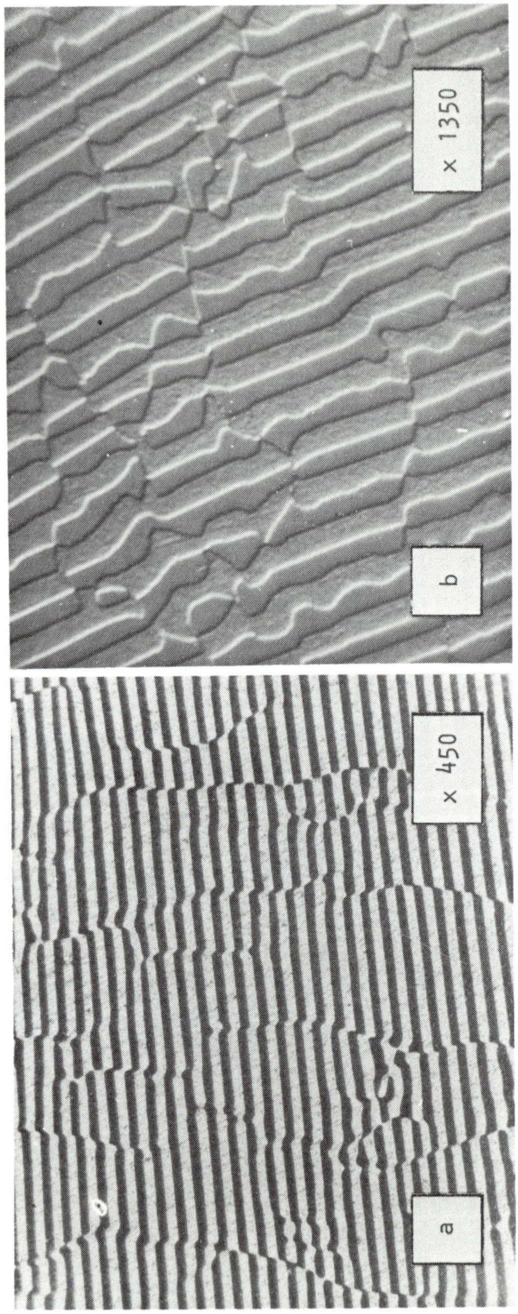


**Figure 3.17 Examples of mismatch surfaces in (a) Al–$Al_2Cu$ and (b) $\gamma$–$\gamma'$–$\delta$**

lamellar faults in a transverse section of the $\gamma$–$\gamma'$–$\delta$ alloy. Although a detailed model has not yet been proposed, faults also play an important role in stabilizing rod eutectic growth. Some fibre eutectics such as Al–$Al_3Ni$ and $\gamma$–$\gamma'$–$Cr_3C_2$ develop with a high density of fibre faults of the type shown in Fig. 3.18, while others such as the Ni–TaC and related alloys seldom if ever exhibit faults. We shall return to this topic when discussing the solidification of eutectics that involve a faceted phase.

## 3.5 OFF-EUTECTIC AND MULTICOMPONENT ALLOYS

When an alloy deviates from eutectic composition it may still be possible to maintain an aligned 'eutectic-like' microstructure if directional solidification is carried out in a sufficiently steep gradient. Two rival explanations for this effect have been proposed, one extending the concept of constitutional supercooling discussed in § 2.3 and the other based on competitive growth of dendritic and eutectic structures; but neither can yet be discounted.

The Jackson–Hunt[6] analysis of eutectic solidification is sufficiently general to account for eutectic-like solidification of a melt of composition $C_0$ that deviates from the eutectic composition $C_E$. (The various compositions are defined in the phase diagram shown in Fig. 3.19.) Assuming that the liquidus boundaries are straight lines with gradients $m_\alpha$ and $m_\beta$, the Jackson–Hunt analysis for lamellar growth shows that, for plane front solidification, the solute concentration in the melt is given by the expression

$$C_L = C_0 + (C_E - C_0)\exp\left(-\frac{V_I x}{D_L}\right) + \sum_{n=1}^{\infty} B_n \cos\left(\frac{2n\pi y}{\lambda_\alpha + \lambda_\beta}\right)\exp\left(-\frac{2n\pi x}{\lambda_\alpha + \lambda_\beta}\right) \tag{3.38}$$

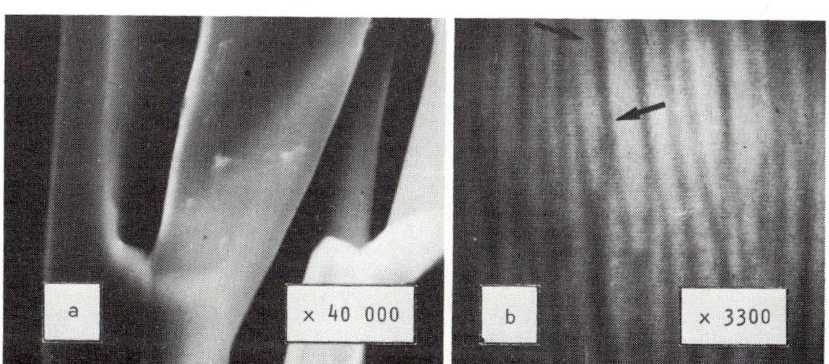


**Figure 3.18 Growth faults in fibre eutectics: (a) $\gamma$–$\gamma'$–$Cr_3C_2$ and (b) Al–$Al_3Ni$**

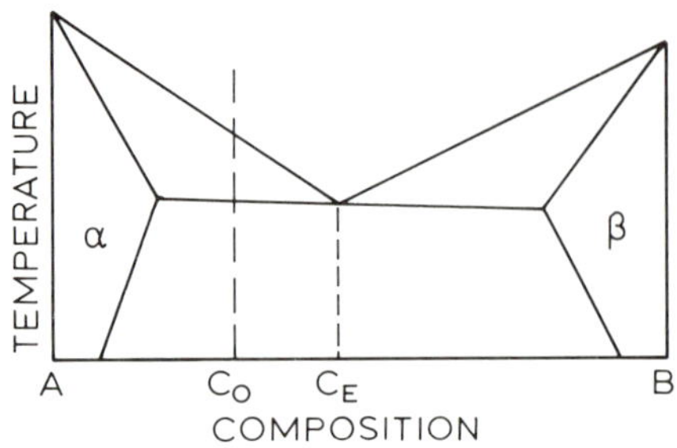


**Figure 3.19 Binary eutectic phase diagram showing relation of off-eutectic composition $C_0$ to the eutectic composition $C_E$**

This is an extension of eqn (3.9). The lateral gradients in solute concentration fall off rapidly with distance normal to the planar solid/liquid interface through the term $\exp[-(2n\pi x)/(\lambda_\alpha+\lambda_\beta)]$ and in addition there is a transversely isotropic solute profile that decays as $\exp(-V_I x/D_L)$. In most practical solidification conditions $V_I/D_L \ll 2\pi/(\lambda_\alpha+\lambda_\beta)$. (Typically, $V_I \sim 10^{-6}$ to $10^{-4}$ m s$^{-1}$, $D_L \sim 10^{-9}$ m$^2$ s$^{-1}$, $\lambda_\alpha+\lambda_\beta \sim 10^{-6}$ to $10^{-7}$ m.) Thus the term modified by $\exp(-V_I x/D_L)$ dominates at distances greater than about the interlamellar spacing from the solid/liquid interface giving the longer-range solute profile in the melt ahead of the solidifying interface

$$C_L^\alpha = C_0\left[1+\frac{C_E-C_0}{C_0}\exp\left(-\frac{V_I x}{D_L}\right)\right] \tag{3.39}$$

This is of exactly the same form as eqn (2.13) which was derived to describe plane front solidification of a single phase. We can follow exactly the same arguments that were used in § 2.3 to show that this solute profile leads to an associated variation in liquidus temperature $T_L$

$$T_L = T_0 + m_\alpha(C_E - C_0)\exp\left(-\frac{V_I x}{D_L}\right) \tag{3.40}$$

Again, as for single-phase solidification, the planar growth front will be stable if the applied temperature gradient $G_L$ exceeds the gradient in $T_L$, otherwise *constitutional supercooling* will occur ahead of the advancing interface and perturbations to the solid/liquid interface will occur. The condition for the maintenance of plane front solidification is

$$\frac{G_L}{V_I} \geqslant \frac{m_\alpha}{D_L}(C_E - C_0) = \frac{\Delta T}{D_L} \tag{3.41}$$

Alternatively this may be expressed as the maximum deviation in solute concentration from the eutectic composition that can give plane front solidification with particular solidification parameters

$$C_0 - C_E = \frac{D_L G_L}{m_\alpha V_I} \tag{3.42}$$

Other authors[16] have proposed that the ranges of compositions giving eutectic-like microstructures can be explained by competitive growth between dendrites of the enriched phase and a coupled eutectic microstructure. It is claimed that the structure, dendritic or eutectic, that forms at the highest temperature will be preferred. Referring to Fig. 3.19, the eutectic structure will only be preferred if the dendrite undercooling $\Delta T_\alpha$ exceeds the sum of the differences between the alloy liquidus and eutectic temperatures $m_\alpha(C_0 - C_E)$ and the eutectic undercooling $\Delta T_e$, i.e.

$$\Delta T_\alpha \geqslant m_\alpha (C_0 - C_E) + \Delta T_e \tag{3.43}$$

We derived an expression for dendrite undercooling in Chapter 2 that predicted a minimum at a particular value of $V_I$ (eqn (2.34)) and the eutectic undercooling is given by eqn (3.22). Substitution into Inequality (3.43) gives an expression for the *maximum* deviation from eutectic composition that will solidify with a plane front for specific solidification parameters

$$C_0 - C_E = \frac{1}{m_\alpha}\left[\frac{G_L D_L}{V_I} + V_I\left(\frac{G_L \lambda^2}{2D_L} + \frac{2\theta(1-k)}{D_L}\right)\right] - 2m_\alpha\sqrt{AB}\,V_I^{1/2} \tag{3.44}$$

where $A$ and $B$ are defined in eqn (3.19). In high temperature gradients the first term in the brackets dominates and (3.44) reduces to exactly the same condition as given by the constitutional supercooling criterion.†

$$C_0 - C_E = \frac{D_L G_L}{m_\alpha V_I}$$

Consequently it is difficult to discriminate between the two possibilities. When solidification occurs in the absence of a temperature gradient, then

$$C_0 - C_E = \frac{2\theta(1-k)}{m_\alpha D_L} V_I - 2m_\alpha \sqrt{AB}\, V_I^{1/2}$$

This indicates that when $G_L = 0$, the composition range giving eutectic-like microstructures increases with increasing $V_I$ as is often observed. Thus dendritic structures are only predicted to occur in an intermediate range of $V_I$.

† This can readily be established by substitution of typical parameters associated with the directional solidification of superalloys: $G_L = 10^4\ \mathrm{K\,m^{-1}}$, $V_I = 8 \times 10^{-4}\ \mathrm{m\,s^{-1}}$, $D_L = 5 \times 10^{-9}\ \mathrm{m^2\,s^{-1}}$.

In alloys containing three or more components, solidification to form a two-phase eutectic structure does not occur at a unique temperature, in general, but takes place over a finite melting range $\Delta T$. This is a direct consequence of the *phase rule* where the introduction of an additional component while retaining the same number of product phases leads to an additional degree of freedom, namely temperature. Whether the extra phases are present as unwanted impurities or as deliberate dopants to optimize certain properties, they will tend to make the attainment of plane front solidification more difficult.

Figure 3.20 shows a schematic isothermal section of a ternary phase diagram in which the locus of the eutectic trough, which is the cusp between the $\alpha$ and $\beta$ liquidus surfaces, is shown as a dashed line. Vertical sections through the cuts $AB$, $A'B'$ and $A''B''$ are shown in Fig. 3.21. In the example shown the eutectic temperature decreases with increasing concentration of the third component C; the melting range is zero for the binary section and increases with increasing $C$. At temperature $T^*$ the alloy of composition $C_0$ starts to freeze, forming solid of compositions $C_{\alpha 0}$ and $C_{\beta 0}$ defined by the intersection of the tie plane through $C_0$ (i.e. the plane defining constant temperature) with the $\alpha$ and $\beta$ phase fields. Following an initial transient, steady state solidification may be set up resulting in a solute concentration gradient in the melt and the formation of the solid phases $C'_{\alpha 0}$ and $C'_{\beta 0}$ at a temperature $T^* - \Delta T$ where $\Delta T$ is the melting range of the alloy $C_0$. Using exactly the same arguments that were invoked to account for plane front solidification in single-phase and binary eutectic alloys, it follows that constitu-

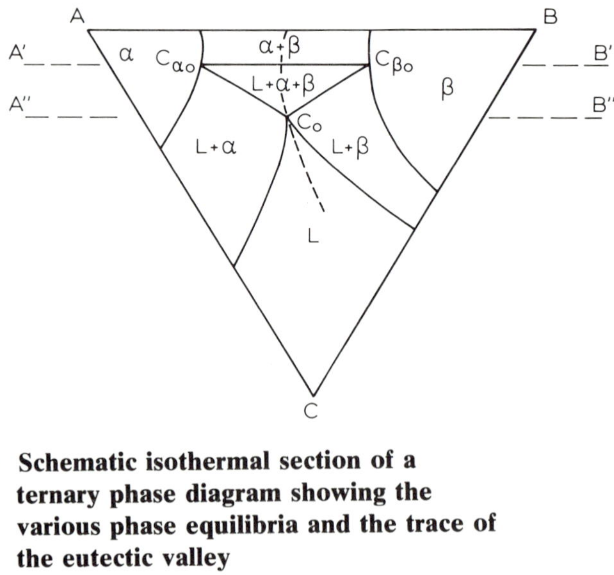


**Figure 3.20 Schematic isothermal section of a ternary phase diagram showing the various phase equilibria and the trace of the eutectic valley**

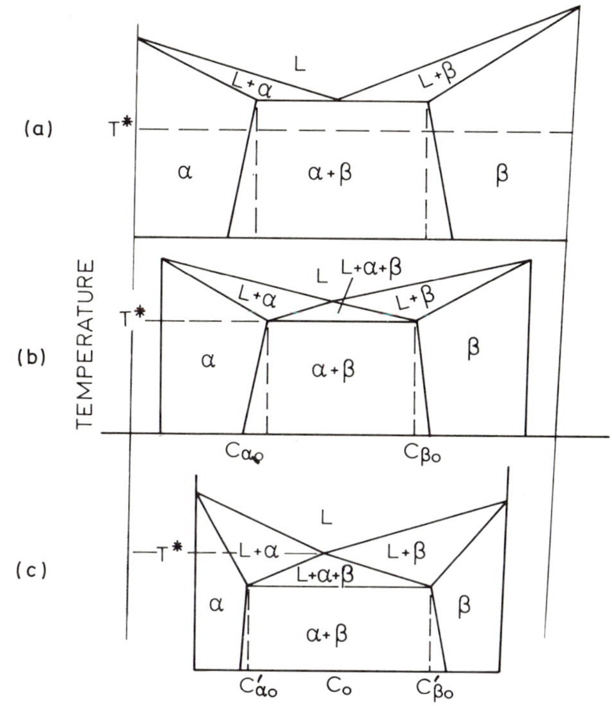


**Figure 3.21 Three sections of the phase diagram shown in Fig. 3.20: (a) *AB* binary; (b) through *A′B′* for constant value of *C*; (c) through *A″B″* for constant value of *C***

tional supercooling will be inhibited and plane front solidification achieved if

$$\frac{G_L}{V_I} \geqslant \frac{\Delta T}{D_L} \tag{3.45}$$

In this case the critical factor in defining $\Delta T$ will be in the slope of the liquidus trough.

The most complete characterization of eutectic microstructure for various values of $G_L$, $V_I$ and solute concentration currently available is that for the Pb–Sn binary system. Hunt[16] has reanalysed the available data to map the boundary between eutectic and dendritic microstructures in the plot of solute concentration versus $G_L/V_I$ shown in Fig. 3.22. The points at high values of $G_L/V_I$, from Mollard and Flemings,[20] are generally compatible with eqn (3.42), showing a proportionate widening in eutectic range with increasing

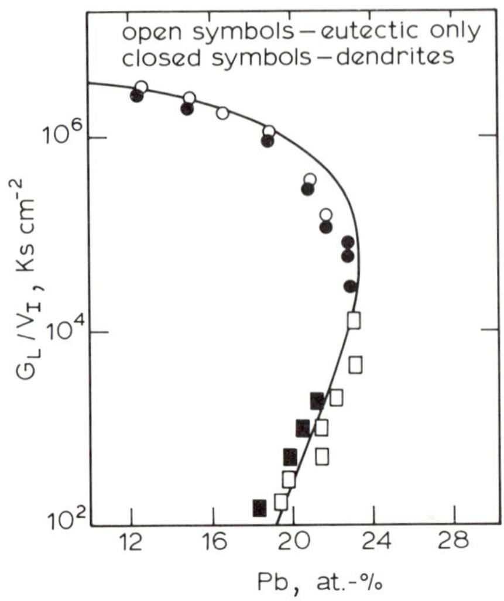


**Figure 3.22 Map of solidification morphologies in Sn–Pb on a plot of $G_L/V_I$ versus Pb composition. The line indicates the boundary between dendritic and eutectic structures [after Hunt[16]]**

$G_L/V_I$. However, inclusion of the high growth rate data of Cline and Livingston[12] suggests that eqn (3.44) may be more appropriate where at high values of $V_I$ the eutectic range increases with increasing $V_I$. The curves shown represent the theoretical eutectic/dendritic boundary calculated from (3.44) using estimated values of the various parameters.

There is less detailed information on the solidification behaviours of the complex alloys of potential engineering importance. Since they are invariably multicomponent alloys having a finite melting range at the eutectic composition, plane front eutectic growth can only occur when Inequality (3.45) is satisfied and this precludes eutectic solidification over a wide range of solidification rates. Indeed one of the recurrent criticisms of *in situ* composites has been the slow processing rates required to obtain well-aligned phases in some systems.

Gigliotti and Henry[21] have pointed out that the long-range solute profile given by eqn (3.39) which can extend to ~10 mm ahead of an interface solidifying at about 6 mm h$^{-1}$ may be significantly modified by various factors such as convective and electromagnetic mixing and that this may lead to significant segregation along the length of the solidified ingot in the manner described for single phases in § 2.7. Figure 3.23 shows the composition profiles calculated for the NITAC 13 alloy along the ingot displayed as functions of

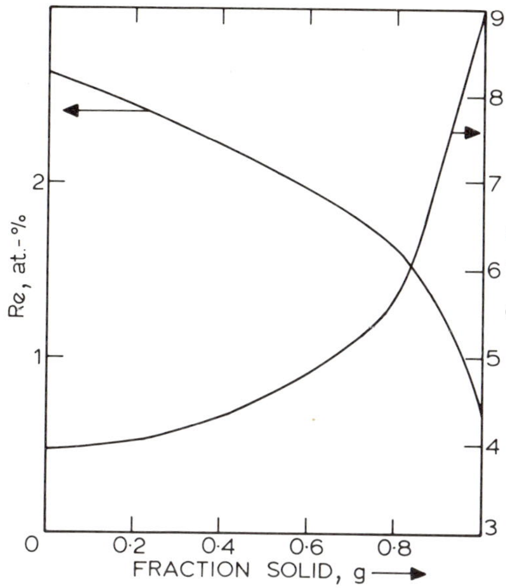


**Figure 3.23 Theoretical variations in concentration of chromium and rhenium as a function of the fraction solidified in NITAC 13 directionally solidified at 6 mm h$^{-1}$ [Gigliotti and Henry[21]]**

the fraction solidified; experimental measurements are in close agreement with these predictions. Clearly this gradation in composition along the solidified material, which provides general support for the constitutional supercooling concepts, will have considerable effect on both the microstructure and associated properties of such alloys. Figure 3.24 shows the microstructures obtained at various points along the length of an ingot of COTAC 74 that has been directionally solidified at 12 mm h$^{-1}$ in a high temperature gradient; the first solid contains primary carbide, the eutectic structure is gradually established with decreasing density of solute-induced growth faults and nickel dendrites occur at the top. Clearly such heterogeneities are detrimental and should be avoided by use of the highest growth rates possible.

The results of Kraft and Thompson[22] from a similar study on Ni-Cr-Al-Nb alloys, however, tend to support the ideas of competitive growth. Typical data for one alloy shown in Fig. 3.25 indicate that the eutectic/dendrite boundary in a plot of $G_L/V_I$ versus composition has a similar form to that shown in Fig. 3.22 although the eutectic zone is subdivided to include both composite and cellular microstructures because of the finite melting range of all of these multicomponent alloys.

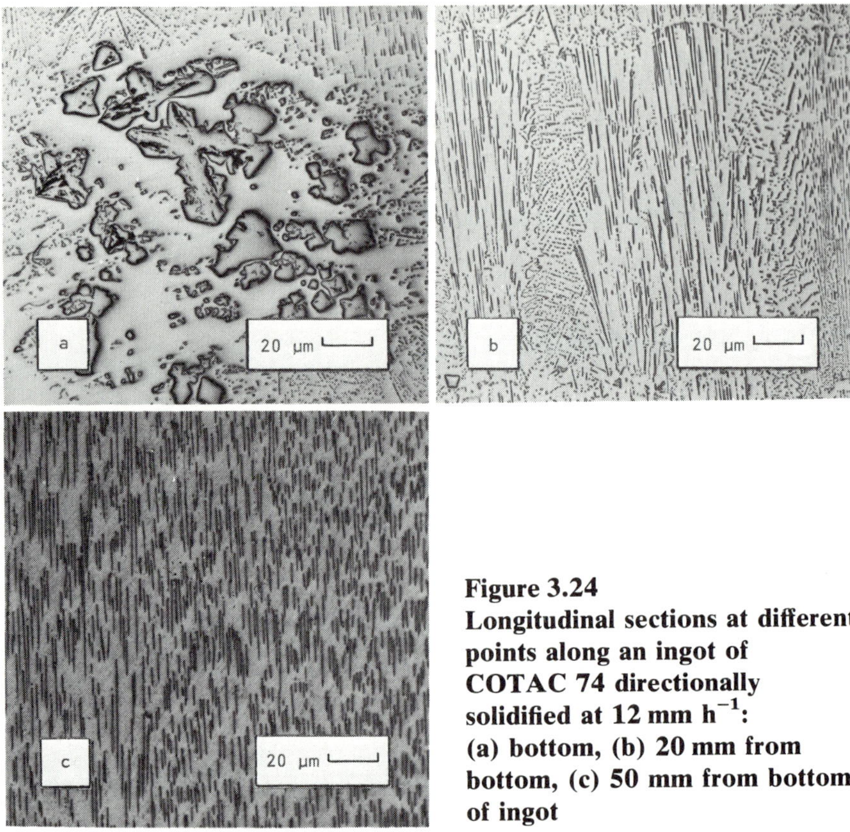


**Figure 3.24**
**Longitudinal sections at different points along an ingot of COTAC 74 directionally solidified at 12 mm h$^{-1}$: (a) bottom, (b) 20 mm from bottom, (c) 50 mm from bottom of ingot**

Young *et al.*[23] have determined as functions of $G_L/V_I$ the lead distances of the primary phases in both nickel- and niobium-rich versions of the $\gamma$–$\gamma'$–$\delta$ alloys by examining the structure of quenched solid/liquid interfaces. A typical example is shown in Fig. 3.26*a*. By plotting the temperature of the protruding cell tips as a function of $G_L/V_I$, as shown in Fig. 3.26*b*, it is possible to determine quite precisely the solidification conditions required to suppress the dendrite formation. In addition to providing strong support for the competitive growth theory, this procedure is a valuable practical tool establishing the minimum processing requirements to achieve well-aligned composite microstructures.

## 3.6 EUTECTICS WITH FACETING PHASES

All the foregoing theory, by ignoring crystallographic growth anisotropy, has only been strictly applicable to eutectics involving two non-faceting phases. However, almost all of the alloys that have been proposed as potential composite materials for use at high temperatures contain at least one highly

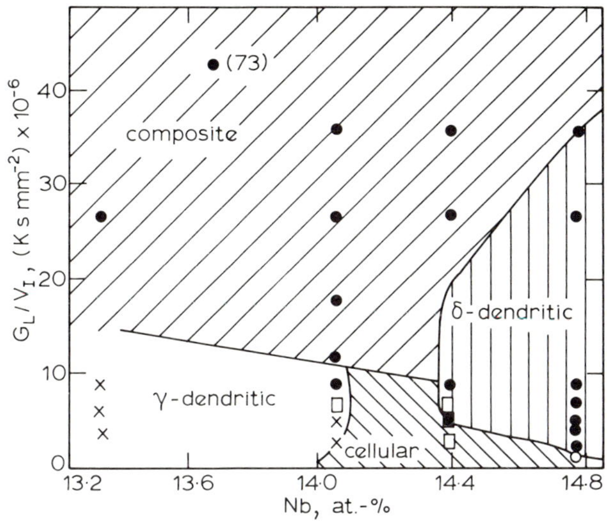


**Figure 3.25 Map indicating the occurrence of various solidification morphologies at various values of $G_L/V_I$ and composition for Ni–Nb–5·7 atom% Al eutectic alloys [Kraft and Thompson[22]]**

faceting intermetallic or non-metallic reinforcing phase (e.g. TaC, NbC, $Ni_3Nb$). In spite of this, the solidification behaviour of many of these materials is well described by the Jackson–Hunt and derivative theories. (These alloys of potential practical importance can solidify with either a very regular morphology typified by the COTAC series of alloys (Fig. 3.27*a*) or a generally aligned but irregularly distributed structure as in the $\gamma$–$\gamma'$–$Cr_3C_2$ alloy (Fig. 3.27*b*).)

Hurle and Hunt[24] have proposed that nf/nf behaviour should occur as long as there is no interference by the facet plane(s) with the equilibrium groove configuration at the intersection of the solid interphase boundaries with the solidification front. When the matrix phase in a rod-eutectic is faceted, the facet can never reach the groove because of the negative curvature as shown in Fig. 3.28*a*; when the minor phase is faceted, or in lamellar eutectics, the facet plane may or may not intersect the groove as is illustrated in Figs. 3.28*b* and *c* respectively. The configurations shown in Figs. 3.28*a* and *b* would be expected to lead to regular structures provided that the appropriate solidification conditions are satisfied; that in Fig. 3.28*c* is likely to lead to a non-isothermal interface and an irregular structure.

The regular f/nf eutectics solidify with a planar interface as shown in Fig. 3.29*a*. Relatively few branches are observed in many such materials.

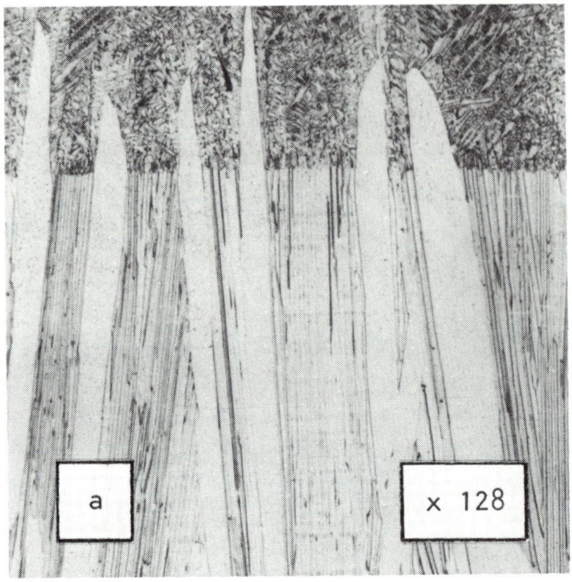


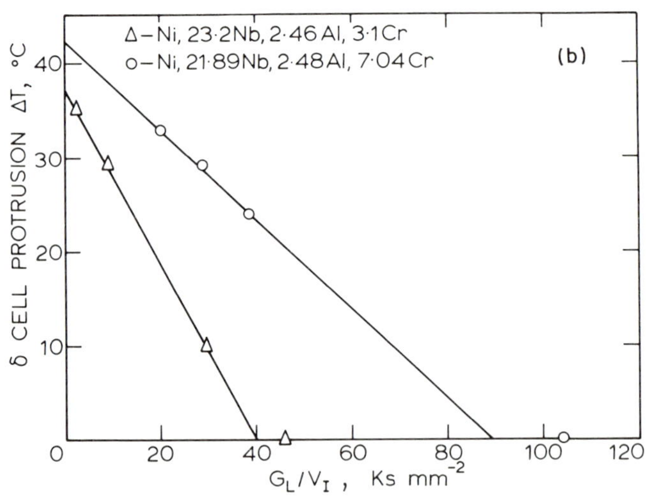


**Figure 3.26** **(a) Longitudinal section of a Ni–23.2Nb–3·1Al–2·5Cr alloy grown with $G_L/V_I = 3 \times 10^4$ K mm$^{-2}$ and quenched**
**(b) Plot of the $\delta$ ($Ni_3Nb$) cell protrusion ahead of the two-phase plane front interface versus $G_L/V_I$ for two $\gamma$–$\gamma'$–$\delta$ alloys [Young *et al.*[23]]**

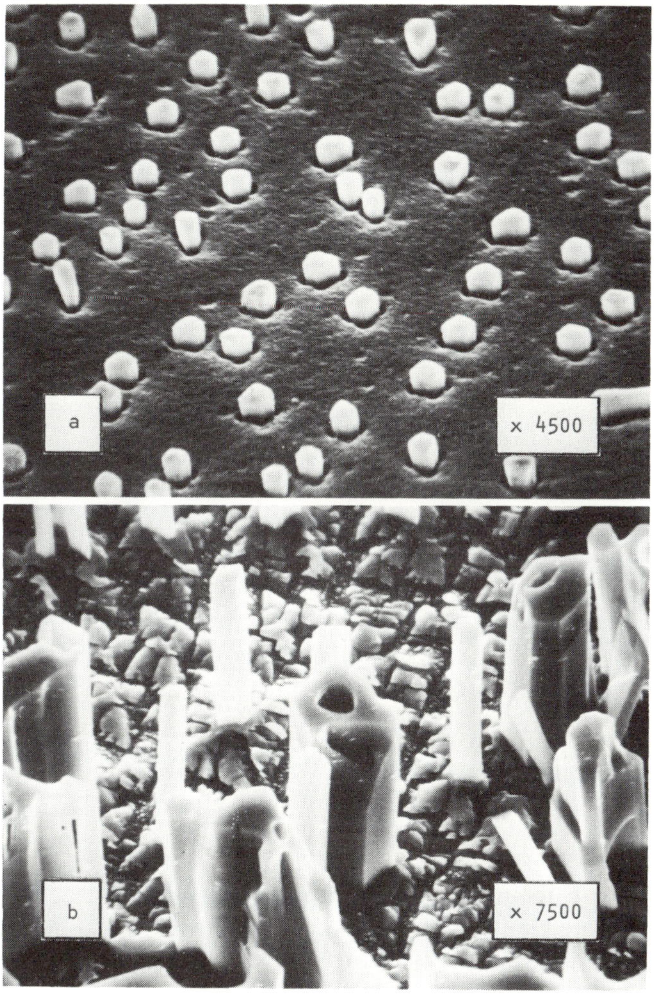


**Figure 3.27 (a) Regular structure of directionally solidified COTAC 74. (b) Irregular structure of directionally solidified $\gamma$–$\gamma'$–$Cr_3C_2$**

Since some method of phase adjustment is necessary to maintain the extremum growth condition, as described in § 3.2, the very slow growth rates required to process alloys like NITAC 13 and the COTAC alloys may be due to the crystallographic constraints that inhibit branching in addition to the melting range criterion (eqn (3.42)). Nevertheless, this category of f/nf eutectic can be solidified with the same morphologies as, and with similar dimensions to, those obtained from nf/nf systems.

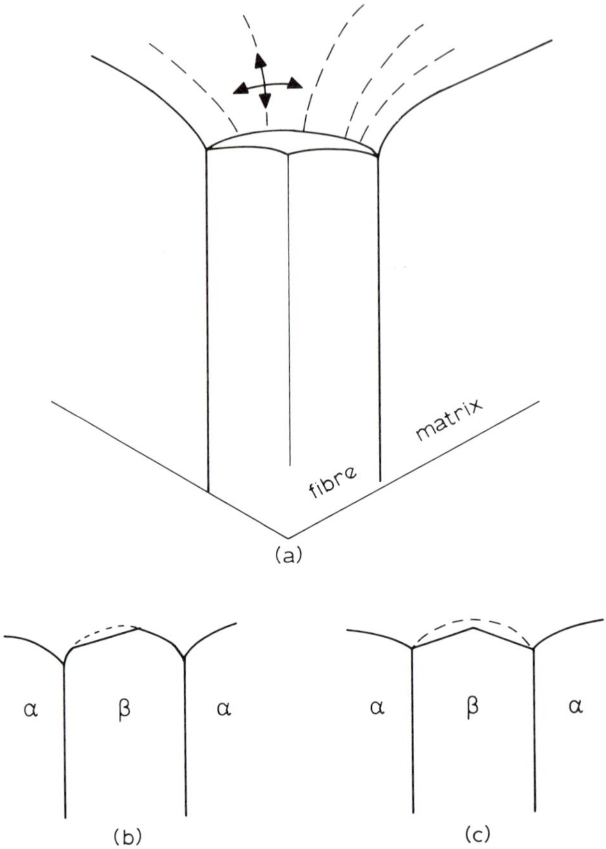


**Figure 3.28 (a) Perspective drawing of a rod eutectic showing the concave component to the curvature of the matrix $r_2$. (b) Faceted rod or lamella where the facet does not interfere with the groove at the $\alpha\beta$ interface. (c) As (b) where the facet does interfere with the growth**

The $\gamma$–$\gamma'$–$Cr_3C_2$ eutectic is an extreme example of alloys that can be directionally solidified with well-aligned but irregularly distributed phases. Such alloys can often be successfully directionally solidified at unexpectedly high rates; in the case of $\gamma$–$\gamma'$–$Cr_3C_2$ solidification rates in excess of $10^3$ mm $h^{-1}$ have been achieved, whereas the regular alloys, like NITAC 13, can often only be aligned when solidified at rates below about 10 mm $h^{-1}$. This amenability to rapid solidification can *not* be attributed to a small melting range as

**Figure 3.29** **Quenched interfaces of f/nf eutectics solidifying with (a) regular and (b) irregular morphologies: (a) (Co,Cr)–$(Co,Cr)_2Zr$, (b) $\gamma$–$\gamma'$–$Cr_3C_2$**

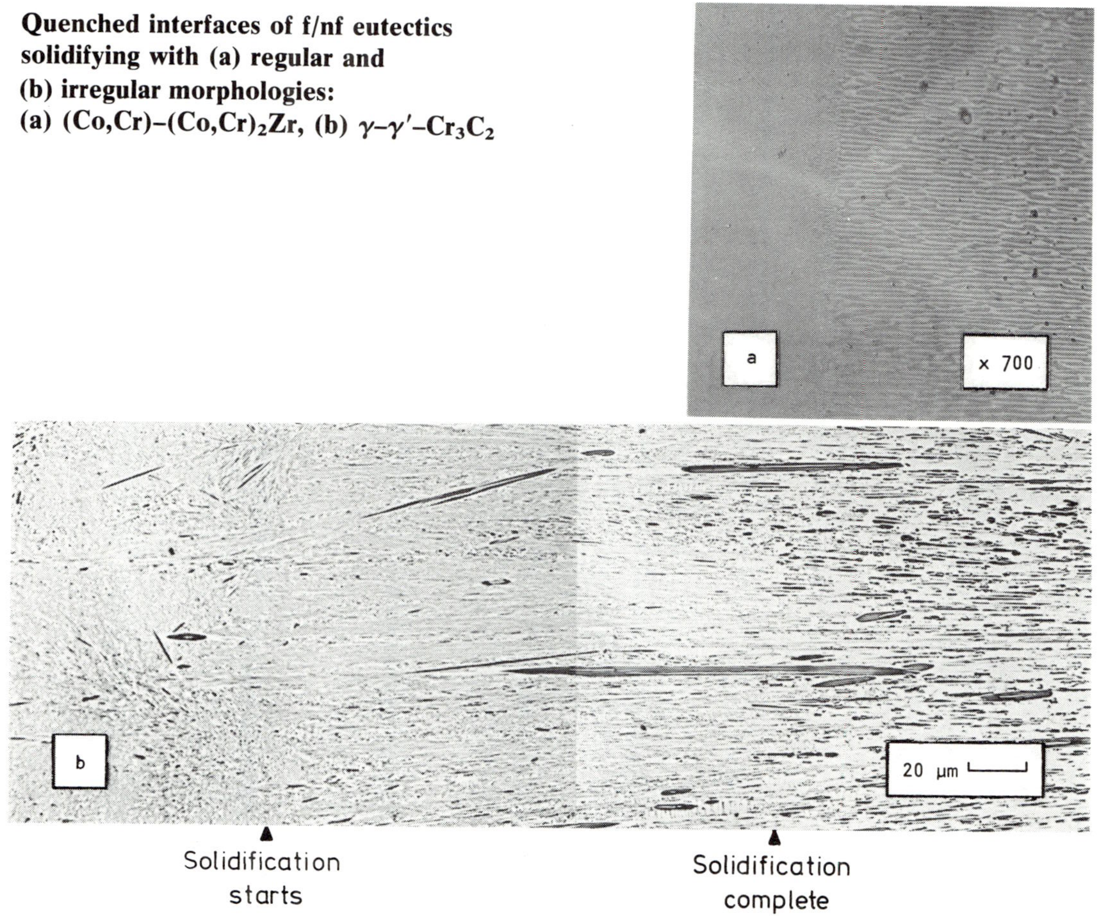

required by the theory of plane front solidification (eqn (3.41)). Indeed, examination of quenched solidification fronts as seen in Figs. 3.29*b* and 3.30, shows them to be far from planar and the principal features of the solidifying structure can be correlated with the results of differential thermal analysis (DTA) which indicate a melting range of ~40 K for $\gamma$–$\gamma'$–$Cr_3C_2$. Figure 3.30 shows the cooling curve for the alloy, corresponding to a cooling rate of 10 K $min^{-1}$, displayed beside a longitudinal section of a specimen quenched from directional solidification at 30 mm $h^{-1}$ in a temperature gradient of 13 K $mm^{-1}$. (This corresponds to a cooling rate of ~6·5 K $mm^{-1}$.) The temperature scale on the cooling curve has been adjusted to represent the temperature distribution along the ingot before quenching. The first minor peak at ~1340°C indicates the initiation of cellular solidification, but the major solidification peak is clearly associated with the planar front at ~1300°C. The broad peak between ~1150 and 1230°C indicates $\gamma'$ precipitation. The results are consistent with the following sequence of events.

(a) Carbide fibres solidify at a higher temperature (~1340°C) than the matrix and project into the melt in a cellular fashion. The good alignment of the fibres is achieved by efficient branching of the carbides. Rapid coarsening of this finely branched structure occurs by liquid state diffusion.
(b) The $\gamma$ matrix solidifies with a near-planar interface at ~1300°C. The carbide fibres are relatively stable at temperatures lower than this due to a discontinuous reduction in the kinetics of coarsening with the transition from liquid- to solid-state diffusion.
(c) $\gamma'$ precipitates from the $\gamma$ matrix in the temperature range from ~1230 to 1150°C in the usual manner.

An important factor in the applicability of this mechanism to $\gamma$–$\gamma'$–$Cr_3C_2$ is that $Cr_3C_2$ readily branches, enabling the carbide dimensions to adjust to the local growth conditions, and has highly anisotropic growth kinetics which maintain a preferred growth direction in spite of irregularities in the solidification front. By contrast, the monocarbides in the NITAC and COTAC alloys appear to have a very low branch density, so that disordered eutectic solidification occurs.

The relationships between the phase dimensions and solidification rates in both regular and irregular eutectics, prepared in a consistent manner, appear to be satisfactorily described by the form of eqn (3.34). However, there have been several reports[14,25] that the microstructural scale of directionally solidified eutectic alloys is sensitive to the temperature gradient $G_L$ in disagreement with the theory. Thus Fig. 3.31 (from Bullock, McLean and Miles[25]) shows the variation in fibre radius with rate of solidification for $\gamma$–$\gamma'$–$Cr_3C_2$ prepared in two quite different temperature gradients; the use of the high temperature gradient leads to significant refinement of the microstructure. Similar observations have been made for other alloys, but the effect of $G_L$ is often less dramatic than in $\gamma$–$\gamma'$–$Cr_3C_2$. It is not clear whether this

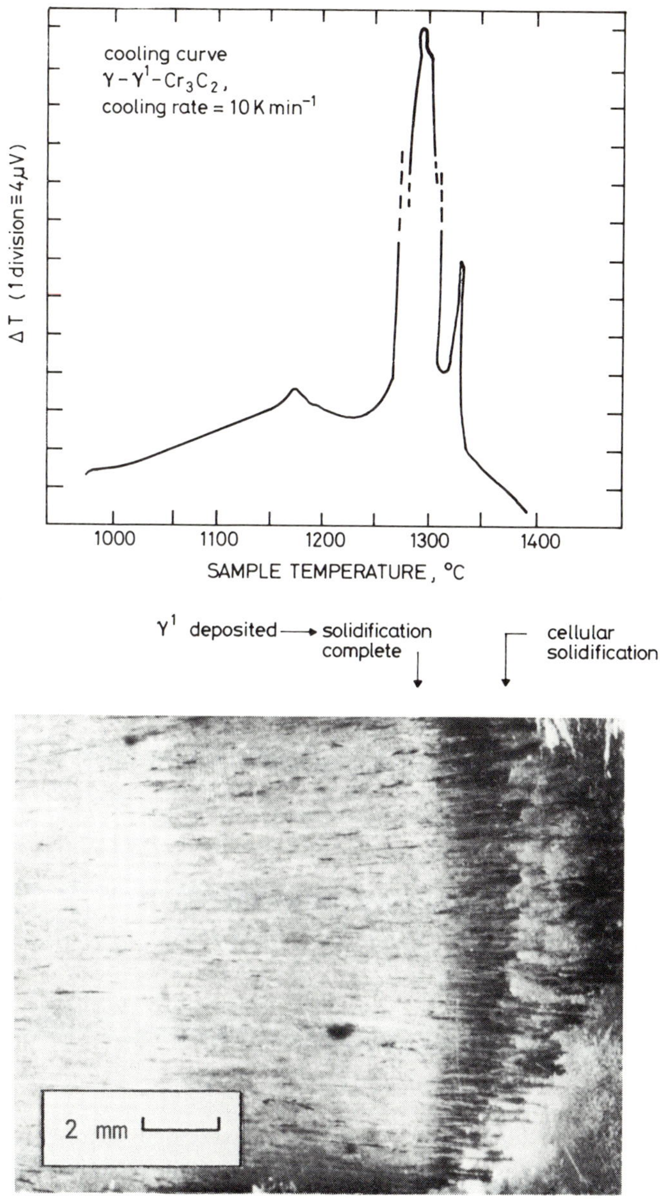


**Figure 3.30 Low magnification photograph of a longitudinal section of $\gamma$–$\gamma'$–$Cr_3C_2$ quenched after directional solidification at 60 mm $h^{-1}$ using liquid metal cooling. Differential thermal analysis is scaled to the same temperature as in the photograph to show the correlation between DTA peaks and solidification features**

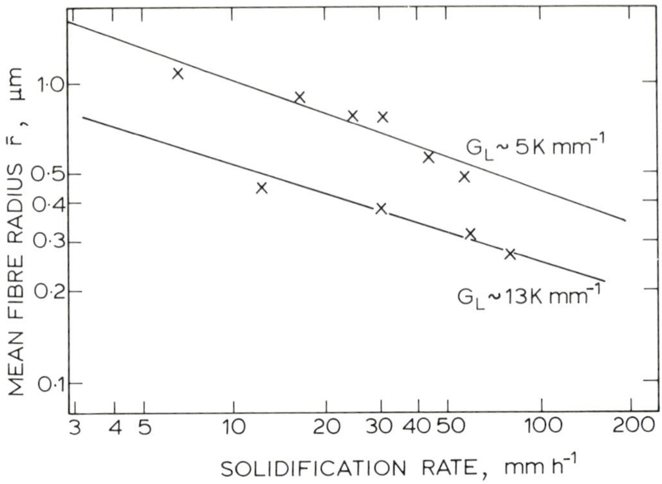


**Figure 3.31 Variation in mean fibre radius with solidification rate for $\gamma$—$\gamma'$–$Cr_3C_2$ directionally solidified in different temperature gradients**

effect is restricted to the irregular eutectics of whether it is simply more apparent in these materials because of the wider range of solidification rates that can be used. Fisher and Kurz[26] have recently proposed a model to account for this temperature gradient effect which envisages high $G_L$ leading to a flattening of the solidification front and to an increased density of fibre branches; available data for the Al–Si and Fe–C eutectics appear to be compatible with the predictions of this model. However, examination of Fig. 3.29 shows that the aligned fibres at the solidification front are very much finer than those occurring further along the ingot that have been cooled at the rate determined by the processing conditions. This clearly implies that, at least for $\gamma$–$\gamma'$–$Cr_3C_2$, the final fibre size is *not* determined at the stage when the carbide fibres solidify but by a subsequent aging process that is also sensitive to the temperature gradient. This is exactly the same effect that was invoked in § 2.5 to explain the coarsening of secondary dendrite arms and of the $\gamma'$ precipitate during post-solidification aging. Correlation with differential thermal analysis suggests that the carbide fibres rapidly grow along ⟨100⟩ directions into the melt in a highly branched but well-aligned structure. In the temperature range between the carbide formation and the matrix solidification, the carbides can coarsen by rapid liquid state diffusion, the extent of coarsening being least in the presence of high temperature gradients.

## 3.7 CRYSTALLOGRAPHY OF *IN SITU* COMPOSITES

The extreme regularity of both the cross-sectional shape and the distribution of phases in many *in situ* composites suggests that there must be preferred orientation relationships between the constituent phases. The existence of such special crystallographic relationships, which have been often assumed rather than proven, have been justified as producing interfaces of low boundary energy between the constituent phases and this has been invoked in explaining the high degree of thermal stability displayed by many *in situ* composites. Cantor[27] has recently reviewed the evidence for the development of special crystallographic relationships during directional solidification of *in situ* composites and has shown that behaviours ranging between strict epitaxial and completely independent growth of the two phases can occur. Moreover, the propensity for preferred crystallographic relationships is not easily related to the microstructural regularity of the phases.

The alloys that have been characterized most extensively from the point of view of crystallography are the lamellar Al–$Al_2Cu$ and fibrous Al–$Al_3Ni$ low temperature model systems which are representative of the two main types of behaviour. In the following discussion we adopt Kraft's[28] notation for specifying the crystallographic relationships between eutectic phases which identifies

(i) the crystallographic axes in each phase that form along the direction of solidification
(ii) the planes of each phase that are parallel to the principal interfaces (or habit planes) that develop between the phases.

This information is written in the form

*growth direction* $\|\langle uvw\rangle\alpha\|\langle uvw\rangle\beta$

*habit planes* $\|\{hkl\}\alpha\|\{hkl\}\beta$

### Al–$Al_2Cu$

During steady state growth a preferred crystallography develops that can be represented approximately by:

*growth direction* $\|\langle 101\rangle \mathrm{Al}\|\langle 210\rangle \mathrm{Al_2Cu}$

*habit plane* $\|\{111\}\mathrm{Al}\|\{211\}\mathrm{Al_2Cu}$

Cantor[27] shows that earlier conflicting reports of crystallographic relationships in this alloy can be attributed to inaccuracies in the measurement techniques and that the various results are generally consistent within the precision of the experimental results. Although there is a clear *tendency* for the phases to adopt this preferred orientation relationship, deviations of up to about 10° from the orientations of both the growth directions and habit planes can occur.

**Al–$Al_3Ni$**

A variety of orientation relationships have been reported for this alloy, mostly based on examination of small areas in single eutectic grains. Cantor and Chadwick[29] have shown that the different crystallographic relationships obtained from different eutectic grains always indicate a ⟨010⟩$Al_3Ni$ growth direction and a series of preferred matrix orientations rather than one unique orientation. The most common relationship is:

| | |
|---|---|
| *growth direction* | ‖⟨110⟩Al‖⟨010⟩$Al_3Ni$ |
| *habit plane* | ‖{331}Al‖{001}$Al_3Ni$ |

The alloys developed as potential structural materials for high temperature application show similar differences in their tendencies to form special orientation relationships between the phases.

**Unique epitaxy**

NiAl–Cr and NiAl–Mo are characteristic of eutectic alloys that consist of phases with similar crystal structures. Such materials often adopt identical growth textures and form a unique orientation relationship between the phases, any lattice parameter mismatch being accommodated by structural dislocations at the interfaces between the phases. Walter and co-workers[30] have particularly extensively studied the NiAl–Cr alloy and various compositional modifications. In the basic alloy the Cr lattice parameter is less than that of NiAl. Additions of Mo, V or W increase and of Fe decrease the Cr lattice parameter; when it exceeds that of NiAl a mixture of faceted rods and lamellae replaces the round rod morphology of the basic alloy. In each case a unique but different crystallography occurs

| | | |
|---|---|---|
| (a) non-faceted | *growth direction* | ‖⟨010⟩NiAl‖⟨010⟩Cr |
| | *habit plane* | ‖{100}NiAl‖{100}Cr |
| (b) faceted | *growth direction* | ‖⟨111⟩NiAl‖⟨111⟩Cr(Mo) |
| | *habit plane* | ‖{211}NiAl‖{211}Cr(Mo) |

**Preferred orientation relationships**

Several alloys that have evolved from the binary Ni–$Ni_3Nb$, or $\gamma$–$\delta$, eutectic alloy have been studied in some detail. Additions of Al lead either to the precipitation of ordered $Ni_3Al$ ($\gamma'$) in the $\gamma$ matrix to form the $\gamma$–$\gamma'$–$\delta$ alloy or, at higher concentrations, to the formation of a lamellar $Ni_3Al$–$Ni_3Nb$ (or $\gamma'$–$\delta$) eutectic. A dominant orientation relationship, satisfied within about 5°, occurs in all of these alloys:[27]

| | |
|---|---|
| *growth direction* | ‖⟨110⟩$\gamma,\gamma'$‖⟨100⟩$\delta$ |
| *lamellar interface* | ‖⟨111}$\gamma,\gamma'$‖{010}$\delta$ |

However, two alternative habit plane relationships have been reported in $\gamma'$–$\delta$ although the $\langle 111\rangle\gamma' \| \langle 100\rangle\delta$ growth direction always occurs, namely:

$$\{113\}\gamma' \| \{031\}\delta \qquad \text{and} \qquad \{111\}\gamma' \| \{012\}\delta$$

Spiller and Smith[31] have examined the irregular $\gamma$–$\gamma'$–$Cr_3C_2$ alloy and shown similar behaviour in that each phase solidifies with preferred growth direction

$$\textit{growth direction} \quad \| \langle 110\rangle\gamma,\ \gamma' \| \langle 001\rangle Cr_3C_2$$

but with no unique misorientation. Two preferred orientations that accounted for over 90% of the fibres were

$$\{100\}Cr_3C_2 \| \{111\}\gamma,\ \gamma' \qquad \text{and} \qquad \{100\}Cr_3C_2 \| \{331\}\gamma,\ \gamma'$$

The interface planes that formed tended to be low index with respect to the orthorhombic fibres but high index relative to the matrix.

**Unrelated orientations**

Many of the *in situ* composites that are of particular interest for high temperature structural applications consist of highly faceting phases in relatively simple Ni-, Co- or Fe-base solid solution matrices. Although there are some conflicting reports, there is a general consensus that during the solidification of these materials the faceting phase grows in preferred directions and adopts a faceted shape that is determined by its own crystallography or anisotropic growth kinetics. There is little evidence of a relationship between the orientations of the faceting and non-faceting phases.

Walter and Cline[32] showed that the fibre orientations that develop during directional solidification of M–TaC composites (where M can be Ni, Co or Fe) can be simply related to the facet, or slow growth directions, observed on primary TaC particles that grow in the melt. Thus, in the basic alloys where {100} facet planes form, the direction of rapid growth would be expected to be parallel to ⟨111⟩ as shown schematically in Fig. 3.32*a* and, indeed, all three alloys directionally solidified with mixtures of trifoils, blades and whiskers with ⟨111⟩ parallel to the solidification direction. Similarly, the changing microstructures obtained when alloy additions were made could always be correlated with the anticipated fast growth directions based on the observed faceting behaviour:

(i) Addition of Cr to Ni–TaC gave {111} facets that led to ⟨100⟩ growth of square section fibres on directional solidification (Fig. 3.32*b*).
(ii) Small additions of Cr to both Co-TaC and Fe–TaC give a mixture of {111} and {100} facets that lead to rapid ⟨110⟩ growth and ⟨110⟩ orientated blades do indeed form on directional solidification (Fig. 3.32*c*).
(iii) Further additions of Cr to Co-TaC lead to the dominance of {111} facets and a transition to rapid ⟨100⟩ growth. Directional solidification gives a mixture of ⟨100⟩ oriented fibres, plates and trifoils (Fig. 3.32*b*).

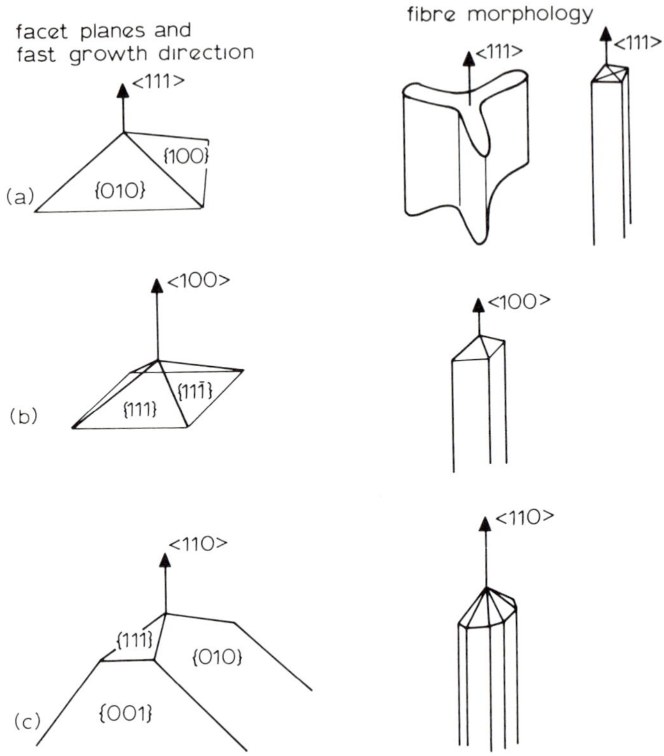


**Figure 3.32 Relationships between the orientation of facet planes and the maximum growth direction: (a) {100} facets, ⟨111⟩ growth; (b) {111} facets, ⟨100⟩ growth; (c) {111} and {100} facets, ⟨110⟩ growth**

The (Co–Cr)-(Co, Cr)$_7$C$_3$ eutectic, which has an irregular morphology, is again characterized by preferred growth directions (⟨211⟩Co, Cr‖⟨0001⟩Cr$_7$C$_3$), but random misorientations.[33]

It is significant that the alloys showing no clear crystallographic interactions generally consist of phases of widely different thermodynamic stabilities. The high enthalpies of formation of the monocarbides such as TaC relative to those of Ni, Co or Fe indicate that the interatomic bond energies associated with the carbides greatly exceed those associated with the metals. Consequently the enthalpy associated with metal/carbide interfaces which crudely can be taken as the sum of the carbide/carbide and metal/metal bond energies less the new carbide/metal bond energies summed across the interface will be dominated by the carbide/carbide contribution. Thus the low energy form in these systems will be determined by the crystallography of the stable phase and will be relatively insensitive to variations in the metal/metal and metal/carbide contributions due to orientation effects. Such interphase ordering will be most important when the phases have similar stabilities and therefore similar bond strengths.

## 3.8 ALTERNATIVE TWO-PHASE REACTIONS

Alloys other than eutectics are capable of solidification by the co-operative growth of two (or more) phases and consequently are capable of producing *in situ* composites. These alternatives (for example monotectics, eutectoids, peritectics) have received rather less attention than have eutectics for a variety of reasons. In particular, in the present context the requirements for high melting points and high temperature strengths militate against most eutectoid and monotectic systems which form their bi-phase structures at relatively low temperatures. Nevertheless these alloys have received some attention and may be of some interest for non-structural applications.

### Eutectoid

The eutectoid reaction is the solid state transformation that is equivalent to eutectic solidification. A schematic phase diagram shown in Fig. 3.33 indicates an alloy of eutectoid composition $C_E$ that transforms from a single phase solid $\gamma$ to two different solid phases $\alpha$ and $\beta$ on cooling through the eutectoid temperature $T_E$. If the transformation is achieved by volume diffusion of the constituent atoms through $\alpha$, then the theory describing the process is identical to that for eutectic solidification. Indeed the theory of eutectoid transformation by Zener[7] pre-dated and inspired subsequent treatments of the eutectic case.

Livingston[24] observed that it is implicit in deriving the relationship between eutectic phase dimensions $\lambda$ and transformation rate $V_I$ that the

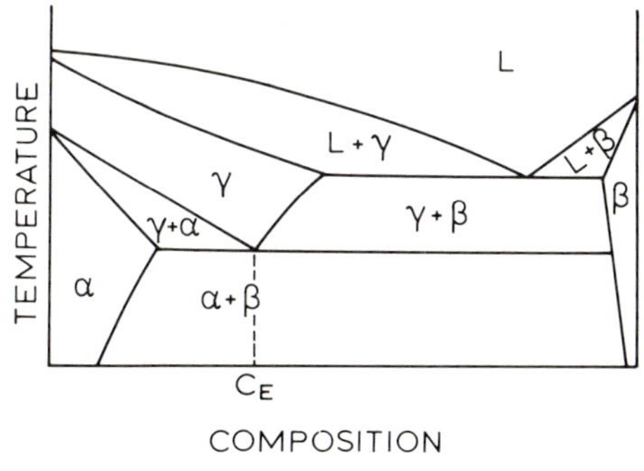


**Figure 3.33 Schematic binary phase diagram showing a eutectoid reaction**

interface temperature is independent of $V_I$ (eqn (3.21))

$$\bar{\lambda}^2 V_I = B/A \tag{3.46}$$

Consequently $B/A$ is independent of solidification conditions principally because of the low undercoolings associated with eutectics ($\leqslant$10 K). In eutectoids, however, where interface undercoolings can be as high as several hundred degrees, the transformation temperature varies strongly with $V_I$ and so therefore do the operative diffusivitities controlling the phase dimensions on transformation. Thus $B/A$ is not a constant for eutectoids but is a function of $V_I$. In attempting to fit $\lambda, V_I$ data empirically to an expression like eqn (3.21), much higher exponents are generally required ($n \sim 3$ to 4)

$$\lambda^n V_I = B'/A' \tag{3.47}$$

Since volume diffusivities through solid phases can be as much as a thousand times less than those in a liquid at the same temperature, this will lead to at least two important differences between eutectoids and eutectics.

(i) In eqn (3.47) $(B'/A')_{\text{eutectoid}} \ll (B/A)_{\text{eutectic}}$, so that for similar processing conditions $\lambda_{\text{eutectoid}} \ll \lambda_{\text{eutectic}}$. The ability to achieve finer microstructures is thought to be advantageous in optimizing mechanical properties, but these structures are likely to be less stable (see Chapter 7).
(ii) The condition for achieving plane front transformation of off-eutectoid compositions given by eqn (3.42) shows that the lower solid state diffusivities will restrict the range of compositions that can be aligned by current processing technology.

**Monotectics**

A typical monotectic reaction is shown schematically in Fig. 3.34. The alloy of monotectic composition $C_M$ solidifies from a single liquid $L_1$ to a mixture of a solid phase $\alpha$ and a liquid $L_2$ that is immiscible in $L_1$. The reaction is very similar in form to eutectic reactions and, since they too are controlled by liquid state diffusion, yields phase distributions of similar dimensions. Since $L_2$ invariably has a much lower volume fraction than $\alpha$, it forms tubes of liquid $L_2$ within a solid $\alpha$ matrix. Livingston and Cline[35] have shown that in the Cu–Pb system $L_2$ can take the form of either continuous rods or aligned strings of spheres depending on the solidification conditions. Hellawell[36] has pointed to two possible reasons for this effect, but the available evidence does not allow either to be confirmed.

(i) The shapes of the reaction interfaces for monotectics and eutectics can be quite different because of differing equilibria between $L_1$, $L_2$ and $\alpha$ for monotectics and $L$, $\alpha$ and $\beta$ for eutectics. In particular, if $\gamma_{L_2\alpha} > \gamma_{L_1L_2} + \gamma_{L_1\alpha}$ (where $\gamma_{ij}$ is the interfacial energy between $i$ and $j$), the cusp angle will be 0 and $L_1$ may completely wet the $L_2$ filament. This could mean that a three-phase mixture would be energetically favoured.

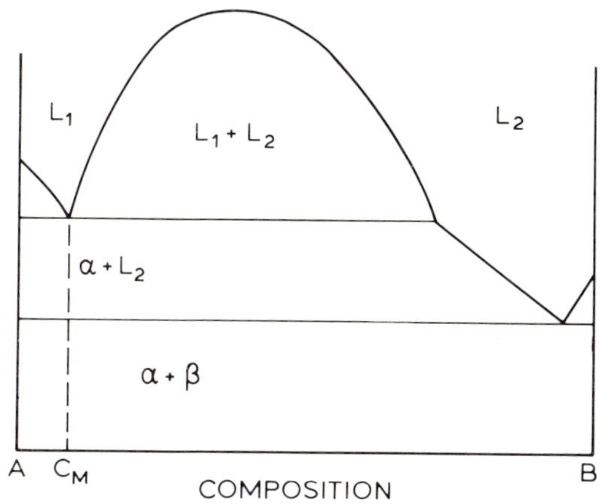


**Figure 3.34 Schematic binary phase diagram showing a monotectic reaction**

(ii) If the cooling rate after the formation of the liquid rods is sufficiently slow, then the rods may spheroidize into a string of liquid spheres by one of the mechanisms described in Chapter 7.

**Peritectics**

The peritectic reaction is rather different from the others discussed above (eutectic, eutectoid, monotectic) in that it involves the combination of a solid ($\alpha$) and liquid ($L$) phase to form a different solid ($\beta$) phase at the peritectic point.

$$\alpha + L \rightarrow \beta$$

An idealized phase diagram containing a peritectic reaction is shown schematically in Fig. 3.35*a*. The reaction indicated, however, is completed only when equilibration can take place. During solidification of and near to the peritectic composition various combinations of $\alpha$ and $\beta$ can be deposited and consequently this provides a possible route for the production of *in situ* composites.

There is as yet no clear agreement on the interpretation of the various microstructures that may be produced on directionally solidifying alloys in systems containing a peritectic reaction. The treatment presented here is based on ideas of Hillert[37] that consider the competitive growth of $\alpha$ and $\beta$ from the melt. The dashed lines in the phase diagram (Fig. 3.35*a*) are the metastable extensions of the $\beta$ liquidus and solidus boundaries.

Conditions for planar solidification of $\alpha$ and $\beta$ can be obtained by inserting the appropriate parameters into eqn (2.17), the condition for constitutional supercooling. For an alloy of solute concentration $C_0$, the boundaries between

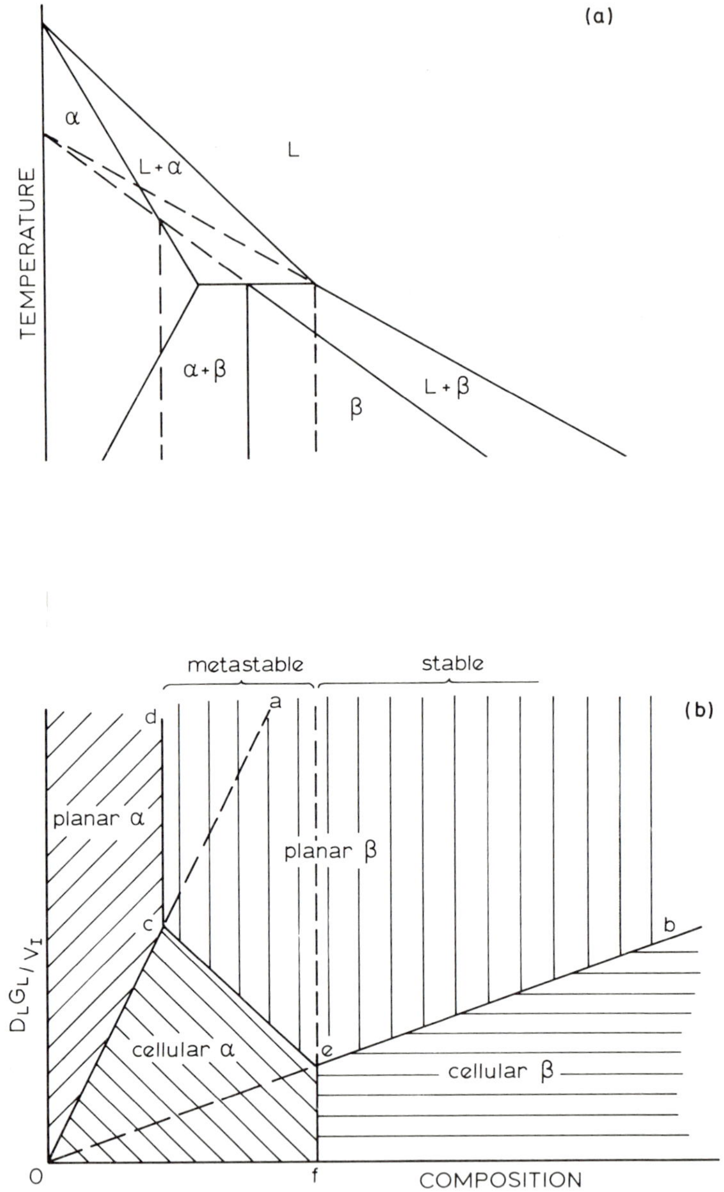


**Figure 3.35** **(a) Schematic binary phase diagram showing a peritectic reaction. (b) Map of different solidification morphologies on plot of $G_L/V_I$ versus composition**

planar and cellular morphologies are given by the expression

$$\text{for } \alpha \quad \frac{D_L G_L}{V_I} = -\frac{m_L^{\alpha} C_0 (1 - k_{\alpha})}{k_{\alpha}}$$

$$\text{for } \beta \quad \frac{D_L G_L}{V_I} = -\frac{m_L^{\beta} C_0 (1 - k_{\beta})}{k_{\beta}}$$

where $m_L^{\alpha}$, $m_L^{\beta}$ are the liquidus slopes and $k_{\alpha}$, $k_{\beta}$ are the partition coefficients. These lines, *Oa* and *Ob*, are drawn in Fig. 3.35*b*. The phase that solidifies at the highest temperature is likely to be the stable form. Thus the intersection between the $\alpha$ and the extrapolated $\beta$ solidus curves marks the boundary between compositions giving stable $\alpha$ and $\beta$ plane front solidification. This boundary *cd* is also included in Fig. 3.35*b*. Similarly, when both $\alpha$ and $\beta$ cells can form, the intersection of the $\alpha$ and $\beta$ liquidus curves, which occurs at the peritectic point, defines the boundary between stable $\alpha$ and $\beta$ cells; this is shown as *ef* in Fig. 3.35*b* which maps the various solidification forms to be expected as functions of composition and solidification conditions.

Where $\alpha$ and $\beta$ cells form they may be surrounded by the complementary phase giving a composite microstructure. Alternatively when compositions to the left of the peritectic point solidify by planar growth of $\beta$ the product is metastable and may subsequently precipitate $\alpha$. At slow-enough solidification rates it may transform by the reaction

$$\beta \rightarrow \alpha + L$$

This is thought to give rise to transverse banding that occurs in directionally solidified peritetic alloys.[38] The banding occurs at relatively low values of $(D_L G_L)/V_I$ under conditions where planar $\beta$ is expected to form, indicated by the shaded area in Figure 3.35*b*, and can be suppressed by increasing $(D_L G_L)/V_I$ to inhibit the degeneration of $\beta$.

## REFERENCES CHAPTER 3

1 R. W. KRAFT: US Patent No. 3129952, 1960

2 W. ALBERS: Proc. Conf. *In Situ* Composites I, National Materials Advisory Board Publication 308-III, Washington, DC, 1973

3 H. WEISS: Proc. Conf. *In Situ* Composites II, Xerox Individualized Publishing, Lexington, Mass., 1976

4 J. D. HUNT AND K. A. JACKSON: *Trans. Metall. Soc. AIME*, 1966, **236**, 843

5 K. A. JACKSON in 'Liquid metals and solidification', p. 174, 1958, Cleveland, Ohio, ASM

6 K. A. JACKSON AND J. D. HUNT: *Trans. Metall. Soc. AIME*, 1966, **236**, 1129

7 C. ZENER: *Trans. Metall. Soc. AIME*, 1946, **167**, 550

8 W. A. TILLER in 'Liquid metals and solidification', pp. 276–318, 1958, Cleveland, Ohio, ASM

9 M. HILLERT: *Jernkontorets Ann.*, 1960, **144**, 520

10 W. W. MULLINS AND J. W. CAHN in 'Decomposition of austenite by diffusional processes,' eds. V. F. Zackay and H. I. Aaronson, p. 123, 1962, New York, Interscience

11 G. E. NASH: *J. Cryst. Growth*, 1977, **38**, 155

12 H. E. CLINE AND J. D. LIVINGSTON: *Trans. Metall. Soc. AIME*, 1969, **245**, 1987

13 J. D. LIVINGSTON, H. E. CLINE, E. F. KOCH AND R. R. RUSSELL: *Acta Metall.*, 1970, **18**, 399

14 G. GARMONG AND T. H. COURTNEY: *Metall. Trans.*, 1975, **6A**, 1945

15 K. D. SHEFFLER, R. H. BARKALOW AND A. YUEN: Proc. Conf. *In Situ* Composites II, Xerox Individualized Publishing, Lexington, Mass., 1976

16 J. D. HUNT: Proc. Conf. *In Situ* Composites II, Xerox Individualized Publishing, Lexington, Mass., 1976

17 M. H. BURDEN AND J. D. HUNT: *J. Cryst. Growth*, 1974, **22**, 99

18 M. TASSA AND J. D. HUNT: *J. Cryst. Growth*, 1976, **34**, 38

19 D. D. DOUBLE: *Mater. Sci. Eng.* 1973, **11**, 325

20 R. MOLLARD AND M. C. FLEMINGS: *Trans. Metall. Soc. AIME*, 1967, **239**, 1526

21 M. F. X. GIGLIOTTI AND M. F. HENRY: Proc. Conf. *In Situ* Composites II, Xerox Individualized Publishing, Lexington, Mass., 1976

22 E. H. KRAFT AND E. R. THOMPSON: Proc. Conf. *In Situ* Composites II, Xerox Individualized Publishing, Lexington, Mass., 1976

23 K. P. YOUNG, B. A. RICKINSON AND M. C. FLEMINGS in 'Solidification and casting of metals', 539, 1979, London, Metals Society

24 D. T. J. HURLE AND J. D. HUNT: 'The solidification of metals', 162, 1968, London, ISI (Pubn 110)

25 E. BULLOCK, M. McLEAN AND D. E. MILES: *Acta Metall.*, 1977, **25**, 332

26 W. KURZ AND D. J. FISHER: *Acta Metall.*, 1981, **29**, 11

27 B. CANTOR in 'Interphase interfaces', ed. G. A. Chadwick and D. A. Smith, Academic Press (to be published)

28 R. W. KRAFT: *Trans. Metall. Soc. AIME*, 1961, **221**, 704

29 B. CANTOR AND G. A. CHADWICK: *J. Cryst. Growth*, 1975, **30**, 109

30 J. L. WALTER, H. E. CLINE AND E. F. KOCH: *Trans. Metall. Soc. AIME*, 1969, **245**, 2073;
H. E. CLINE AND J. L. WALTER: *Metall. Trans.*, 1970, **1**, 2907

31 G. SPILLER AND D. A. SMITH: *J. Cryst. Growth*, 1980, **50**, 445

32 J. L. WALTER AND H. E. CLINE: Proc. Conf. *In Situ* Composites I, National Materials Advisory Board Publication 308-I, 1973;
J. L. WALTER: Proc. Conf. *In Situ* Composites II, Xerox Individualized Publishing, Lexington, Mass., 1976

33 E. R. THOMPSON AND F. D. LEMKEY: *Metall. Trans.*, 1970, **1**, 2799

34 J. D. LIVINGSTON: Proc. Conf. *In Situ* Composites I, National Materials Advisory Board Publication 308-I, 1973

35 J. D. LIVINGSTON AND H. E. CLINE: *Trans. Metall. Soc. AIME*, 1969, **245**, 351

36 A. HELLAWELL: Proc. Conf. *In Situ* Composites III, Ginn and Company, Lexington, Mass., 1979

**REFERENCES CHAPTER 3**

37 M. HILLERT: 'Solidification and casting of metals', 81, 1979, London, Metals Society

38 H. D. BRODY AND S. A. DAVID: 'Solidification and casting of metals', 144, 1979, London, Metals Society

39 A. HELLAWELL: *Met. Mater.*, 1967, **11**, 361

# 4
# Practice of directional solidification

## 4.1 INTRODUCTION

The previous two chapters have dealt with the relationships between directional solidification conditions and the microstructures of superalloys and eutectic composites. This section will consider the problems associated with the production of useful components, having a desired microstructure, by various approaches to directional solidification. For clarity the evolution of methods of controlling heat flow to obtain optimum solidification characteristics and of fabricating components will be described separately. Historically, of course, these two aspects have developed in parallel having been motivated both by a drive for more economic production rates and by a need to overcome particular difficulties encountered in directionally solidifying certain new alloys, such as *in situ* composites. However, in the following we shall indicate the principal heat transfer characteristics associated with different directional solidification processes in relation to a simple cylindrical casting geometry before discussing the problems of preparing components of complex shape.

Analytical solutions to the appropriate heat flow equations are only available for a few combinations of simple geometries of both the casting and the furnace where a single simple heat transfer mechanism is operative (e.g. cylindrical castings and axial conduction). More usually, the heat flow analyses that are available have been obtained by numerical solution of the appropriate equations. No attempt will be made here to describe the details of these analyses. Rather, the contributions of various cooling mechanisms will be indicated by simple physical arguments and, where appropriate, the characteristics of various directional solidification rigs in terms of the principal processing variables will be cited without full proof. Indeed, it is likely that rigorous heat transfer analyses have been most useful in optimizing approaches to directional solidification that have been initially devised by fairly qualitative considerations.

## 4.2 HEAT FLOW CONTROL

It was shown in Chapters 2 and 3 that directional solidification depends on the presence of a temperature gradient and that the microstructures of directionally solidified materials depend on the magnitudes of the temperature gradient in the melt at the solidification front $G_L$ and on the rate of solidification $V_I$. Generally, developments in directional solidification processing technology have been directed to

(a) suppression of variations in $G_L$ during directional solidification in order to maintain a consistent microstructure
(b) increasing $G_L$ to permit faster $V_I$ in order to obtain finer microstructures with improved mechanical properties, and
(c) independent control of $G_L$ and $V_I$ to both maintain the required microstructures in components of complex shape and effect economies of operation.

There is some arbitrariness in categorizing the various types of apparatus that have been applied to directional solidification. Here we roughly follow the chronological sequence by grouping systems according to the disposition of the casting relative to the heating source.

### 4.2.1. Stationary casting and stationary furnace

The simplest method of imposing some directionality during freezing is to cast into a mould that has efficient heat extraction at one end and good thermal insulation elsewhere. Figure 4.1 represents a cylindrical mould with a cooled baseplate that is maintained at a constant temperature $T_0$ and ideally insulated sides through which *no* heat can flow. When molten alloy at its melting temperature $T_m$ is poured into the mould, heat can only be extracted through the base so that heat flow is unidirectional along the cylinder axis and solidification proceeds from bottom to top. The only temperature variation is in the solid as indicated in Fig. 4.1 since the liquid is at the melting temperature $T_m$. The general differential equation describing linear heat flow in a static system is given by Carlslaw and Jaeger:[1]

$$\alpha \frac{d^2T}{dx^2} = \frac{dT}{dt} \tag{4.1}$$

where $\alpha = K/\rho C$, $\rho$ = density, $C$ = specific heat and $K$ = thermal conductivity of the alloy (all may be different in the solid and liquid states which are represented by the subscripts s and L respectively). Equation (4.1) can only be solved analytically for the simple cases where there is perfect thermal contact at the cooling plate (i.e. the casting in contact with the cooling plate always has the temperature $T_0$), there is no melt superheat (i.e. the liquid metal is cast at its freezing point) and there is no melting range (i.e. pure

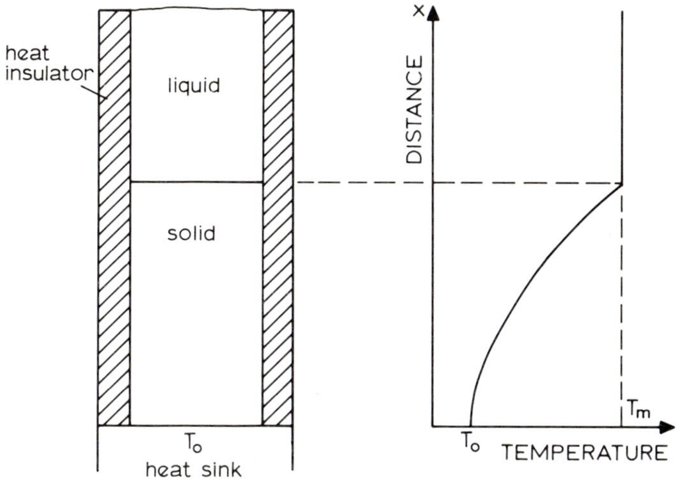


**Figure 4.1 Schematic illustration of the temperature distribution during solidification of an alloy cast at its melting point $T_m$ into a mould with perfectly insulating sides and a cooled base at temperature $T_0$**

metal or eutectic alloy). Then

$$T = A + B \operatorname{erf}\left(\frac{x}{2\sqrt{\alpha_s t}}\right) \tag{4.2}$$

where erf $(y)$ is the Error Function and the constants $A$ and $B$ are determined by the following boundary conditions:

(i) $T = T_0$ at $x = 0$ for all $t$
(ii) $T = T_m$ at $x = X$, where $X$ is the length of solidified ingot
(iii) $K_s(dT/dx)_{x=X} = \rho L_f V_I$, where $L_f$ is the latent heat of fusion.

Condition (iii) expresses the need to remove the heat generated on freezing. Putting condition (ii) into eqn (4.2) we have

$$T_m = A + B \operatorname{erf}\left(\frac{X}{2\sqrt{\alpha_s t}}\right) \tag{4.3}$$

which requires that the argument of the Error Function, $X/2\sqrt{\alpha_s t}$, be constant ($=\phi$, say). Then

$$X = 2\phi\sqrt{\alpha_s t} \tag{4.4}$$

indicating that the solidification front proceeds as a parabolic function of time.

Differentiation of eqn (4.4) gives an expression for $V_I$:

$$V_I = \frac{dX}{dt} = \frac{2\alpha_s\phi^2}{X} \tag{4.5}$$

This indicates that the rate of solidification progressively decreases as freezing proceeds. Putting the boundary conditions (i) and (ii) successively into eqn (4.2) gives expressions for $A$ and $B$:

$$A = T_0 \qquad B = \frac{T_m - T_0}{\operatorname{erf} \phi}$$

Before applying the heat flux conservation condition (iii), eqn (4.2) must be differentiated.

$$\frac{dT}{dx} = \frac{B}{\sqrt{\pi\alpha_s t}} \exp\left(-\frac{x^2}{4\alpha_s t}\right) \tag{4.6}$$

Then

$$\left(\frac{dT}{dx}\right)_{x=X} = \frac{2(T_m - T_0)}{\sqrt{\pi}X} \frac{\phi}{\operatorname{erf} \phi \exp \phi^2} = \frac{\rho L_f V_I}{K_s} \tag{4.7}$$

i.e.

$$\sqrt{\pi}\phi \operatorname{erf} \phi \exp \phi^2 = \frac{(T_m - T_0)K_s}{\alpha_s \rho L_f} \tag{4.8}$$

Equation (4.8) can be solved by iterative methods to give $\phi$ as a function of the parameters on the right hand side of the expression. The result is shown in Fig. 4.2.[2] Clearly the boundary temperatures $T_0$ and $T_m$ and the physical constants of the alloy completely determine both the rate of solidification $V_I$ and the temperature gradient in the solid $G_s$ both of which vary during casting, i.e.

$$V_I \propto \frac{1}{X}, \qquad G_s \propto \frac{1}{X} \quad \text{and} \quad G_L = 0.$$

In the absence of superheat there is no temperature gradient in the liquid metal. The temperature in the solid rises from $T_0$ at the cooling plate to the casting melting point $T_m$ over the thickness of the solidified zone $X$ and is constant ($=T_m$) in the remaining melt as is shown schematically in Fig. 4.1. When the temperature of the melt is initially above $T_m$ ($=T_s$, say) then the temperature distribution, shown schematically in Fig. 4.3*a*, involves temperature gradients in both the solid and liquid. Consequently, the boundary condition indicating heat flux conservation at the solidification front must be generalized to include a contribution from conduction in the melt. The heat

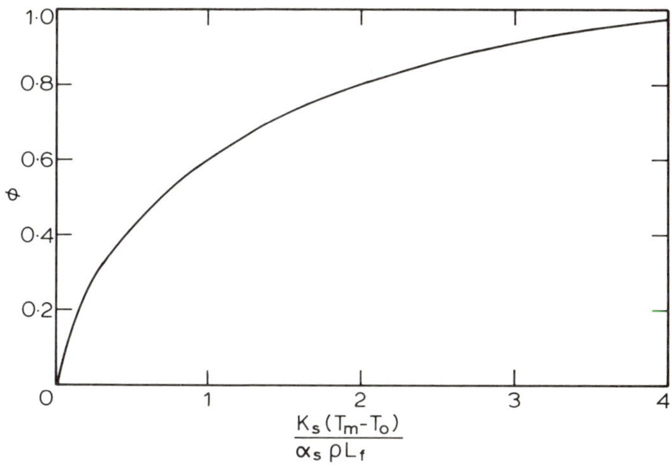


**Figure 4.2** **Variation in $\phi = x/(2\sqrt{\alpha_s t})$ with $(T_m - T_0)K_s/\alpha_s \rho L_f$ for the casting configuration shown in Fig. 4.1 [Ref. 2]**

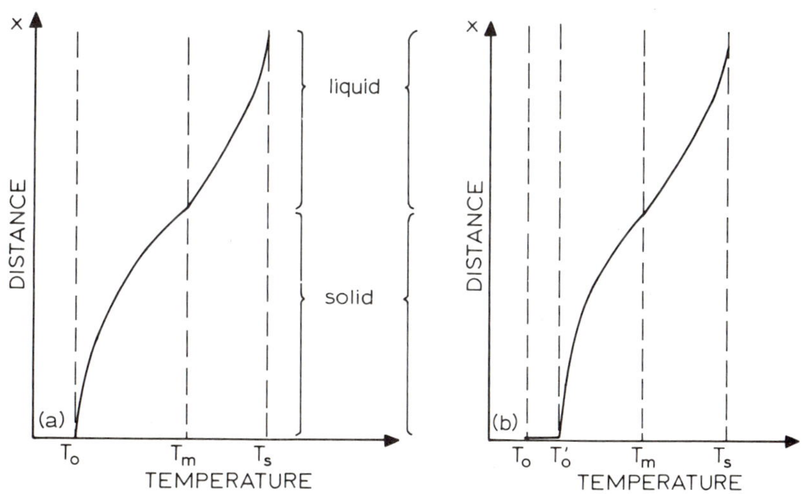


**Figure 4.3** **Schematic illustrations of temperature distributions in the directional casting configuration shown in Fig. 4.1 for (a) a higher melt temperature $T_s$ than the melting point, and (b) resistance to heat transfer across the melt/cooling plate interface**

generated during solidification is therefore the difference in heat conducted through the liquid and solid phases.

(iv) $$K_s\left(\frac{dT}{dx}\right)_{x=X}^{\text{solid}}-K_L\left(\frac{dT}{dx}\right)_{x=X}^{\text{liquid}}=\rho L_f V_I$$

Thus the temperature distribution in the casting is not a smooth curve but there is a sudden change in slope on crossing the solid/liquid interface as indicated in Fig. 4.3*a*.

Also, there will normally be a finite resistance to heat flow between the casting and the cooling plate so that the temperature of the ingot adjacent to the chill plate will not be constant but will vary during solidification of the casting. The temperature of the first solid portion $T'_0$ adjusts so that the heat flux $q$ to the cold sink $T_0$ equals that being conducted to the end of the ingot. To a first approximation $q$ may be described by the Newtonian heat transfer equation

(v) $$q=h(T'_0-T_0)=K_s\left(\frac{dT}{dx}\right)_{x=0}$$

where $h$ is a constant referred to as the heat transfer coefficient.

Hills, Malhotra and Moore[3] have considered the effects of both superheat and Newtonian heat transfer at the chill during directional casting of congruently melting alloys (i.e. $\Delta T=0$). They have solved eqn (4.1) numerically subject to the boundary conditions (i), (ii), (iv) and (v) and identify four different cooling and solidification modes that may occur during the freezing of a finite casting. The different stages, not all of which need occur in a given sequence and which are illustrated schematically in Fig. 4.4, are arbitrarily categorized by (a) whether the entire melt changes in temperature on cooling or if only a transition zone does so, and (b) whether or not solid is present. The liquid progressively cools from the chilled end until the temperature is low enough for solid to form. The results of their calculations are compared with experimental measurements on the lead–tin eutectic in Fig. 4.5. The variation in thickness of solid eutectic $X$ with time for various melt temperatures (Fig. 4.5*a*) shows that the parabolic growth rate at zero superheat, which is predicted by both eqn (4.4) and the numerical calculations, does not occur at higher degrees of superheat; rather, the curves become slightly S-shaped although they are generally parallel to each other, indicating that control of superheat has little influence on $V_I$. The temperature of the ingot adjacent to the chill is shown as a function of time in Fig. 4.5*b*; those parts of the curves that lie below the eutectic temperatures are roughly parallel, being displaced by about the amount of superheat ($T_s-T_m$). The calculations do not show the temperature distributions along the castings so that $G_L$, the temperature gradients in the liquid at the solid/liquid interfaces, are not available. However, the similarity in shapes of the curves in Figs. 4.5*a* and

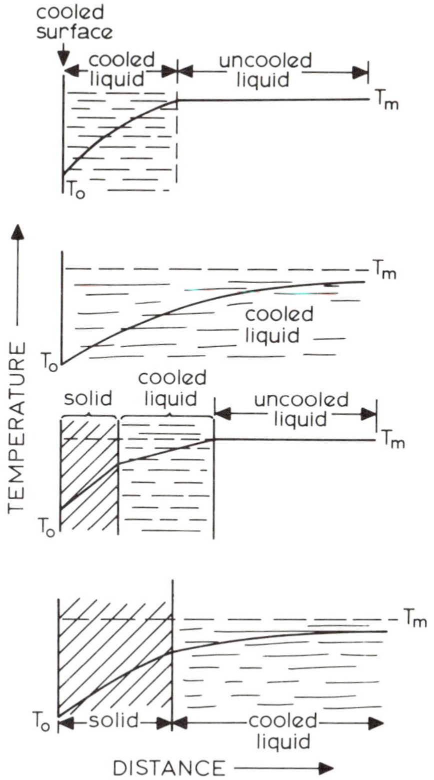


**Figure 4.4 Various types of temperature distributions that can occur during the solidification of castings of congruently melting alloys of finite size [Ref. 3]**

*b*, which are simply mutually displaced, suggests that the general temperature distributions at the moving solidification fronts are insensitive to the level of superheat which essentially defines a delay period during which pre-solidification cooling occurs.

Solidification during the directional casting described above is completely predetermined by the size and geometry of the component, by certain physical properties of the alloy (e.g. latent heat, specific heat, density), by the thermal characteristics of alloy, mould and cooling plate and by the temperatures of the initial melt and of the heat sink. Once cast, there is no control either over the local solidification rate or of the temperature gradients. Consequently, since the microstructures obtained will vary between different parts of the casting, the approach cannot be used in this simple form.

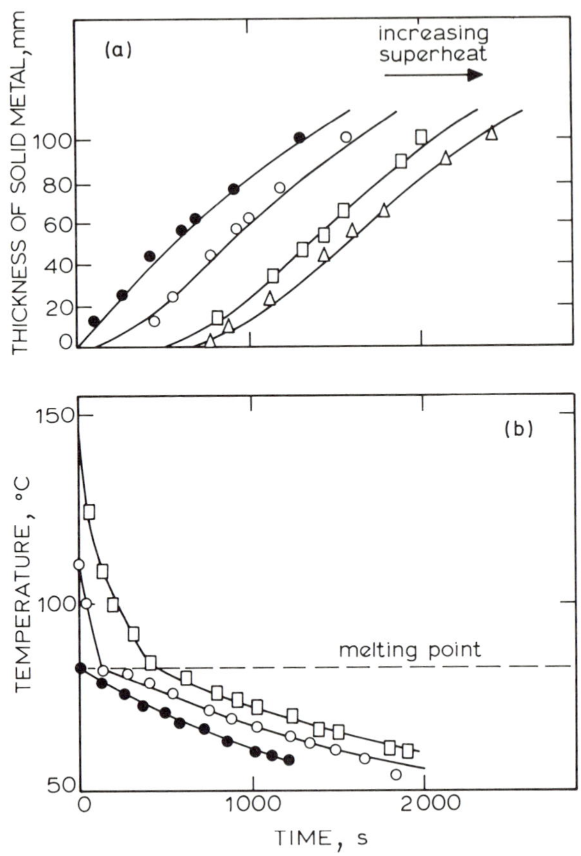


**Figure 4.5** **(a) Comparison of the calculated and observed times at which solidification occurs during directional casting as a function of distance from the chill for various superheats of a Pb–Sn eutectic melt**
**(b) Temperatures in melt adjacent to chill [Ref. 3]**

If the melt is poured into a heated, rather than cold, mould at a temperature above the melting point of the alloy and the temperature gradually reduced in a predetermined manner, then the velocity of the solidification front is dependent on the difference between the heat input from the heating source $q_{\mathrm{IN}}$ and the heat output to the chill plate $q_{\mathrm{OUT}}$ which can be related to the latent heat of fusion. This gives a new boundary condition for the heat flow equation (4.1).

(vi) $\quad q_{\mathrm{OUT}} - q_{\mathrm{IN}} = \rho L_{\mathrm{f}} V_{\mathrm{I}}$

Thus, whereas simple directional casting leads to a progressive decrease in $V_I$, it is possible in principle to maintain a constant $V_I$ by controlling the heat input. However, this approach gives little control over the other important solidification parameter, $G_L$. Nevertheless, the principle of programmed heat input was the basis of the first commercial application of directional solidification to the manufacture of turbine blades for aero-gas turbines.[4] Several modifications of this approach have been used.

**(a) Exothermic moulds**

The early experimental work of VerSnyder and Guard[5] made use of exothermic compounds packed around the sides of the mould and of a water-cooled base in order to control the cooling rate. This process leads to relatively low temperature gradients and cannot be controlled after the charge has been poured into the mould. The variability of the process makes it unsuitable for the manufacture of large components or when there is a demand for premium quality castings. However, the low capital cost has led to renewed interest for applications where small batches of components are needed.[6,7] Figure 4.6 shows a schematic diagram of an investment casting mould incorporating exothermic compound that has recently been designed for the production of components for small engines.

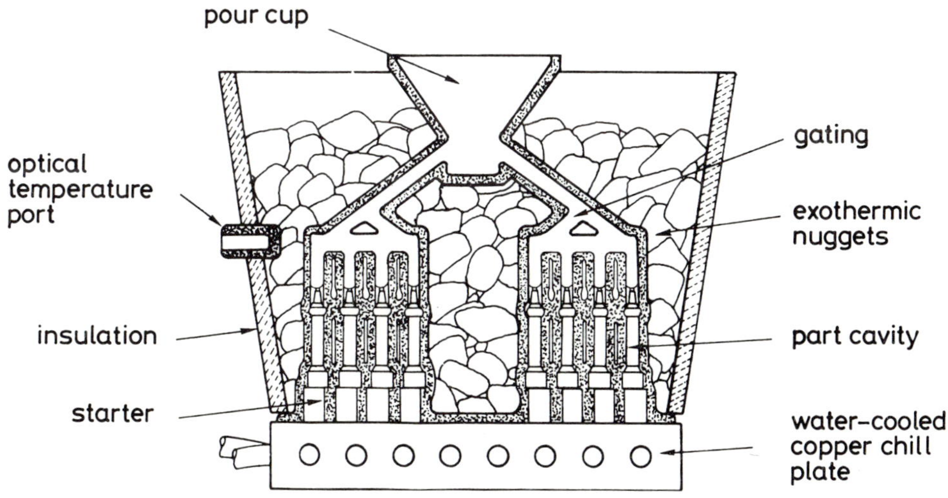


**Figure 4.6 Schematic cross-section of mould and chill for the production of directionally solidified turbine blades by the exothermic process [Ref. 5]**

### (b) Power-down process

In this approach the power control is via a furnace that is external to the casting. Consequently the casting temperature can be monitored and the rate of power reduction modified as required to give a steady solidification rate. The background to the development of the process has been described in some detail by VerSnyder *et al.*[8] Initially the mould was heated resistively via two independent molybdenum windings that surrounded the ceramic mould as is shown schematically in Fig. 4.7. Operation of the process, in particular the preparation of moulds, was simplified by the use of induction heating directly coupling into conventional investment moulds as illustrated in Fig. 4.8. By careful design of the winding configuration and programming the power input into different sections of the coil, quite regular cooling rates can be achieved as illustrated by the plot shown in Fig. 4.9 of times when the solidus and liquidus temperatures occur along the length of a directionally solidifying ingot.

In all of these static solidification processes the distance between the dendritic growth front and the heat sink gradually increases with a concomitant decrease in temperature gradient. This is illustrated by the heat transfer calculations of Monteith and Piwonka[9] for the solidification of idealized blade-like shapes of MarM200 by the power-down process. Their results, shown in Fig. 4.10, plot the temperature distribution along the casting after

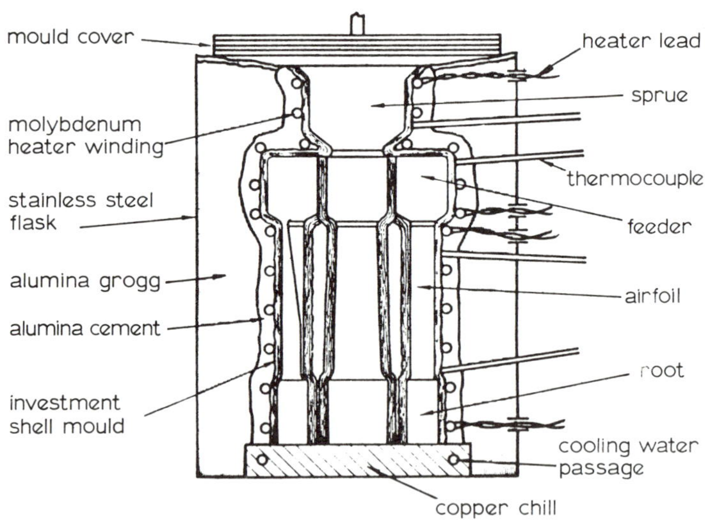


**Figure 4.7 Schematic cross-section of resistance heated mould assembly for casting directionally solidified turbine blades by the power-down process [Ref. 7]**

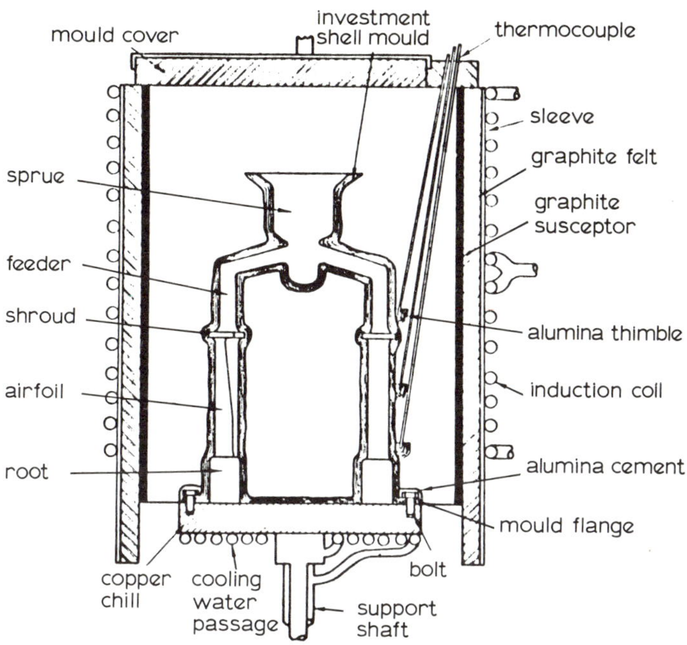


**Figure 4.8 Schematic cross-section of inductively heated mould assembly for directional solidification by the power-down process [Ref. 7]**

various periods from the start of solidification; this indicates a progressive reduction in temperature gradient from $\sim 15\ \mathrm{K\,mm^{-1}}$ at the root to $\sim 2\ \mathrm{K\,mm^{-1}}$ at the tip of the casting. Giamei and VerSnyder[10] are more pessimistic about the temperature profile obtained by this method, suggesting that a typical temperature gradient 100 mm from the chill would be $1\ \mathrm{K\,mm^{-1}}$ using the highest practical levels of superheat (up to ~1600°C). Under these conditions the time required to directionally solidify a turbine blade ~200 mm in length is approximately 2 h. Attempts to increase the solidification rate by reducing the power input more rapidly simply lead to a reduction in temperature gradient. We have seen in Chapter 2 that the ratio $G_L/V_I$ is important in determining the solidification morphology; when this ratio falls below a critical value then the structure reverts from columnar to equiaxed and at an even earlier stage freckles may form. Consequently there is a maximum practical length of ingot that can be maintained with a columnar microstructure. Modifications of the process that lead to higher temperature gradients will have the dual benefits of allowing higher solidification rates that are more

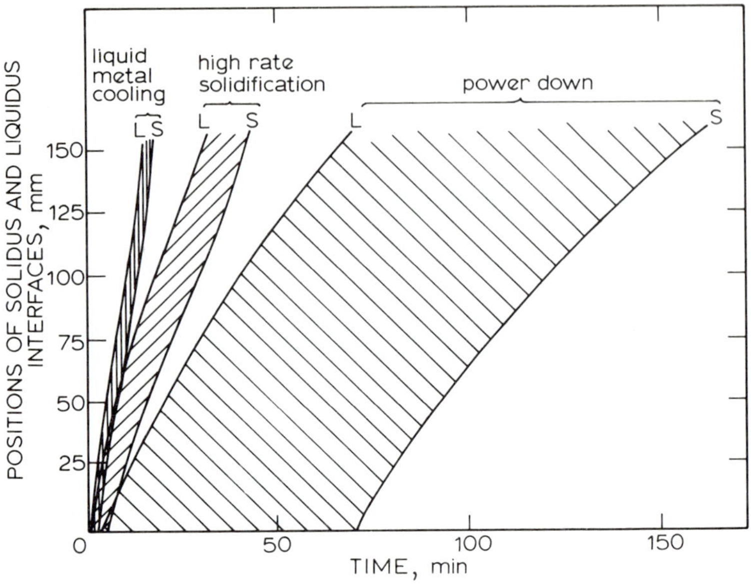


**Figure 4.9 Comparison of the solidus and liquidus temperature isotherms along an ingot as a function of time for MarM200 directionally solidified by different processes [Ref. 13]**

commercially attractive and of allowing aligned structures to be maintained over longer ingots.

### 4.2.2 Moving furnace or ingot

Most current laboratory and commercial processes are variants of this method, often designated the HRS (high rate solidification) process, which is a derivative of the Bridgman–Stockbarger crystal growing technique. The casting is removed at a controlled withdrawal rate from the furnace (or vice versa) with the latter maintained at full power.

Unidirectional heat flow along a solid moving with a velocity $V_I$ is described by Carlslaw and Jaeger[1] as

$$\alpha \frac{d^2T}{dx^2} - V_I \frac{dT}{dx} = \frac{dT}{dt} \tag{4.9}$$

This is of the same form as eqn (4.1) with an additional term to account for the heat removed due to translation of the ingot. Consider the case of an infinitely long cylinder of radius $r$ moving from a heating zone with a temperature $T_s$ into a cooling zone of temperature $T_0$. The $x$-axis is along the

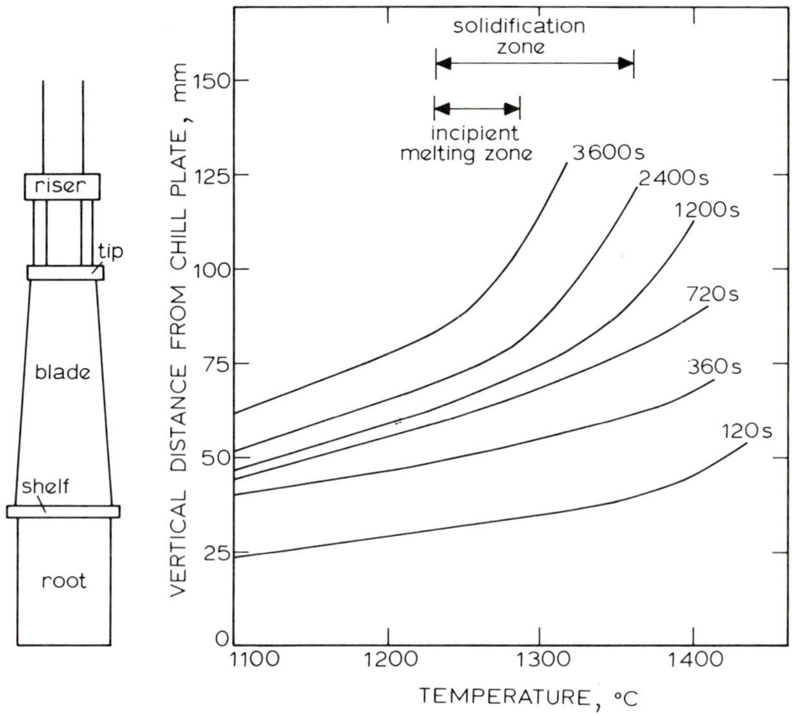


**Figure 4.10 Calculated plots of the temperature distributions that occur at various times during directional solidification of MarM200 in a typical blade shape by the power-down process [Ref. 9]**

cylinder axis and the origin is at the junction between the heating and cooling zones as illustrated in Fig. 4.11. Clearly the solution to eqn (4.9) must satisfy the boundary conditions

(vii) $T \to T_s$ as $x \to +\infty$

$T \to T_0$ as $x \to -\infty$

In addition there must be conservation of heat flux at the solid/liquid interface represented by boundary condition (iv).

A schematic temperature distribution is indicated in Fig. 4.11 showing the solid/liquid interface positioned at a distance $X$ above the junction between the heating and cooling zones. As $V_I$ is increased $X$ will become progressively smaller. Since cooling invariably occurs by way of the ingot surface and therefore involves radial heat flow when $x < 0$, axial solidification

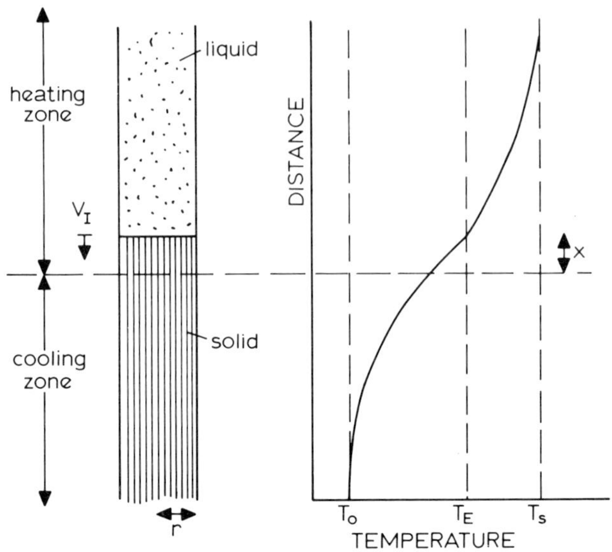


**Figure 4.11 Schematic illustration of directional solidification of a rod by moving it relative to fixed heating source and cold sink and the steady state temperature profile**

can only occur as long as $X>0$ when there is longitudinal heat flow at the solid/liquid interface by conduction along the solid ingot. The *maximum* practicable velocity $(V_I)_{max}$ therefore occurs when $X=0$. Livingston *et al.*[11] have shown that when a steady state is established (i.e. $dT/dt=0$) and there is Newtonian heat transfer across the ingot surface (i.e. $q=h(T-T_0)$)

$$(V_I)_{max}=\frac{1}{C}\left(\frac{2hK_s}{r}\right)(Z^{1/2}-Z^{-1/2}) \tag{4.10}$$

where

$$Z=\frac{L_f+C(T_s-T_0)}{L_f+C(T_s-T_m)}$$ and $C$ is the specific heat.

Thus $(V_I)_{max}$ can be increased by increasing the cooling rate (i.e. by increasing $h$ or by decreasing $r$ or $T_0$) or by decreasing the melt superheat ($T_s$).

Heat is dissipated by a combination of conduction along the ingot to the chilled base (which dominates at the beginning of the process) and of radiation from the mould surface to the colder surroundings outside the furnace, which becomes increasingly important as solidification proceeds. It is essential to minimize the radiant heat loss from the furnace to the cooling zone in order to maximize the temperature gradient. In particular, radiation shields form an efficient baffle at the furnace base when designed with the smallest opening compatible with the casting geometry in order to encourage directional heat flow normal to the component cross-section when the solid/liquid interface is maintained above the level of the baffle. The baffle essentially separates the heat source (furnace) from the heat sink (exterior) allowing near steady state conditions to be approached. Schubert[12] has noted that there is an optimum distance of this baffle from the furnace to ensure that the solidification front is maintained on the furnace side of the baffle. For rod shaped specimens, this 'cooling distance' should exceed $(K_s V_I)/C$.

At the beginning of directional solidification by HRS, the major source of heat loss is by conduction to the chill through the solidified portion of the ingot, but this contribution decreases linearly with the length of ingot solidified $X$. Following Giamei and Tschinkel[13] the heat loss $q$ from a portion of the ingot at temperature $T$ may be expressed

$$q = (h_C + h_R)(T - T_0) \tag{4.11}$$

where $h_C$ and $h_R$ are the conductive and radiative heat transfer coefficients:

$$h_C = \frac{K_s}{X}$$

and

$$h_R = \frac{\sigma(\varepsilon T^4 - \alpha T_0^4)}{(T - T_0)} \tag{4.12}$$

($T_0$ is the temperature of the heat sink, $\sigma$ is the Stefan–Boltzmann coefficient, $\varepsilon$ is the emissivity of the ingot and $\alpha$ is the absorbivity of the surroundings.) Clearly $h_C$ progressively decreases with increasing $X$ whereas $h_R$ remains constant. Taking $\varepsilon = \alpha = 0{\cdot}5$, $\sigma = 1.88 \times 10^{-7}\ \mathrm{J\ m^{-2}\ s^{-1}\ K^{-4}}$, $T = 1590\ \mathrm{K}$ and $T_0 = 0$,

$$h_R \sim 150\ \mathrm{J\ m^{-2}\ s^{-1}\ K^{-1}}.$$

Assuming a typical value of $K_s = 24{\cdot}5\ \mathrm{J\ m^{-1}\ s^{-1}\ K^{-1}}$ for superalloys

$$h_C = 2450\ \mathrm{J\ m^{-2}\ s^{-1}\ K^{-1}} \quad \text{when } X = 10\ \mathrm{mm}.$$

Thus initially $h_C \gg h_R$ but the two terms become equal when $X \sim 160$ mm. After this length is solidified a steady state is rapidly established. This is shown in the plot of the positions of the solidus and liquidus surfaces with time shown in Fig. 4.9. Comparison with the curve for the power-down process (Fig. 4.9) shows that the duration of the solidification cycle has been cut by a factor of four by using high rate solidification.

When steady state with dominant radiative heat loss has been established, the factors governing the temperature distribution in the ingot can be estimated. The difference between the heat flowing through cross-sections of the ingot at distance $-x$ and $-(x+\mathrm{d}x)$ below the solidification front may be equated to the heat loss by radiation through the edge of the volume element $q_{\mathrm{R}}$, i.e.

$$q_{-x} - q_{-(x+\mathrm{d}x)} = q_{\mathrm{R}} \tag{4.13}$$

For a cylindrical ingot of radius $r$,

$$\pi r^2 K_{\mathrm{s}} \left(\frac{\mathrm{d}T}{\mathrm{d}x}\right)_{-x} - \pi r^2 K_{\mathrm{s}} \left(\frac{\mathrm{d}T}{\mathrm{d}x}\right)_{-(x+\mathrm{d}x)} = 2\pi r \sigma \varepsilon (T^4 - T_0^4)\,\mathrm{d}x$$

or

$$\frac{\mathrm{d}^2T}{\mathrm{d}x^2} = \frac{2\sigma\varepsilon}{K_{\mathrm{s}} r}(T^4 - T_0^4) \tag{4.14}$$

But at steady state, when $\mathrm{d}T/\mathrm{d}t = 0$, eqn (4.9) may be written

$$\alpha \frac{\mathrm{d}^2T}{\mathrm{d}x^2} - V_{\mathrm{I}} \frac{\mathrm{d}T}{\mathrm{d}x} = 0 \tag{4.15}$$

Thus

$$\left(\frac{\mathrm{d}T}{\mathrm{d}x}\right)_{x=X}^{\text{solid}} = \frac{2\alpha\sigma\varepsilon}{V_{\mathrm{I}} K_{\mathrm{s}} r}(T_{\mathrm{m}}^4 - T_0^4) = G_{\mathrm{s}} \tag{4.16}$$

The temperature gradient in the liquid is given by boundary condition (iv)

$$\left(\frac{\mathrm{d}T}{\mathrm{d}x}\right)_{x=X}^{\text{liquid}} = \frac{1}{K_{\mathrm{L}}}\left[K_{\mathrm{s}} \left(\frac{\mathrm{d}T}{\mathrm{d}x}\right)_{x=X}^{\text{solid}} - \rho L V_{\mathrm{I}}\right] = G_{\mathrm{L}} \tag{4.17}$$

Clearly $G_{\mathrm{L}}$ is sensitive to both $V_{\mathrm{I}}$ and $r$, thin ingots and slow solidification rates leading to higher temperature gradients. However, eqn (4.17) shows that if a critical value of $V_{\mathrm{I}}$ is exceeded then $G_{\mathrm{L}}$ becomes negative

$$(V_{\mathrm{I}})_{\mathrm{crit}} = \left[\frac{2\alpha\sigma\varepsilon}{r\rho L}(T_{\mathrm{m}}^4 - T_0^4)\right]^{1/2} \tag{4.18}$$

The *maximum* steady state temperature gradient that can be achieved by the HRS process is completely determined by the radiation characteristics included in eqn (4.16) and the dimensions of the casting. Any major improvement depends on replacing the radiative cooling step by a more effective heat transfer mode.

### 4.2.3 Methods of using improved cooling

Enhanced cooling has been achieved on the laboratory scale by a variety of approaches such as tightly fitting cooling rings,[14] water sprays,[15] liquid metal cooling[16] and fluidized bed cooling.[17] All are designed to give higher heat transfer coefficients than can be obtained by radiation and involve the maintenance of close contact between the cooling medium and the solidifying ingot or component. Some of the techniques that have been applied successfully to simple cylindrical ingots on a laboratory scale are less appropriate to the production of complex shaped components (e.g. cooling rings and water sprays). The most versatile methods have used fluid cooling media that can accommodate quite complex changes in section (e.g. liquid metal and fluidized bed cooling).

Tschinkel *et al.*[18] have developed the use of liquid metal cooling during directional solidification, using tin as the quenching medium, into a viable commercial process capable of directionally solidifying quite complex clusters of turbine blades. An advanced form of this commercial rig is shown schematically in Fig. 4.12.

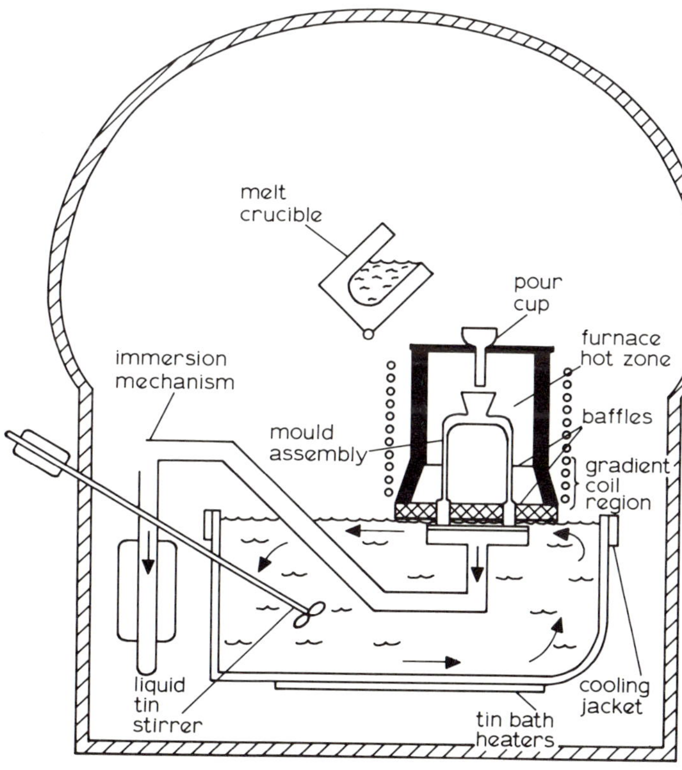


**Figure 4.12 Schematic illustration of a high temperature gradient liquid metal cooling furnace [Ref. 18]**

Curran *et al.*[19] have described the results of heat transfer calculations of directional solidification involving liquid tin cooling. Figure 4.13 shows the effects of the temperatures of the heating zone and of the top surface of the cooling medium on the temperature gradient. For relatively low hot zone temperatures (<1500°C) the coolant temperature has little effect on the temperature gradient; at higher melt temperatures there is a clear advantage in reducing the coolant temperature. One method of doing so is by agitation of the cooling fluid to maintain a uniform temperature throughout the quenchant. Combination of forced agitation and reduction in coolant temperature leads to the efficiency of the process being limited by the heat input capability. As with the HRS process, the design of heat shields and baffles between the heating and cooling zones is crucial to the optimization of the process. In particular, concentration of the heat input near the base of the furnace adjacent to the coolant bath, through the use of inductive concentrating devices or shaped furnace elements, is particularly effective in increasing the temperature gradient. The progressive changes in hot zone profiles obtained by Curran *et al.*[19] through modification of the baffle design are illustrated in Fig. 4.14. This

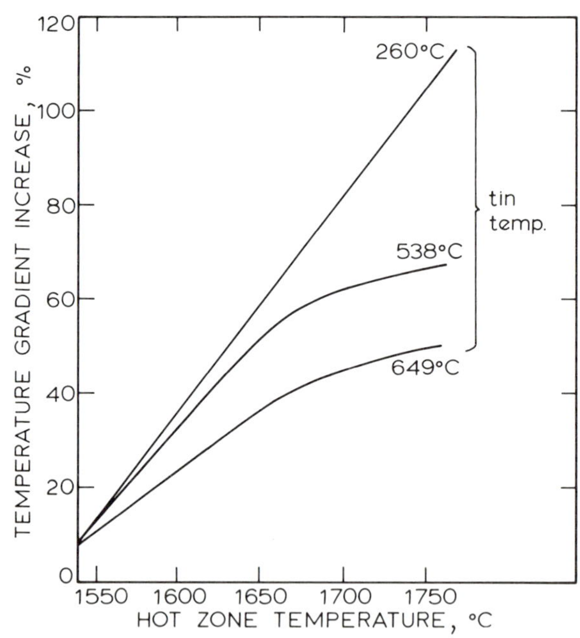


**Figure 4.13 Effect of hot zone temperature and cooling bath temperature on the temperature gradient for a solidification rate of 51 mm h$^{-1}$ [Ref. 19]**

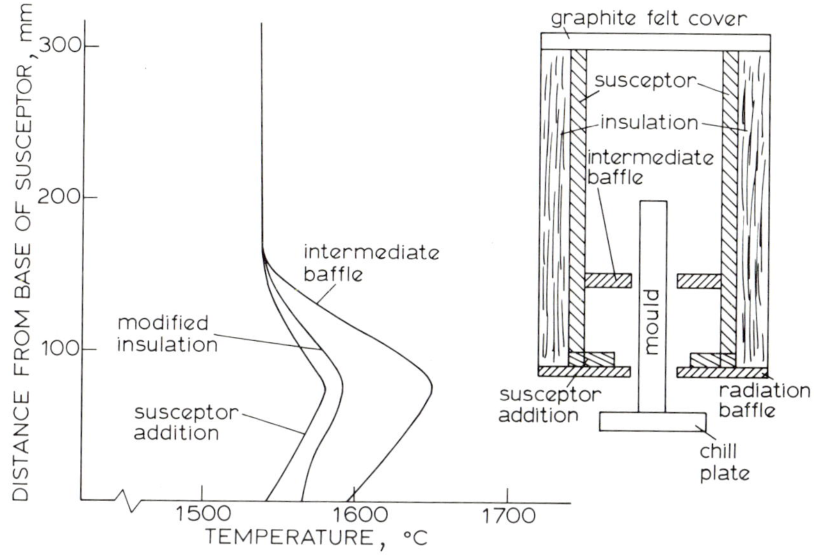


**Figure 4.14 Effects of baffle, insulation and susceptor modification on the hot zone profiles of a directional solidification rig incorporating liquid metal cooling [Ref. 19]**

can lead to an approximate doubling of the temperature gradient from $\sim 10\ \mathrm{K\ mm^{-1}}$ for the original stagnant liquid metal coolant to $\sim 20\ \mathrm{K\ mm^{-1}}$ for the optimized process (for ~13 mm diameter rods). Rabinovitch and Hauser[20] have made use of the liquid tin to enhance both the heat input and heat output in the apparatus illustrated in Fig. 4.15 for which they claim a temperature gradient of $15\ \mathrm{K\ mm^{-1}}$. This technique makes the further departure from the previous liquid metal cooling techniques of using prealloyed powder in thin pure alumina moulds rather than by casting the melt into conventional investment casting shell moulds.

By changing the configuration of withdrawal apparatus (either HRS or liquid metal cooling) to allow the heat input to be concentrated in a shallow pool immediately above the solidification front, Flemings and co-workers[21] have achieved very high temperature gradients on a laboratory scale ($\sim 100\ \mathrm{K\ mm^{-1}}$). One of several rigs developed by Flemings is shown in Fig. 4.16; all of them produce thin directionally solidified rod in a semi-continuous production rather than in component form.

Some further improvement may be possible through the use of cooling media that remain fluid at much lower temperatures. Bullock *et al.*[22] have

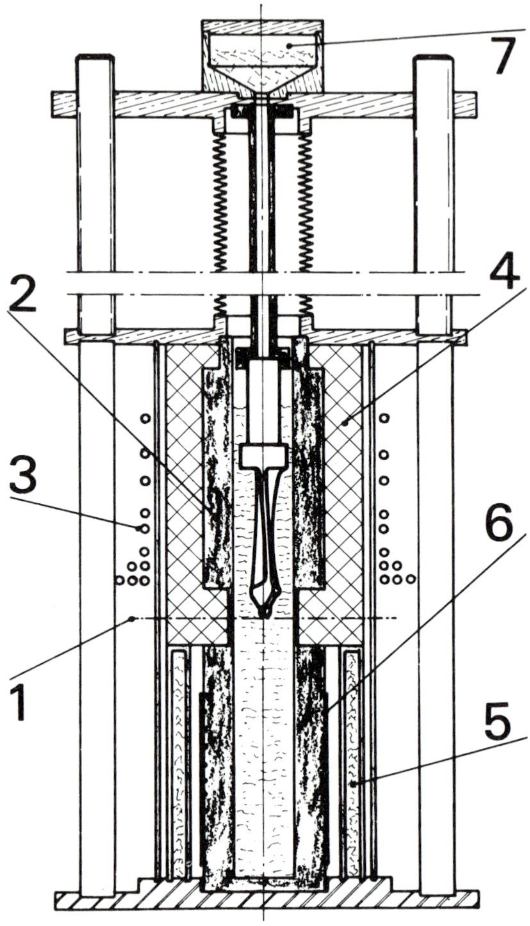


1 *solidification front*
2 *liquid tin column in graphite container*
3 *induction heating coil*
4 *insulation*
5 *cold sink*
6 *heater*
7 *pre-alloyed powder feed*

**Figure 4.15 Directional solidification rig for processing *in situ* composites that uses the good heat transfer characteristics of liquid tin to enhance both heating and cooling [Ref. 20]**

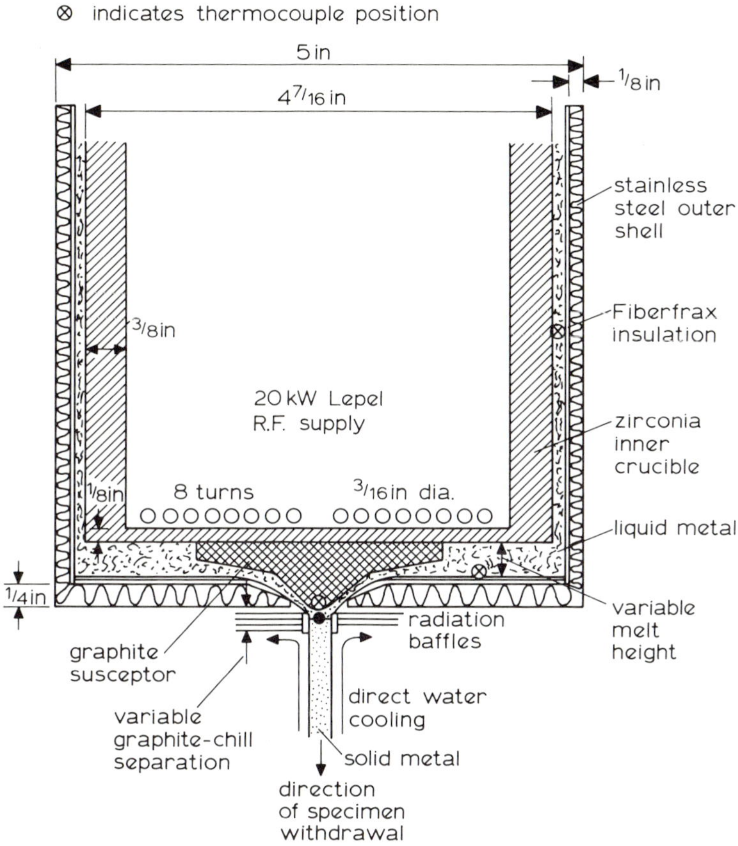


**Figure 4.16 Experimental directional solidification process which concentrates heat in a thin zone above the solid/melt interface [Ref. 21]**

described the use of low melting point metals (~70°C) on the laboratory scale in both the stagnant and agitated modes. These rigs, which are shown schematically in Fig. 4.17, are estimated to give temperature gradients of ~15 and ~25 K $mm^{-1}$ respectively for 12·5 mm diameter ingots and have been successfully applied to the directional solidification of both eutectics and superalloys. However, the compositions of such low melting point alloys, which contain such potentially noxious elements as bismuth and lead, make them unattractive for commercial application. Similarly, gallium and gallium/indium mixtures are unlikely to be used as coolants in component production because of the high cost and the possibility of liquid metal embrittlement of the product. In

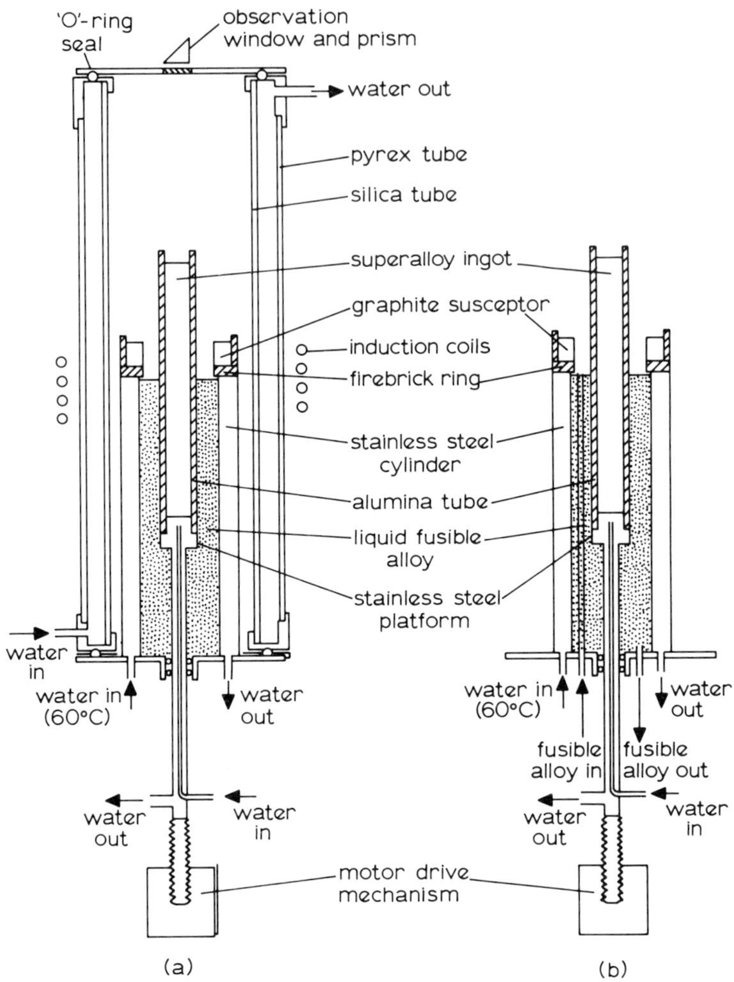


**Figure 4.17 Schematic diagram of modifications of a laboratory directional solidification process: (a) stagnant liquid metal cooling, and (b) pumped liquid metal cooling**

order to avoid these objections, Quested and Northwood[17] have developed the use of fluidized beds of stable ceramic powders (e.g. alumina) suspended in an inert gas (e.g. argon or helium) as quenching media and have applied this method, shown schematically in Fig. 4.18, to both eutectics and superalloys. It is estimated that temperature gradients approaching those achieved

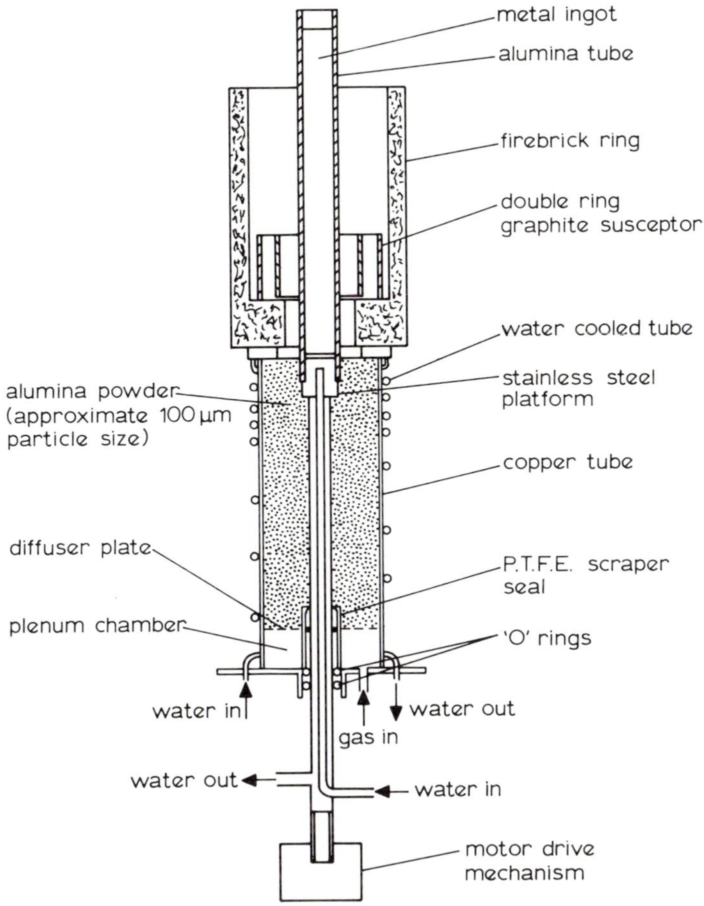


**Figure 4.18 Schematic illustration of a laboratory rig for directional solidification using fluidized bed cooling [Ref. 17]**

by liquid metal cooling can be obtained without the dangers of contamination that are inherent in most liquid metal coolants that have been proposed.

### 4.2.4 Miscellaneous methods

In all of the processes described above the entire charge is initially molten. Containment of certain alloys, particularly those containing reactive elements such as yttrium and titanium, in available ceramics can be a severe problem. Moreover, the attainable heat transfer coefficients are limited by the conductivities of the mould materials.

Zone melting has been used on the laboratory scale to produce small directionally solidified ingots. The theory and practice of this method has been fully described by Pfann.[23] Where relatively large ingots are produced, the charge must be contained within a ceramic sheath, as during the normal freezing procedure, but the period of contact between melt and mould is considerably shortened, thereby reducing any interaction between ceramic and melt. Floating zone techniques requiring no containment can only be used for ingot diameters less than about 10 mm. An additional advantage of this approach is that a steady state composition can be maintained; the narrow molten zone being fed by solid of controlled composition reduces the effects of segregation that can be significant during normal freezing of materials, such as the COTAC and NITAC *in situ* composites, that contain heavy elements. The major disadvantage is that component shapes cannot be produced. Consequently, the method is unsuitable for the production of engineering components such as turbine blades but may be of value in the manufacture of material required for their physical properties (e.g. electrical, magnetic).

Hellawell[24] has pointed out that electro-slag, or electro-flux, remelting is essentially a directional solidification process that is akin to zone-melting since a relatively narrow molten pool is continuously fed with solid bar stock. The process is represented schematically in Fig. 4.19. Electrical resistance heating causes melting of the pool which is fed through a floating molten slag

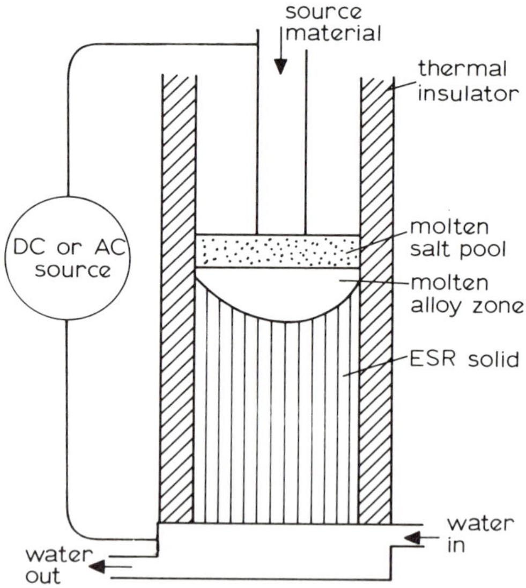


**Figure 4.19 Schematic illustration of a longitudinal section through a rod during electroslag remelting**

(or flux) and heat loss is by conduction along the ingot. The method can be applied to a large range of materials. However, it again produces the processed alloy in ingot rather than component form and consequently has not been considered seriously as a directional solidification technique. Johnson and Hellawell[25] and Quested *et al.*[26] have attempted to apply the technique to the directional solidification of eutectic alloys. Although some phase alignment can be achieved, as shown in Fig. 4.20 for the (Fe, Cr, Al, Y)–$Cr_7C_3$ *in situ* composite, lack of control over the process leads to microstructural variations. The main attraction of the technique is in offering relatively cheap bulk ingots in a directionally solidified form that could be used for less sophisticated applications (e.g. bolts) than at present.

Edge-defined film-fed growth (EFG), which is a derivative of the Czochralski crystal growth technique, has been investigated as a method of producing either long tape or aerofoil shapes of directionally solidified eutectics.[27,28] The method, illustrated schematically in Fig. 4.21, depends on the capillary rise of the melt up a die which is not wet by the liquid charge. A seed crystal makes contact with the melt emerging from the die and is withdrawn at a predetermined rate. Relatively large temperature gradients of ~40–60 K $mm^{-1}$ have been claimed for this method since the small section sizes enhance heat transfer. However, there appear to be problems associated with

**Figure 4.20 Longitudinal section of a (Fe, Cr, Al, Y)–$Cr_7C_3$ *in situ* composite after electroslag remelting**

the durability of the available die materials (e.g. TaC, TiN and HfC) which tend to interact with the melts of the potentially useful *in situ* composites such as $\gamma$–$\gamma'$–$\delta$ and the COTAC alloys. Nevertheless, well-aligned $\gamma$–$\gamma'$–$\delta$ has been produced by this route in the form of ribbons although attempts to prepare aerofoil-shaped hollow bar gave cellular $\gamma$–$\gamma'$–$\delta$.

## 4.3 COMPONENT MANUFACTURE

In principle, components could be machined from directionally solidified ingots or near-shape castings. Although this approach is often used to produce testpieces on a laboratory scale and has been adopted to facilitate early engine trials of materials that are difficult to directionally solidify, its high cost militates against its being exploited commercially. Rather, the technology that has been developed to prepare conventionally cast components such as gas turbine blades has been extended to allow fully shaped parts, often with complex internal channels and requiring minimal finishing, to be cast directly in the directionally solidified form. The process of directional solidification requires better melt/mould compatibility than in conventional casting and also controlled heat flow to ensure that solidification can occur continuously along the

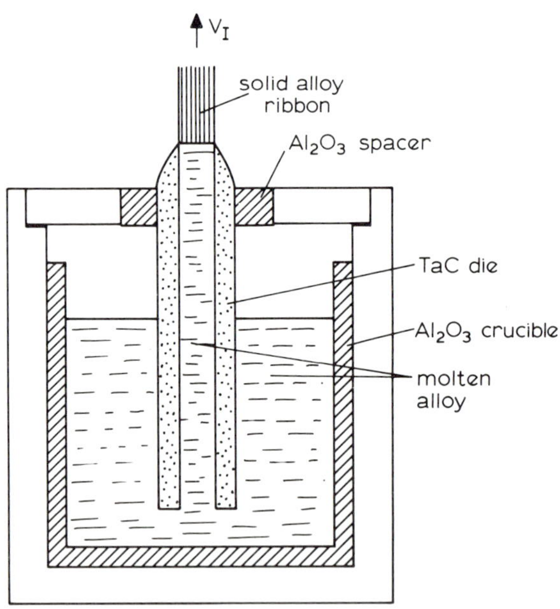


**Figure 4.21 Schematic illustration of the configuration for directional solidification by edge-defined film-fed growth [Refs. 27, 28]**

length of the component; these conditions are often difficult to satisfy, particularly for some highly reactive eutectic alloys which require very slow solidification rates to achieve plane front solidification and very tight compositional control. In such cases it may be necessary to consider radical changes in either the moulding materials or the approach to component manufacture.

The investment casting or 'lost-wax' technique has been utilized to produce cast shapes for many centuries (certainly dating to 3000 BC in Mesopotamia). The various stages in application to gas turbine components from superalloys are illustrated schematically in Fig. 4.22. They are as follows:

(a) Wax models of the components are prepared and arranged in clusters connected by wax replicas of runners and risers to enable multiple parts to be produced in a single casting. The size of the arrays currently used to produce directionally solidified turbine blades varies between one and twenty depending on the complexity of the blade and the magnitude of the temperature gradient required to successfully process particular alloys. Thus directional solidification of complex air-cooled blades requires more careful heat flow control, and therefore smaller clusters, than solid blades and single blade processing has been adopted to give sufficiently high

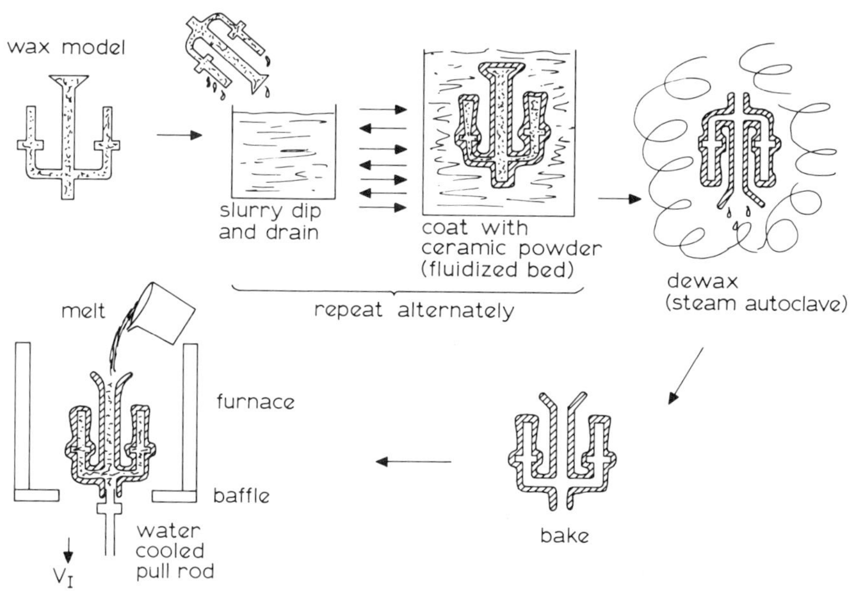


**Figure 4.22 Schematic illustration of the stages during the manufacture and use of investment moulds for directional solidification**

temperature gradients to ensure plane front solidification of certain *in situ* composites.

(b) The moulds are produced by coating these waxes with several layers of ceramic powders held together by a suitable binder. There are various ways of building up such a shell, but the 'shell-moulding' procedure described below is typical. The early and some current production of directionally solidified superalloys have used the same ceramics, silica and zircon, employed in conventional casting. The mould is deposited on the wax pattern by a succession of dips into various ceramic slurries; the precise procedure and nature of the slurries varies between foundries and is often proprietary. However, typically the first few coatings which will eventually be in contact with the melt consist of finely divided zircon ($ZrSiO_4$), alumina ($Al_2O_3$) or silica ($SiO_2$) bound by either colloidal silica or ethyl silicate while the main support for the mould may be provided by subsequent stuccoing by larger particles of zircon, silica, mullite or fused alumina. High temperature strength of the mould is important to prevent deformation due to the weight of the liquid metal charge.

(c) When the ceramic structure has been fully built up the mould is dried, dewaxed (usually using a steam autoclave) and fired in order to completely remove residual wax and to consolidate and strengthen the ceramic. The action of the binder can be quite complex, progressing through various chemical states during the drying and firing of the mould and leading to a progressive increase in strength. For example, the ethyl silicate binder first reacts with water to form a polymerous fluosilic acid sol:

$$Si(OC_2H_5)_4 + 4H_2O \rightarrow Si(OH)_4 + 4C_2H_5OH$$

When this dries, the sol decomposes imparting 'green' strength to the mould

$$3Si(OH)_4 \rightarrow H_2Si_2O_5 + H_2SiO_3 + 4H_2O$$

During the baking stage, further elimination of water leads to the formation of an $SiO_2$ cement of high strength

$$H_2Si_2O_5 + H_2SiO_3 \rightarrow 3SiO_2 + 2H_2O$$

If complex internal channels are required, as in air-cooled turbine blades, a ceramic core in the shape of the desired channels, that can be dissolved chemically after the casting is completed, is inserted in the mould. The core materials must be sufficiently stable to prevent interaction with the melt during casting, but must be able to be leached efficiently from the solid component. For conventional casting and directional solidification of relatively non-reactive alloys, silica is a suitable core material; however, when directionally solidifying reactive alloys, such as some eutectic composites, such coring is inadequate and this has necessitated the development of either new ceramics or of novel casting approaches that do not depend on the use of cores.

(d) The alloy charge is melted and cast into the mould normally in either vacuum or an inert atmosphere. For directional solidification the mould is removed at a controlled rate from a furnace designed from the criteria discussed in § 4.2. During conventional casting the mould remains at temperatures considerably below that of the melt and is in contact with molten metal for a relatively short time. The variation in mould and melt temperature with time after casting shown in Fig. 4.23*a* is fairly typical of turbine blade castings in nickel- or cobalt-base alloys. Thus the mould need only retain its integrity for a minute or so. For directional solidification, however, the mould is first heated to above the alloy solidus temperature to minimize the danger of mould failure on casting due to thermal shock and is subjected to thermal stresses because the mould base is artificially cooled to maintain a temperature gradient. The superheated charge is poured into the preheated mould which is then withdrawn from the furnace at a relatively slow rate so that liquid metal remains in

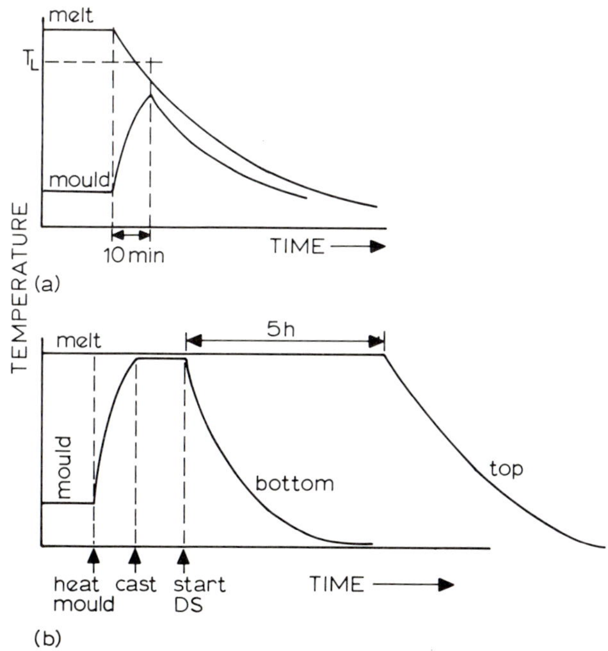


**Figure 4.23 Comparison of the changes with time of the melt and mould temperatures during (a) conventional casting and (b) directional solidification**

contact with some of the mould throughout the directional solidification procedure which can last from about 1 h for superalloys to about 24 h for some *in situ* composites. The variations in metal and mould temperatures during directional solidification are shown schematically in Fig. 4.23*b*. Clearly the directional solidification procedure places much greater demands on the chemical and mechanical integrity of moulds than does conventional casting.

Turbine blades of three different sizes and configurations for aero-gas turbine engines, made by this process, are shown in Fig. 4.24; all have been etched to show the grain structure.

**Figure 4.24 Directionally solidified turbine blades etched to show the grain structure [Rolls-Royce Ltd]**

Melt containment in the conditions required to successfully directionally solidify some advanced *in situ* composite materials, such as the NITAC series of alloys, has necessitated the development of new $Al_2O_3$-based moulding materials. There are two particular problems associated with some *in situ* composites that prevent the use of conventional zircon- or silica-based ceramics.

(a) Much higher superheat temperatures (~1750°C for NITAC 13) than for superalloys (~1500°C) are used in order to ensure high temperature gradients. At such temperatures the $SiO_2$ binder softens, or even melts, leading to sagging or rupture of the mould.
(b) These alloys contain much higher proportions than do superalloys of the reactive elements C and Al which can reduce the $SiO_2$ or $ZrSiO_4$ moulding materials leading to penetration of melt into the ceramic with deterioration of surface finish and to serious decarburization of the melts with associated loss of the near eutectic compositions that are essential to obtain composite microstructures.

Figure 4.25 compares the microstructures of the $\gamma$–$\gamma'$–$Cr_3C_2$ *in situ* composite after directional solidification in moulds of recrystallized alumina and of

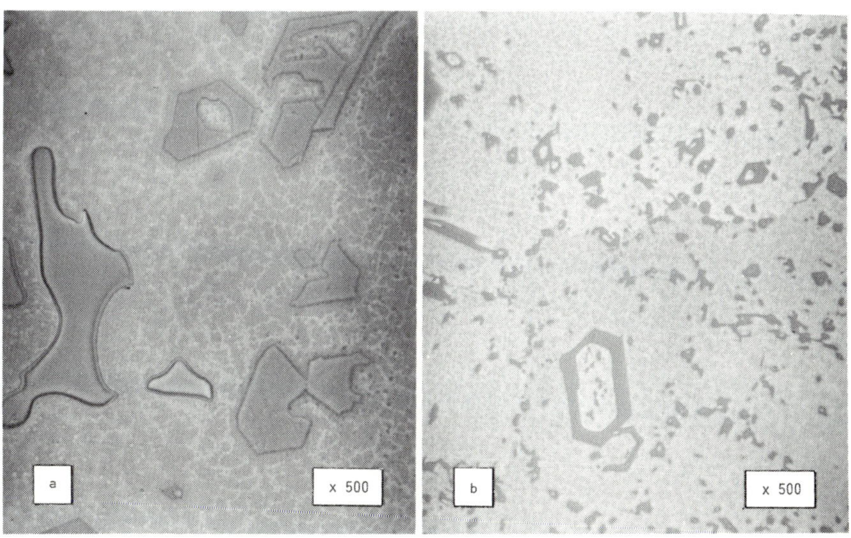


**Figure 4.25 Comparison of microstructure of the $\gamma$–$\gamma'$–$Cr_3C_2$ *in situ* composite after directional solidification in (a) zircon/silica investment mould and (b) recrystallized alumina**

zircon/silica; the latter shows the presence of a large proportion of an additional phase, probably ZrC, due to interaction with the mould that modifies the chemistry throughout the thickness of the casting causing a large deviation from eutectic composition.

Shell moulds of mixed $Al_2O_3/SiO_2$ composition (88 wt% $Al_2O_3$ + 12 wt% $SiO_2$) that are suitable for high-carbon eutectic alloys have been developed by General Electric; their preparation and properties are described by Greskovich *et al.*[29] and by Gigliotti and Greskovich.[30] The procedure was to dip the wax form into a suspension of fine $Al_2O_3$ powder (of specified particle size) and colloidal $SiO_2$ (36 wt% $SiO_2$, 64 wt% $H_2O$, 0·15 wt% $Na_2O$), drain and then dip into a fluidized bed of coarser powder (~300 μm). After drying the entire procedure was repeated several times with increasing $Al_2O_3$ particle sizes at the stuccoing stage until a rigid mould was obtained. On heat treatment above 1600°C but below 1890°C the $Al_2O_3$ and $SiO_2$ in the binder react to form mullite, $2Al_2O_3.SiO_2$, which has better thermal shock resistance, is less reactive, and is stable to higher temperatures than $SiO_2$. The reactivity between $Al_2O_3$ and $SiO_2$ to form mullite depends strongly on both the $Al_2O_3$ particle size and temperature in accord with reaction rate theory. Thus the best mould properties were obtained using the finest $Al_2O_3$ particle size considered (~3·5 μm). This was manifested as a greater proportion of mullite in the binder and a higher sag temperature of the moulds relative to the mould prepared with greater $Al_2O_3$ particle sizes.

The reactivity of such moulds to NITAC melts was shown to be very sensitive to the furnace atmosphere.[30] When directional solidification was carried out in an argon atmosphere there was extensive metal/mould reaction leading to penetration of the mould by metal, poor surface finish and decarburization of the alloys, all very similar to that found with more traditional mould materials. However, when an argon plus carbon monoxide mixture was used there was no discernible mould/metal interaction; this was attributed to the formation of a protective $Al_2O_3$ layer at the melt/mould interface. By control of the furnace atmosphere it has been possible to manufacture turbine blades of the NITAC alloys at rates as low as 6 mm $h^{-1}$.

An alternative approach to the containment of reactive eutectic alloys has been to use pure alumina moulds produced by plasma spraying. In this technique, pioneered by the ONERA group, the wax pattern is first coated by copper in order to dissipate the heat and then rotated in a plasma flame. After the wax is removed the mould is fired at 1800°C in order to achieve maximum densification. The main advantages of this approach are that metal/mould interactions are eliminated and that the moulds are extremely thin (~0·8 mm) giving better heat transfer and therefore higher temperature gradients during directional solidification. On the debit side the moulds are much more fragile than conventional shell moulds and are particularly vulnerable to thermal shock. Rabinovitch and Hauser[20] have overcome this problem by feeding the mould with prealloyed powder during directional solidification rather than by casting superheated melt into it.

Clearly $SiO_2$-based core materials will be inadequate for use with reactive melts to produce hollow components. Some success has been achieved with $Al_2O_3$ cores prepared by the Koralox process which appears to combine good metal mould compatibility with an ability to yield under cooling stresses and to be removed from the solid casting.[31] The secret appears to be deliberately to aim for incomplete densification of the ceramic compact which then collapses when stressed and to arrange for elongated grain boundaries to facilitate leaching from the solid blade. An alternative approach has been to avoid the use of cores by casting the component in two halves that can be joined by diffusion bonding to provide a hollow blade. This technique, which was developed by Phipps and Hayes[32] and is illustrated schematically in Fig. 4.26, has the advantage that the internal channels can be inspected but, of course, depends on a good bond being made between the two halves. In addition, attempts have been made to produce cooling channels in turbine blades by post-solidification drilling by a variety of techniques (chemical, mechanical, electron beam etc.), but none appear to be commercially viable.

Alexander and Graham[33] have speculated on how turbine blades might be fabricated from materials that are difficult to directionally solidify and some of the more plausible approaches are illustrated in Fig. 4.27. Figure 4.27*b* envisages the aerofoil and root sections being solidified separately and then being joined by brazing or diffusion bonding; with this method the high $G_L/V_I$ required to align the phases in some eutectic alloys could be achieved for the narrow aerofoil section where the improved high temperature capability is required while simpler and perhaps more appropriate materials could be used for the root. In a variant of this method (4.27*d*), entire blades might

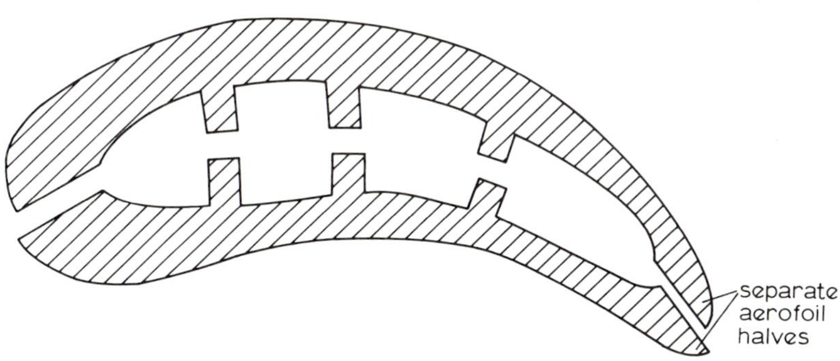


**Figure 4.26 Schematic illustration of hollow cooled turbine blades produced by diffusion bonding of directionally solidified segments [after Phipps and Hayes[32]]**

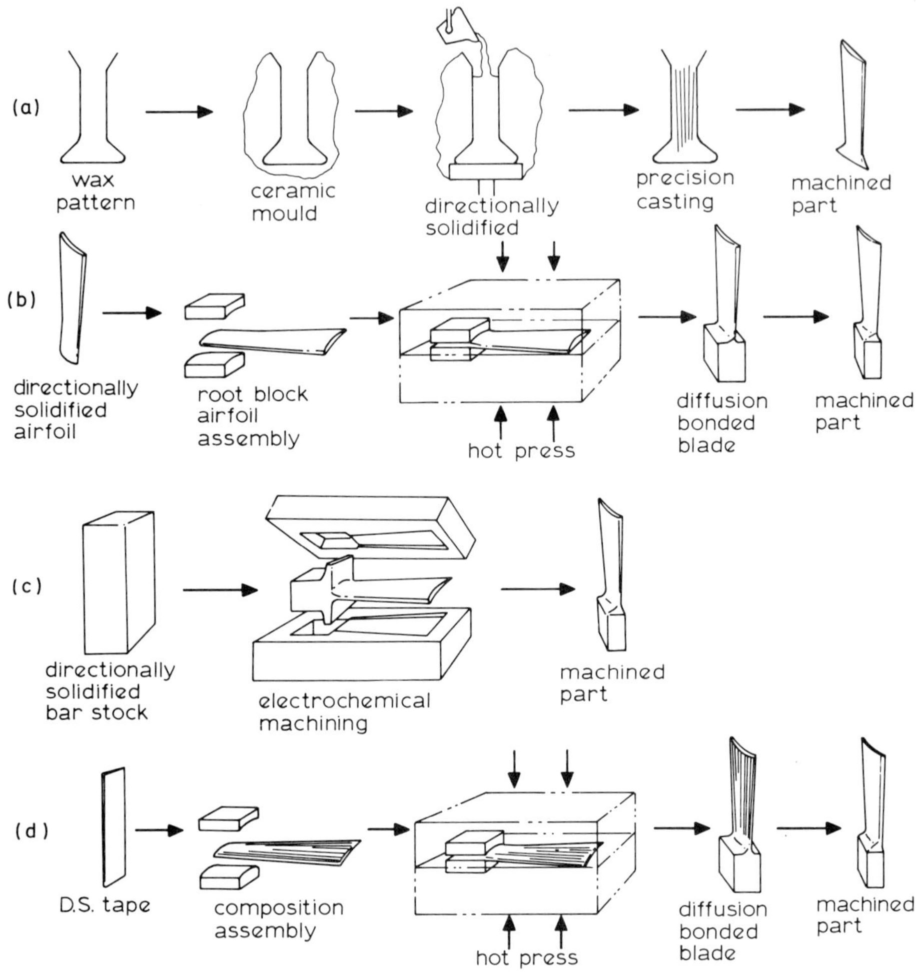


**Figure 4.27 Schematic illustration of different methods of producing components which consist entirely or partially of directionally solidified material: (a) investment casting of entire component; (b) diffusion bonding of directionally solidified and conventionally cast segments; (c) machining from directionally solidified block; (d) assembly using thin directionally solidified tape [after Alexander and Graham[33]]**

be fabricated by layering and bonding of directionally solidified tapes prepared by the method of Edge-defined Film Growth. Neither of these approaches appears to have been seriously tested. However, it has been demonstrated that strong eutectic/eutectic and eutectic/superalloy bonds can be prepared, by either brazing or diffusion bonding, that fail in one or other of the metals rather than at the junction between them. Examples of diffusion bonds for $\gamma$–$\gamma'$–$Cr_3C_2$ joined both to itself and to the superalloy MarM246 are shown in Fig. 4.28. Blades could also be formed or machined from blocks of *in situ* composite (Fig. 4.27*c*); although this would eliminate problems of surface roughness due to metal/mould interactions, (i) it would be extremely wasteful in material, (ii) it would be difficult to control microstructures in large section sizes and (iii) internal cooling channels would be difficult to insert.

## 4.4 EFFECTS OF SECTION SHAPE

The solidification morphologies obtained on directionally solidifying both superalloys and eutectic composites, discussed in Chapters 2 and 3, depend on

(i) the form of the advancing solid/liquid interface (planar, curved, rough etc.)
(ii) the temperature gradient at the solid/liquid interface
(iii) the rate of solidification.

All of these factors are closely related to the efficiency of heat extraction and the local directions of heat flow in the castings. The aim in designing directionally solidified components must be to ensure that heat can flow efficiently along the direction of withdrawal of the mould and that, in the case of *in situ* composites, a high temperature gradient is maintained. This clearly introduces

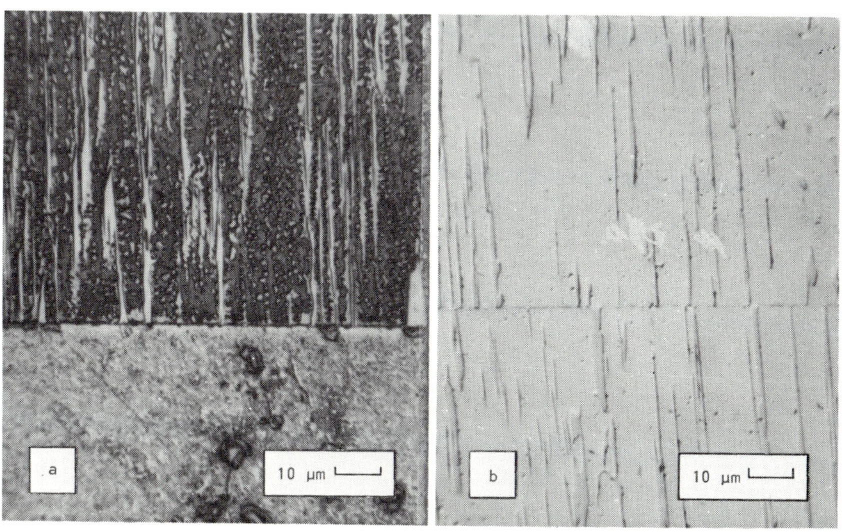


**Figure 4.28 Sections through diffusion bonds between similar and dissimilar alloys: (a) $\gamma$–$\gamma'$–$Cr_3C_2$ to MarM246 (b) $\gamma$–$\gamma'$–$Cr_3C_2$ to $\gamma$–$\gamma'$–$Cr_3C_2$**

an additional constraint to the design of castings and consequently the components designed for conventional casting will not necessarily be appropriate to directional solidification.

The effects of changes in section during directional solidification have been considered theoretically and tested with low melting point model systems by Farag and co-workers.[34,35] The simplest situations are illustrated schematically in Fig. 4.29 where there is a sharp (a) decrease and (b) increase in casting diameter during directional solidification. In the first case (decrease) the change will be associated with an increase in temperature gradient and solidification in the reduced section can proceed without nucleation of new solid phase. However, when the section increases (Fig. 4.29*b*) the *reduction* in temperature gradient is also associated with both (i) a distortion of the longitudinal heat flow directions along the ingot and therefore of the directions of the dendrites and reinforcing fibres and (ii) the need to establish the optimum growth morphology over a finite distance at the parts of the ingots that do not adjoin solid metal. If a common microstructural scale is desired, it will be necessary to programme the withdrawal rate to match the local temperature gradient, and this has been used notably by Sahm and Schubert[36] to maintain suitable aligned microstructures across the root/aerofoil transition in very large turbine blades of both superalloys and the 73C eutectic intended for use in industrial gas turbines. Clearly, it is preferable to solidify the large

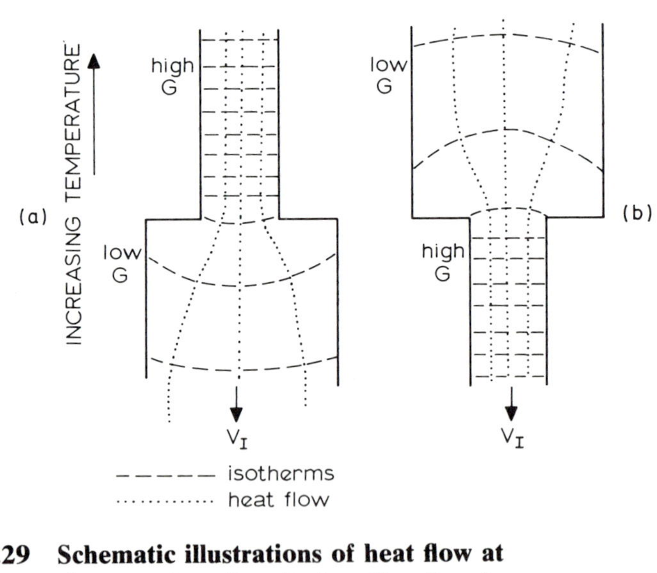


**Figure 4.29 Schematic illustrations of heat flow at changes of section during directional solidification: (a) decrease in section; (b) increase in section**

sections first in order to avoid the 'sort-out' zone in establishing aligned structures; where this is not possible, changes in section should be graduated rather than sharp.

The principal features are illustrated in the work of Rabinovitch and Hauser[20] on the alloy COTAC 74 which is illustrated in Fig. 4.30. The microstructure is quite well aligned and the only effect the sudden reduction in section has is the formation of a nickel-dendrite at the outside corner; this may be a consequence of the final solidification transient (cf. Fig. 2.4) since solidification terminates there. The effect at the increasing section is more

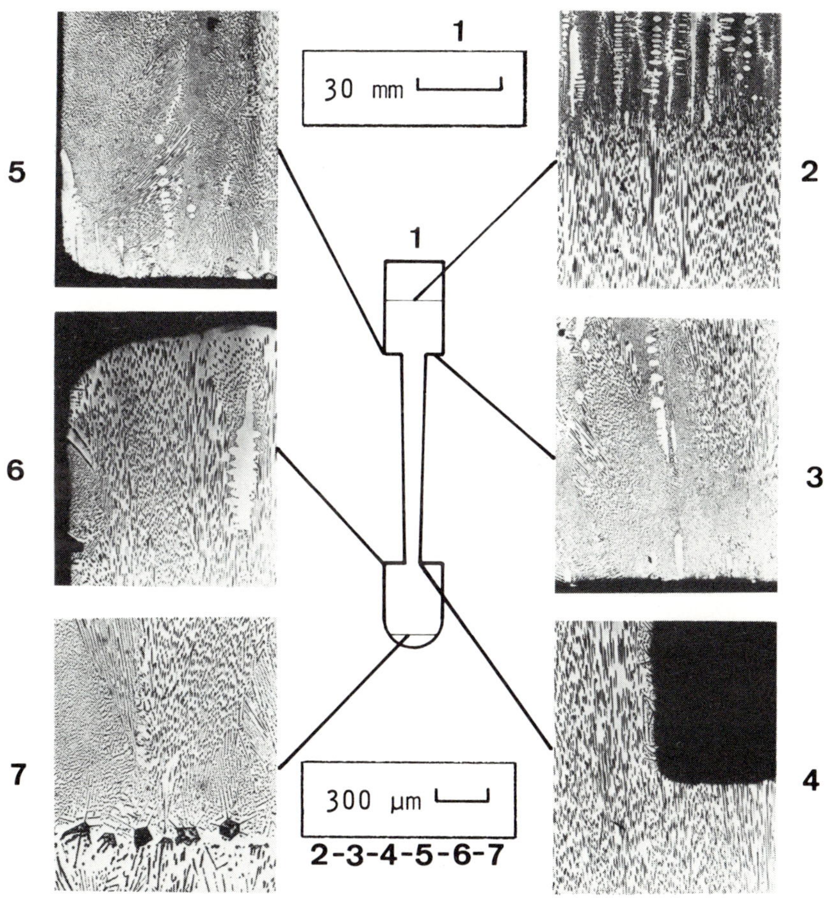


**Figure 4.30 Microstructure occurring at different places during directional solidification of COTAC 74 in the form of a shaped testpiece [Ref. 20]**

severe where extensive formation of dendrites shows that the reduced temperature gradient cannot satisfy the $G_L/V_I$ condition required to maintain the composite microstructure, and where the areas of eutectic structure are badly aligned.

## 4.5 SINGLE CRYSTALS

During directional solidification of superalloys a ⟨001⟩ texture develops along the direction of withdrawal by a process of competitive growth (§ 2.10). In the first solid portion there is a random distribution of orientations of the crystals formed; those crystals with ⟨001⟩ directions parallel to the solidification direction solidify more rapidly than, and eventually eliminate, the others by an overgrowth mechanism indicated schematically in Fig. 4.31. Consequently there is an initial transient region during which the density of elongated grains increases until a stable density is established where all the columnar grains are misorientated by rotations about ⟨001⟩. Figure 4.32 shows a longitudinal section of a directionally solidified plate, demonstrating the variation in grain structure. It is necessary, therefore, to ensure that any casting has a disposable starter so that the desired texture is established before solidification of the component commences.

After the ⟨001⟩ texture has been established, there is no driving force for further increase in grain size. However, Piearcey[37] showed that single

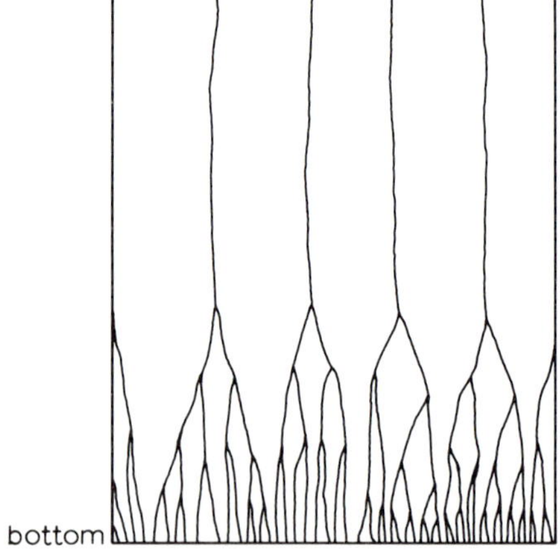


**Figure 4.31 Schematic illustration of the variation in grain structure along the length of a directionally solidified ingot**

**Figure 4.32 Longitudinal etched surfaces of directionally solidified IN738LC showing change in density of grains as solidification proceeds**

crystal turbine blades could be produced by inserting a constriction in the mould, after the initial transient, in order to select a single ⟨001⟩ crystal for the top part of the casting. Various types of constriction have been used successfully as grain selectors (e.g. spirals, right angles, kinks); one typical configuration is shown schematically in Fig. 4.33. All appear to operate effectively and subsequent growth of the single crystal appears to be no more difficult than normal directional solidification. Thus the plant designed for simple directional solidification can also be used to produce ⟨001⟩ single crystal components by only a simple change in the mould configuration.

Other crystal orientations can be produced by seeding with existing crystals of the desired orientation. This has been used extensively in laboratory studies of the anisotropic properties of superalloy single crystals, but clearly has disadvantages on an industrial scale.

**Figure 4.33 Configuration of casting to produce single crystal components**

*(courtesy of J. E. Northwood, NGTE)*

## REFERENCES CHAPTER 4


1 H. S. CARLSLAW AND J. C. JAEGER: 'Conduction of heat in solids', 1959, Oxford University Press

2 A. GARCIA AND M. PRATES: *Metall. Trans.*, 1978, **9B**, 449

3 A. W. D. HILLS, S. L. MALHOTRA AND M. R. MOORE: *Metall. Trans.*, 1975, **6B**, 131

4 F. L. VERSNYDER: US Patent No. 3 260 505, 1966

5 F. L. VERSNYDER AND R. W. GUARD: *Trans. ASM*, 1960, **52**, 485

6. D. H. COMEY: 'Cost benefit analysis of advanced materials technologies for small aircraft turbine engines,' NASA report CR-135265, September 1977

7 G. S. HOPPIN AND L. W. SINK: AIME Fall Meeting, October 1977

8 F. L. VERSNYDER, R. B. BARLOW, L. W. SINK AND B. J. PIEARCEY: *Mod. Cast.*, December 1967, p. 68

9 B. G. MONTEITH AND T. S. PIWONKA: *J. Vac. Sci. Technol.*, 1970, **7**, S126

10 A. F. GIAMEI AND F. L. VERSNYDER: Proc. 4th Bolton Landing Conference, June 1974

11 J. D. LIVINGSTON, H. E. CLINE, E. F. KOCH AND R. R. RUSSELL: *Acta Metall.*, 1970, **18**, 399

12 F. SCHUBERT: Thyssen Giesserei AG, Bochum, Germany, Final Report of COST 50 Programme D10/2, February 1976

13 A. F. GIAMEI AND J. G. TSCHINKEL: *Metall. Trans.*, 1976, **7A**, 1427

14 H. E. CLINE AND J. D. LIVINGSTON: *Trans. AIME*, 1969, **245**, 1987

15 F. D. LEMKEY: NASA Contractor Report NASA CR-2278, November 1973

16 P. R. SAHM AND M. LORENZ: *J. Mater. Sci.*, 1972, **7**, 793

17 P. N. QUESTED AND J. E. NORTHWOOD: British Patent No. 7943193, 1980

18 J. G. TSCHINKEL, A. F. GIAMEI AND B. H. KEAR: US Patent No. 3 763 926, October 1973

19 P. M. CURRAN, L. F. SCHULMEISTER, J. S. ERICKSON AND A. F. GIAMEI: Proc. Conf. *In Situ* Composites II, (eds. M. R. Jackson, J. L. Walter, F. D. Lemkey, R. W. Hertzberg), Xerox Individualized Publishing, Lexington, Mass., 1976

20 M. RABINOVITCH AND J. M. HAUSER: Proc. Conf. *In Situ* Composites III (eds. J. L. Walter, M. F. Gigliotti, B. F. Oliver, H. Bibring), Ginn and Company, Lexington, Mass., 1979

21 M. C. FLEMINGS, D. S. LEE, M. A. NEFF, K. P. YOUNG AND B. A. RICKINSON: Proc. Conf. *In Situ* Composites III (eds. J. L. Walter, M. F. Gigliotti, B. F. Oliver, H. Bibring), Ginn and Company, Lexington, Mass., 1979

22 E. BULLOCK, P. N. QUESTED AND M. MCLEAN: Proc. Conf. *In Situ* Composites III (eds. J. L. Walter, M. F. Gigliotti, B. F. Oliver, H. Bibring), Ginn and Company, Lexington, Mass., 1979

23 W. G. PFANN: 'Zone melting', 1966, New York, Wiley

24 A. HELLAWELL: Proc. AGARD Specialists Meeting on Directionally Solidified *In Situ* Composites, AGARD-CP-156, Washington, DC, 1974

25 A. J. D. JOHNSON AND A. HELLAWELL: *Metall. Trans.*, 1972, **3**, 1016

26 P. N. QUESTED, E. BULLOCK AND M. McLEAN: 'Solidification and casting of metals', 466, 1979, London, Metals Society

27 F. H. COCKS, J. T. A. POLLOCK AND J. S. BAILEY: Proc. Conf. *In Situ* Composites I, National Materials Advisory Board Publication 308-I, Lakeville, Conn., 1972

28 G. F. HURLEY AND N. W. MARR: Proc. Conf. *In Situ* Composites II, Xerox Individualized Publishing, Lexington, Mass., 1976

29 C. GRESKOVITCH, M. F. X. GIGLIOTTI AND P. SVEC: *Trans. J. Brit. Ceram. Soc.*, 1978, **77**, 98

30 M. F. X. GIGLIOTTI AND C. GRESKOVITCH: *Metall. Trans.*, 1980, **11A**, 319

31 *Aviation Week and Space Technology*, 11 August 1980, p. 15

32 C. M. PHIPPS AND D. R. HAYES: US Patent No. 3 965 963, 1976

33 J. A. ALEXANDER AND L. D. GRAHAM: Proc. AGARD Specialists Meeting on Directionally Solidified *In Situ* Composites, AGARD-CP-156, Washington, DC, 1974

34 N. A. EL-MAHALLAWY AND M. M. FARAG: 'Solidification and casting of metals', 106, 1979, London, Metals Society
35 N. A. EL-MAHALLAWY AND M. M. FARAG: Proc. Conf. *In Situ* Composites III, Ginn and Company, Lexington, Mass., 1979
36 P. R. SAHM AND F. SCHUBERT: Proc. Conf. *In Situ* Composites III, Ginn and Company, Lexington, Mass., 1979
37 B. J. PIEARCEY: US Patent No. 3 494 709, 1970

# 5
# Mechanical behaviour: superalloys

## 5.1 HISTORICAL PERSPECTIVE

Directional solidification was developed as a viable method of processing engineering alloys in response to a specific problem that emerged during the evolution of superalloys for application in gas turbines. The demand for alloys that retain their strength and stability to ever increasing temperatures was largely satisfied by the nickel-base superalloys that derived their strength from a dispersion of ordered $\gamma'$ particles. The approach to extending the temperature range of these materials has been to progressively increase the temperature at which the $\gamma'$ particles go into solution in the $\gamma$-nickel matrix; as the gap between the $\gamma'$ solvus and the alloy liquidus temperatures was narrowed it became impracticable to produce the alloys by thermomechanical processing (wrought superalloys) and components were prepared by precision casting. However, this change was accompanied by a reduction in creep elongation that was associated with considerable scatter in mechanical properties and often with limitations in thermal fatigue behaviour. It was recognized that fracture of such polycrystalline materials at high temperatures was generally initiated at grain boundaries lying transverse to the applied stress. VerSnyder and colleagues[1] reasoned that if the density of these detrimental transverse grain boundaries could be reduced by producing columnar grains by directional solidification, or even eliminated by growing single crystal components, then the fracture process should be suppressed and certain mechanical properties, particularly ductility, should be improved and indeed the benefits obtained in VerSnyder's early work on the alloy MarM200 were dramatic. However, as we discussed in Chapter 2, directional solidification affects several aspects of the microstructure, such as giving a preferred crystallography and reducing microporosity, in addition to altering the grain boundary morphology and the improvements in mechanical behaviour obtained through directional solidification are likely to result from a combination of the various microstructural characteristics.

The extent of improvement to creep behaviour that can be obtained is illustrated by the creep curves for the advanced nickel-base superalloy MarM200 in the conventionally cast, directionally solidified and single crystal forms shown in Fig. 5.1.[2] There is little difference in the early stages of creep deformations of the three types of materials, but whereas the conventionally cast specimen fails after 2–3% elongation, the columnar grained and single crystal alloys exhibit dramatic increases in the strain at rupture to ~20 to 30%; this is accompanied by substantially increased rupture lives relative to the conventionally cast alloy. However, such large improvements in life on directional solidification are not exhibited by all alloys and test conditions. The major benefits are reported for alloys which have low intrinsic rupture ductilities in the conventionally cast condition and during tests at relatively low temperatures. This is illustrated by the work of Piearcey and Terkelson,[3] shown in Fig. 5.2, which compares the temperatures giving lives of 100 h at stresses of 551·6 and 137·9 MPa of several alloys in the conventionally cast and directionally solidified conditions. Clearly the largest differences are at the higher stress (i.e. lower temperature) for the low ductility alloys MarM200 and IN100. This general trend has been confirmed by Northwood and Homewood[4] who show that the rupture lives of the alloys MarM246 and IN100 are extended by directional solidification whereas the alloys IN713 and IN738,

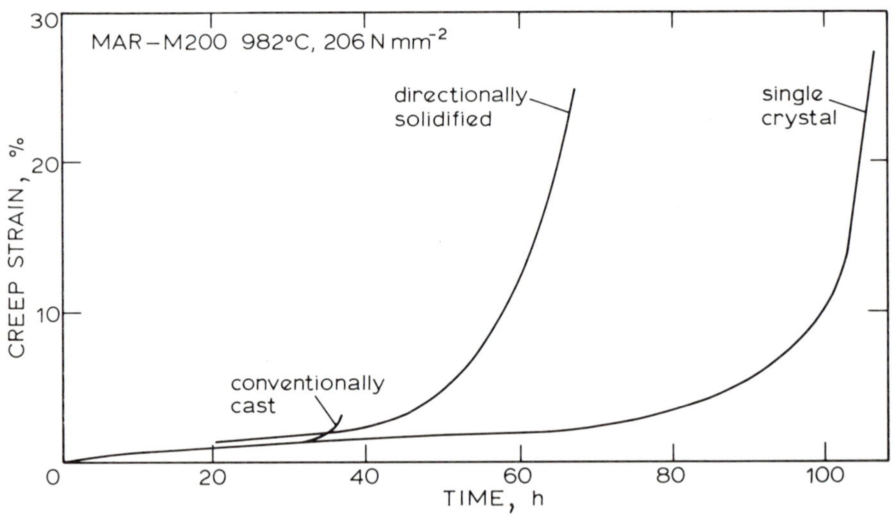


**Figure 5.1** **Comparison of creep curves for MarM200 tested at 1255 K and 206·8 MPa in (a) conventionally cast, (b) directionally solidified and (c) single crystal forms [after Kear and Piearcey[2]]**

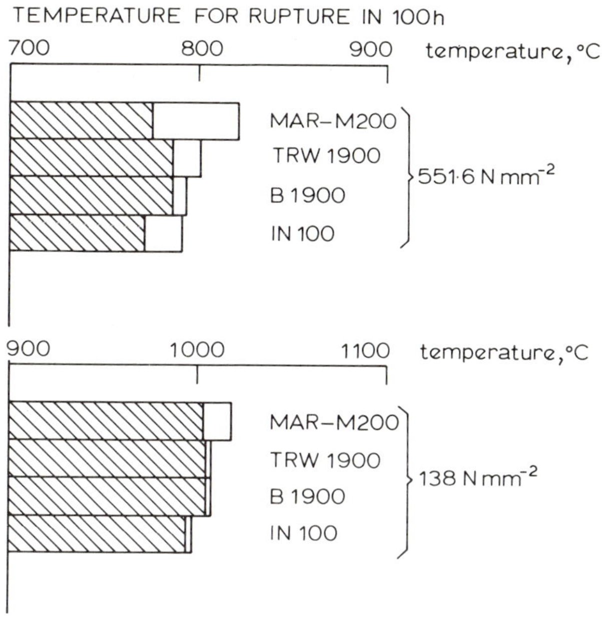


**Figure 5.2 Temperatures to produce rupture in 100 h at 551·6 and 137·9 MPa for various alloys in the conventionally cast and directionally solidified forms (hatched and open bars respectively) [after Piearcey and Terkelson[3]]**

which have higher ductility in the conventionally cast condition, apparently derive no such benefit. In further works on the alloy IN738 Woodford and Frawley[5] also reported no extension in life through directional solidification, but Sahm and Schubert[6] demonstrated a clear advantage in very long term tests as shown in Fig. 5.3.

The two alloys deriving the greatest advantage from directional solidification, MarM200 and MarM246, have very similar compositions differing only in the refractory element contents. (Alloy compositions are listed in the Appendix.) They have relatively low chromium levels and high volume fractions and solution temperatures of the $\gamma'$ precipitate phase. On the other hand, the alloys showing a marginal or no improvement in rupture life, such as IN713 and IN738, tend to have high chromium content and low $\gamma'$ volume fractions and solution temperatures. It is not clear from this evidence if these high chromium alloys are inherently less amenable to improvement by directional solidification or if it is relatively more difficult to achieve the

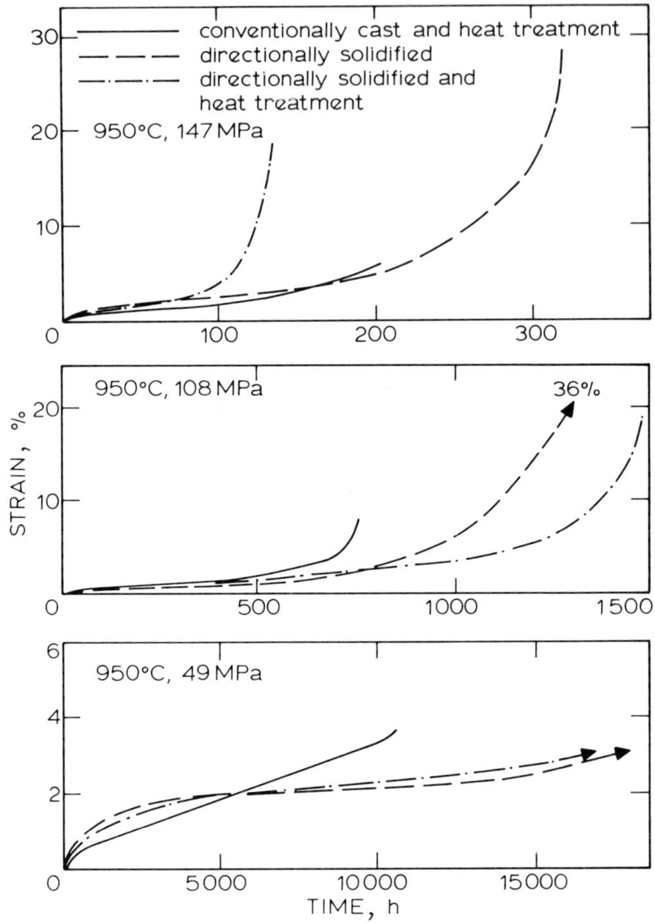


**Figure 5.3 Comparison of creep curves at 950°C with various stresses for IN738LC in the directionally solidified and conventionally cast forms [after Sahm and Schubert[6]]**

solidification conditions required to obtain the microstructures that lead to improved creep performance. The theme of the relationship of solidification microstructure to mechanical behaviour will be taken up later in this chapter.

In spite of the variability of rupture life improvement with alloy and testing conditions, all nickel-base superalloys have shown substantially increased creep rupture strains after directional solidification. Figure 5.4 indicates the band of elongations at fracture for superalloys in the conventionally cast and directionally solidified conditions; the columnar castings have

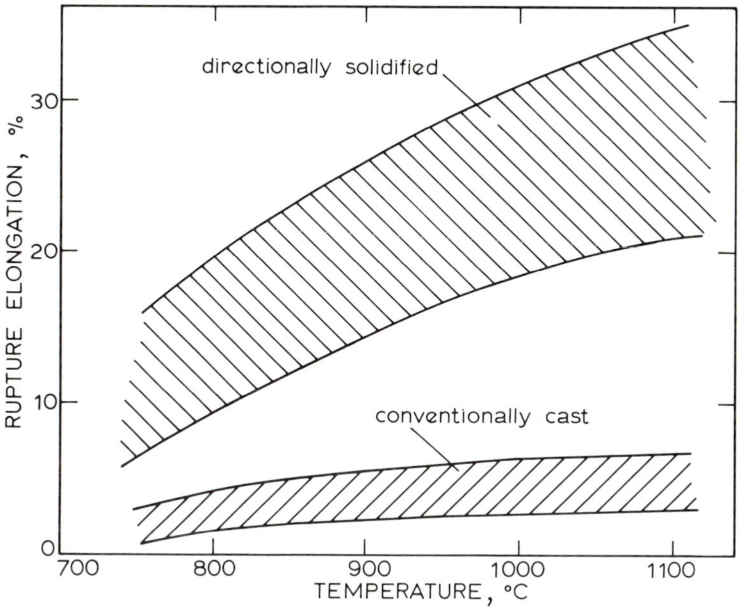


**Figure 5.4 Comparison of the rupture ductilities of nickel-base superalloys in the conventionally cast and directionally solidified conditions [after Piearcey and Terkelson[3]]**

considerably greater elongations at failure than the conventionally cast material. These improved ductilities allow relaxation of the stresses that develop during changing temperatures and this is manifested as a greatly improved resistance to thermal fatigue of the directionally solidified over conventionally cast alloy. Indeed, fracture of turbine blades in service, particularly of air cooled blades in aero-engines, normally results from thermal fatigue rather than creep damage. The extent of the improvement[7,8] to thermal fatigue resistance by directional solidification, illustrated in Fig. 5.5, again is greatest for those alloys with low inherent ductility in the conventionally cast condition (e.g. MarM200), but all alloys appear to derive a substantial benefit. This is perhaps the single most important reason for the commercial exploitation of directionally solidified superalloys.

It has been possible to gain further property improvements by progressing from columnar grains to single crystals with relatively little extension of the processing technology. The rate at which directional solidification has progressed from the laboratory to industrial practice has been relatively rapid and clear commercial benefits have been obtained. However, scientific under-

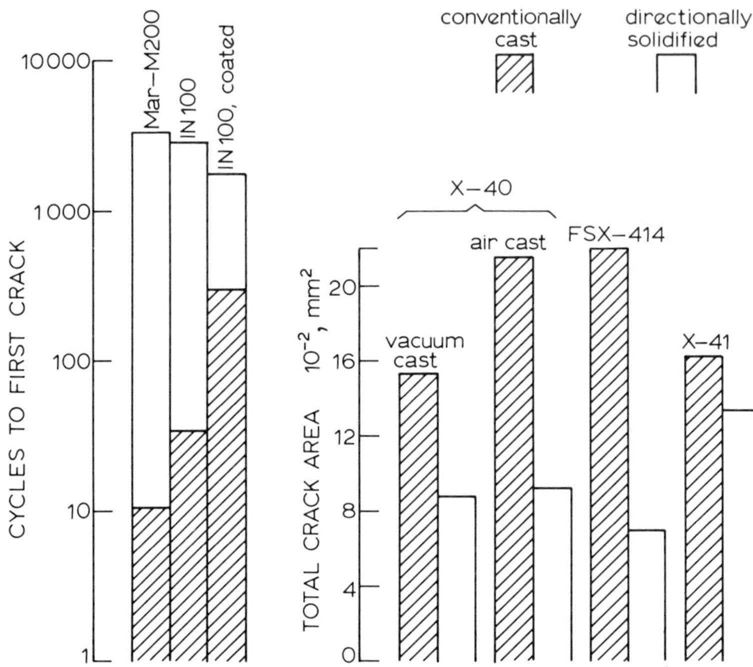


**Figure 5.5 Dependence of thermal fatigue behaviour of superalloys in the conventionally cast and directionally solidified conditions**

standing of the factors leading to the improved mechanical behaviour of directionally solidified structures has tended to lag behind the technological advances. A primary aim of this chapter will be to review the most recent work on directionally solidified superalloys with particular regard to the role of the various microstructures that can be obtained through control of the solidification conditions and to discuss appropriate models describing this material behaviour. It is hoped that this will both point to process modifications that may lead to further property advantages and provide some basis for property prediction and extrapolation.

## 5.2 ELASTIC ANISOTROPY

On the macroscopic scale conventionally cast alloys have isotropic elastic and plastic properties provided that the constituent grains have a random distribution of crystallographic orientations. In the single crystal form the mechanical behaviour of even cubic materials strongly depends on the orientation of the applied stress and reflects the full crystallographic symmetry. The columnar grain structure produced by directional solidification which has a well-defined growth direction (⟨100⟩ for superalloys) but random rotational orientations about this direction will have an intermediate type of behaviour displaying differences between the solidification and transverse directions but being isotropic in the transverse directions. Figure 5.6 indicates schematically the differences in structure and anisotropy associated with equiaxed and columnar grain structures and with single crystals. It is important that the designer be able to account for the elastic and yield response that these materials are likely to encounter in service.

Consider a cubic single crystal with orthogonal axes $0xyz$ along the ⟨100⟩ directions. The elastic behaviour of an anisotropic cubic material is fully defined by three elastic constants $C_{11}$, $C_{12}$ and $C_{44}$.

$$\left.\begin{aligned}\sigma_{11}&=C_{11}\varepsilon_{11}+C_{12}\varepsilon_{22}+C_{12}\varepsilon_{33}\\ \sigma_{22}&=C_{12}\varepsilon_{11}+C_{11}\varepsilon_{22}+C_{12}\varepsilon_{33}\\ \sigma_{33}&=C_{12}\varepsilon_{11}+C_{12}\varepsilon_{22}+C_{11}\varepsilon_{33}\\ \sigma_{23}&=2C_{44}\varepsilon_{23}\\ \sigma_{31}&=2C_{44}\varepsilon_{31}\\ \sigma_{12}&=2C_{44}\varepsilon_{12}\end{aligned}\right\}\tag{5.1}$$

Alternatively the strain components can be expressed in terms of stress through the elastic compliance coefficients.

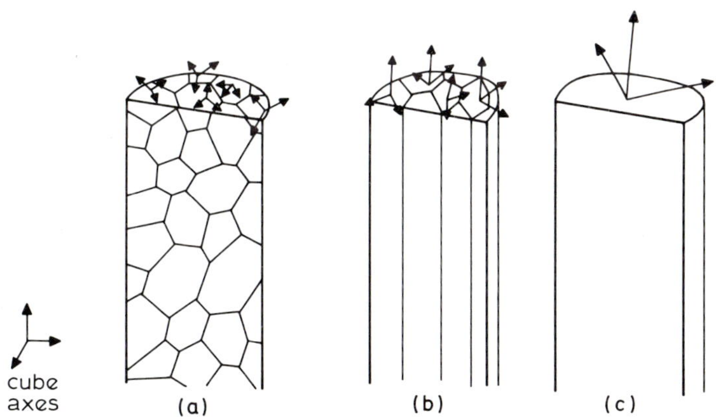


**Figure 5.6 Schematic illustration of the different grain structures and orientation relationships associated with (a) conventionally cast, (b) directionally solidified and (c) single crystal superalloys**

$$\left.\begin{aligned}
\varepsilon_{11} &= S_{11}\sigma_{11} + S_{12}\sigma_{22} + S_{12}\sigma_{33} \\
\varepsilon_{22} &= S_{12}\sigma_{11} + S_{11}\sigma_{22} + S_{12}\sigma_{33} \\
\varepsilon_{33} &= S_{12}\sigma_{11} + S_{12}\sigma_{22} + S_{11}\sigma_{33} \\
\varepsilon_{23} &= \tfrac{1}{2}S_{44}\sigma_{23} \\
\varepsilon_{31} &= \tfrac{1}{2}S_{44}\sigma_{31} \\
\varepsilon_{12} &= \tfrac{1}{2}S_{44}\sigma_{12}
\end{aligned}\right\} \tag{5.2}$$

where

$$S_{11} = \frac{C_{11} + C_{12}}{(C_{11} - C_{12})(C_{11} + 2C_{12})},$$

$$S_{12} = -\frac{C_{12}}{(C_{11} - C_{12})(C_{11} + 2C_{12})} \quad \text{and} \quad S_{44} = \frac{1}{C_{44}}$$

Hirth and Lothe[9] have discussed the circumstances under which either the elastic constants $C_{ij}$ or elastic compliances $S_{ij}$ should be averaged in order to determine the mean elastic behaviour of a polycrystal. The extremes of behaviour are represented by the microstructures shown schematically in Fig. 5.7. For columnar grains (5.7*a*) stressed parallel to their long axes, all crystals are geometrically required to be in the same state of strain and this is achieved by redistribution of stresses between the various crystals; in this case the

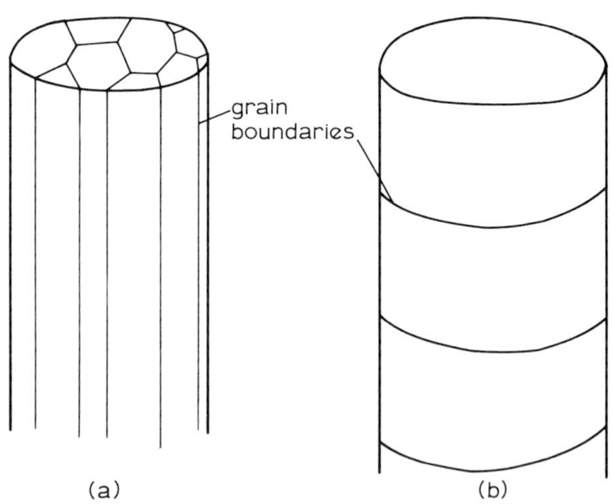


**Figure 5.7 Extremes of grain morphologies leading to (a) equal strain and (b) equal stress on different grains**

elastic constants $C_{ij}$ are averaged (the Voigt average). This approach would be relevant to the calculation of the elasticity of arrays of columnar grains having different growth axes. Although that situation does not commonly occur, it would be appropriate to calculating the change in elastic constants due to normal deviations (of ~10°) from the desired ⟨100⟩ texture in directionally solidified superalloys. In the situation illustrated in Fig. 5.7*b* each crystal must support the full stress and in general the strains are different. Here it is appropriate to average the compliances (Reuss average). This configuration would occur in wires having a 'bamboo' crystal structure. However, it is also approached for columnar grains stressed in a transverse direction.

Wells[10] has calculated the differences in the elastic constants of columnar grained superalloys in the transverse and longitudinal directions. The Young's modulus $E$ along the cube solidification axis is simply derived from eqns (5.2) by setting $\sigma_{22}=\sigma_{33}=0$. Then

$$E_{100}=\frac{\sigma_{11}}{\varepsilon_{11}}=\frac{1}{S_{11}} \tag{5.3}$$

Consider a stress $\sigma_\theta$ in an arbitrary direction in the $yz$ plane, which is normal to the solidification direction, that makes an angle $\theta$ with $0y$. The components of stress referred to the cube axes are

$$\left.\begin{aligned}\sigma_{22}&=\sigma_\theta \cos^2\theta\\ \sigma_{33}&=\sigma_\theta \sin^2\theta\\ \sigma_{23}&=\sigma_\theta \sin\theta\cos\theta=\sigma_{32}\end{aligned}\right\} \tag{5.4}$$

The appropriate strains obtained by substitution in eqns (5.2) can be resolved to give the tensile strain in the direction of the original applied stress which after algebraic manipulation yields

$$\varepsilon_\theta=\sigma_\theta\{S_{11}+\sin^2\theta\cos^2\theta(2S_{12}+S_{44}-2S_{11})\} \tag{5.5}$$

Thus, the Young's modulus in this direction

$$E_\theta=\frac{\sigma_\theta}{\varepsilon_\theta}=\{S_{11}+\sin^2\theta\cos^2\theta(2S_{12}+S_{44}-2S_{11})\}^{-1} \tag{5.6}$$

The average transverse value $E_{\mathrm{T}}$ is obtained by integrating $E_\theta$ over all independent values of $\theta$.

$$\begin{aligned}E_{\mathrm{T}}&=\frac{\int_0^{\pi/4}E_\theta\,\mathrm{d}\theta}{\int_0^{\pi/4}\mathrm{d}\theta}\\ &=\frac{4}{\pi}\int_0^{\pi/4}\frac{\mathrm{d}\theta}{a^2-b^2\sin^2 2\theta}\end{aligned} \tag{5.7}$$

where $a^2 = S_{11}$ and $b^2 = \frac{1}{4}(2S_{11} - S_{44} - 2S_{12})$. This is a standard integral[11] with a solution

$$E_T = 2[S_{11}(2S_{11} + S_{44} + 2S_{12})]^{-1/2} \tag{5.8}$$

Taking the data for the elastic compliances of nickel listed in Table 5.1 as typical of nickel-base superalloys we estimate that

$$E_{100} = 1{\cdot}36 \times 10^5 \text{ MPa}$$

$$E_T = 1{\cdot}78 \times 10^5 \text{ MPa}$$

It is more difficult to calculate the average modulus over all possible orientations rather than for the ⟨100⟩ zone described above. Moreover, the situation in a real polycrystal is likely to be intermediate between those appropriate to the Voigt and Reuss averaging procedures. For the case of nickel, the data given by Hirth and Lothe[9] indicate

$$E_{\text{Voigt}} = 1{\cdot}99 \times 10^5 \text{ MPa}$$

$$E_{\text{Reuss}} = 2{\cdot}04 \times 10^5 \text{ MPa}$$

This predicts that ⟨100⟩ oriented single crystals and columnar grains should have a 30% lower Young's modulus than the equiaxed material. This is confirmed by experiment; Fig. 5.8 is typical of the differences in $E$ for directionally solidified and conventionally cast superalloys that are observed over a wide range of temperatures.[12]

Such elastic anisotropy should have considerable implications to designs involving directionally solidified components. Wells[10] points out that, in addition to the greater longitudinal and transverse elastic strains resulting from a given stress, the directionally solidified superalloy will have less flexural rigidity than the isotropic material about both transverse and longitudinal axes (~62 and 82% of the equiaxed values respectively). However, the torsional rigidity of the directionally solidified alloy will be greater than that of the conventional casting by about 45%. In addition, the thermally induced stresses $\Delta\sigma$, due to thermal expansion and contraction when the temperature changes $\Delta T$, are

**Table 5.1 Elastic constants and compliances of nickel**

| $10^{-5}$ MPa$^{-1}$ | | | $10^4$ MPa | | |
|---|---|---|---|---|---|
| $S_{11}$ | $-S_{12}$ | $S_{44}$ | $C_{11}$ | $C_{12}$ | $C_{44}$ |
| 0·734 | 0·274 | 0·802 | 24·65 | 14·73 | 12·47 |

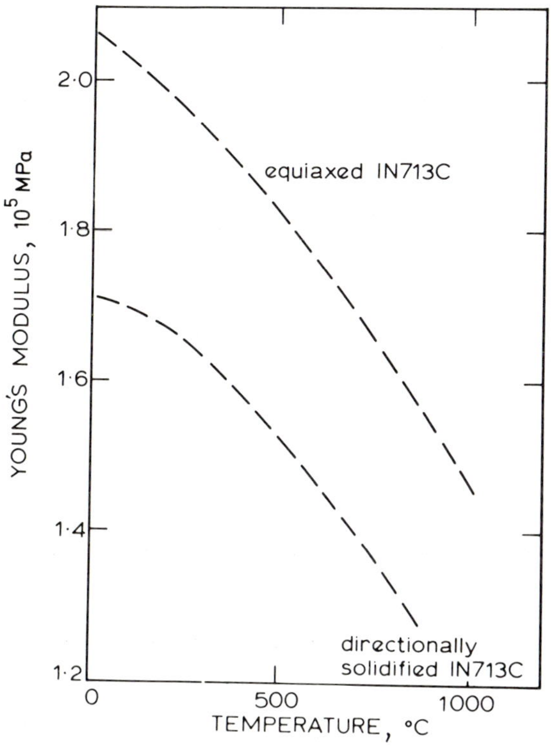


**Figure 5.8 Variation in Young's modulus *E* with temperature for the superalloy IN713LC in the conventionally cast and directionally solidified forms [Ref. 12]**

considerably less in the directionally solidified than in the conventionally cast alloy and, as we shall discuss in § 5.7, this contributes to the impressive thermal fatigue resistance of directionally solidified alloys.

$$\Delta\sigma \sim E\alpha\Delta T \tag{5.9}$$

(where $\alpha$ = coefficient of linear expansion).

The value of $E$ for an arbitrary orientation in a cubic crystal, given by Nye,[13] can be expressed as

$$E(\theta, \phi) = [S_{11} - 2(S_{11} - S_{12} - \tfrac{1}{2}S_{44}) \sin^2\theta\,(\cos^2\theta + \tfrac{1}{4}\sin^2\theta \sin^2 2\phi)]^{-1}$$

where $\theta$, $\phi$ define the orientation of the crystal axis relative to the cube directions as shown in Fig. 5.9. Taking the values of the elastic compliances for pure nickel the orientation dependence of $E$ is shown in Fig. 5.10. This shows that there is very considerable variation in elastic behaviour with crystal

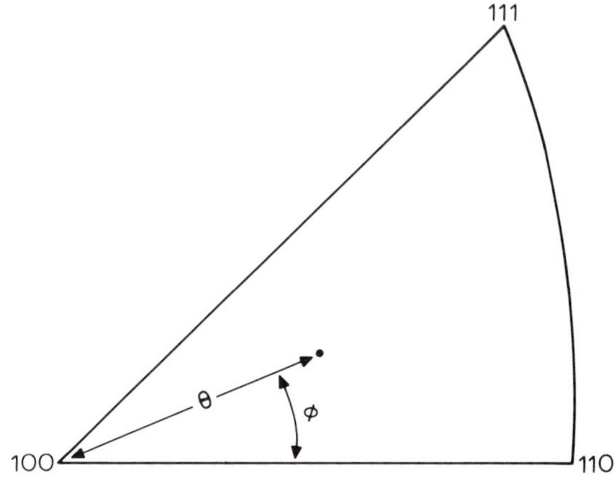


**Figure 5.9 Definition of the orientation parameters $\theta$ and $\phi$ describing the location of an arbitrary direction relative to the cubic crystallography**

direction $E$ having its minimum value along ⟨100⟩ and being maximum along ⟨111⟩. This is an important factor in determining the optimum orientation for single crystal components.

## 5.3 ANISOTROPY OF PLASTICITY

Tensile tests on single crystal superalloys of different orientations show a large anisotropy in the deformation behaviour at temperatures below about 760°C, but virtual isotropy at higher temperatures. The data of Kear and Piearcey[2] on single crystals of MarM200, shown in Fig. 5.11, indicate that at 760°C and below those crystals stressed parallel to ⟨100⟩ and ⟨111⟩ strain harden, while those with less rational orientations yield without hardening. These differences are reflected in the changes in specimen cross-sections, those of the ⟨100⟩ and ⟨111⟩ oriented crystals remaining circular whereas those of other orientations become elliptical. In addition, while the ⟨100⟩ and ⟨111⟩ crystals retain their original orientation, the orientations of the other crystals change during deformation in the manner shown in Fig. 5.12*a*. At higher temperatures (>980°C), however, there are virtually no differences in the various tensile curves (Fig. 5.11*d*) all of which show a degree of strain hardening and are characterized by the retention of the original specimen shape and orientation. The orientation changes that occur during creep deformation of single crystal superalloys differ in detail from those that occur during tensile testing; the rotations observed by McKay *et al.*[14] during stress rupture testing of MarM247 at 774°C are shown in Fig. 5.12*c*.

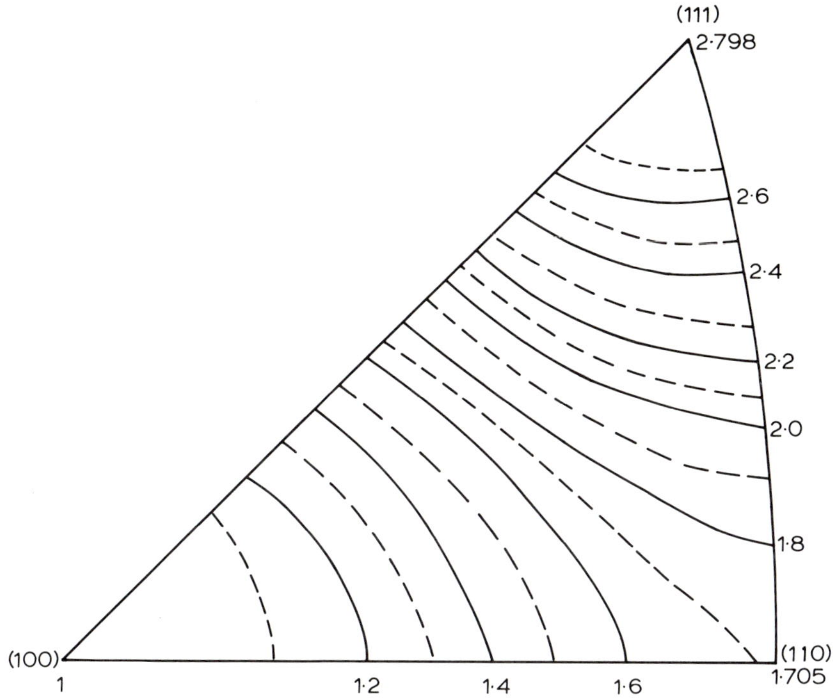


**Figure 5.10 Calculated relative variation in Young's modulus *E* with crystallographic orientation for nickel**

The lattice rotations, anisotropic reduction in area and lack of work hardening associated with crystals with orientations other than ⟨100⟩ and ⟨111⟩ that are observed at the lowest temperatures have been correlated with the occurrence of easy glide on a single slip system. Work hardening and the retention of the original specimen shape and orientation occurs for those crystals that are symmetrically oriented with respect to several slip systems so that dislocation cross-slip and interaction can occur; the dramatic change to isotropic deformation of all orientations at temperatures above 980°C implies that additional slip systems are activated at high temperatures. The different deformation behaviours can be rationalized on the basis of Schmid's law which expresses the critical resolved shear stress (CRSS) $\tau$ for slip in a specified direction ⟨*pqr*⟩ on the glide plane {*hkl*} in terms of the angles made by the direction of applied tensile stress with the surface normal to the slip plane $\phi$ and with the slip direction $\lambda$.

$$\tau = \sigma_y \cos\phi \cos\lambda \qquad (5.10)$$

where $\sigma_y$ is the yield stress.

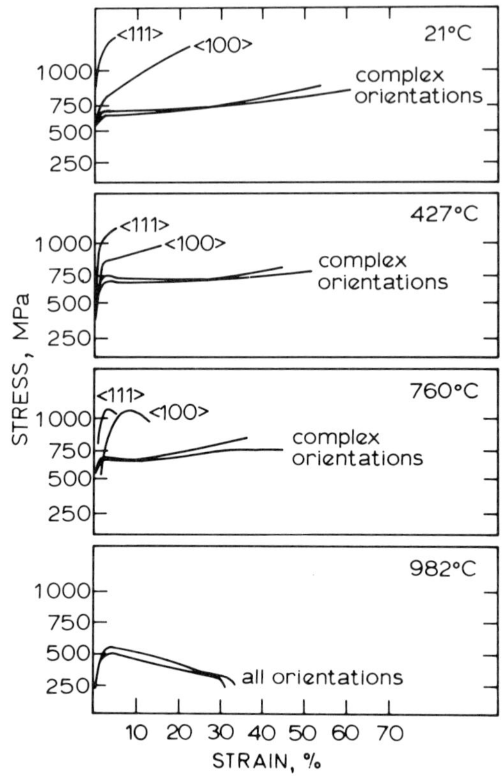


**Figure 5.11 Tensile curves for single crystals of MarM200 of different orientations tested at different temperatures [Ref. 2]**

Figure 5.13 shows the contours of constant Schmid factor (i.e. $\cos\phi \cos\lambda$) for the $\{111\}\langle\bar{1}01\rangle$ and $\{111\}\langle 112\rangle$ slip systems. During plastic deformation, the single crystals rotate towards the operative slip direction and the directions of rotation associated with those slip systems are indicated in Fig. 5.14. Comparison with the experimental data of Fig. 5.12 suggests that $\{111\}\langle\bar{1}01\rangle$ slip occurs during tensile testing whereas $\{111\}\langle 112\rangle$ slip dominates under creep conditions (both at the lower temperatures). It has been suggested by Kear and Piearcey[2] that the onset of multiple slip at temperatures above ~980°C is due to the activation of slip on cube planes in addition to the octahedral slip that occurs at the lower temperatures. Careful transmission electron microscopy and dislocation analysis by Kear and co-workers[15] has largely confirmed these deductions. Further support comes from the work of McKay *et al.*[14] who show that data from tensile testing of MarM247 indicate a CRSS of 400 MPa for slip on $\{111\}\langle\bar{1}01\rangle$ that is independent of orientation and temperature between 20 and 760°C.

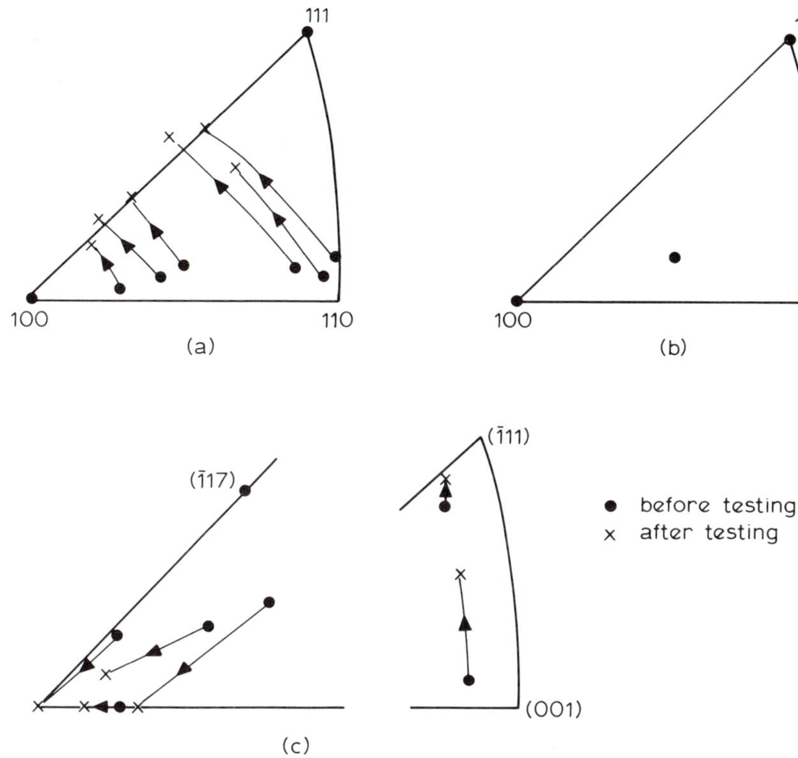


**Figure 5.12 Changes in orientation that occur during mechanical deformation of superalloys:**
**(a) MarM200 tensile testing at 21 to 760°C**
**(b) MarM200 tensile testing at 980°C**
**(c) MarM247 creep testing at 774°C**
**[Refs. 2 and 14]**

The detailed mechanisms during creep of single crystals of nickel-base superalloys are considerably more complex than those operating during tensile testing. The transition from primary to secondary (or steady state) creep cannot occur until there is sufficient intersecting cross-slip to enable strain hardening to occur and, for the {111} ⟨112⟩ slip system, this cannot occur until the crystals have rotated to the ⟨001⟩-*Y* and *X*-⟨$\bar{1}$11⟩ boundaries shown in Fig. 5.14*b*. The smaller the rotation required to reach these boundaries and achieve cross-slip, the smaller is the amount of primary creep and the sooner is secondary creep established. McKay *et al.*[14] have shown that this also leads to longer rupture lives for crystals oriented closest to directions of

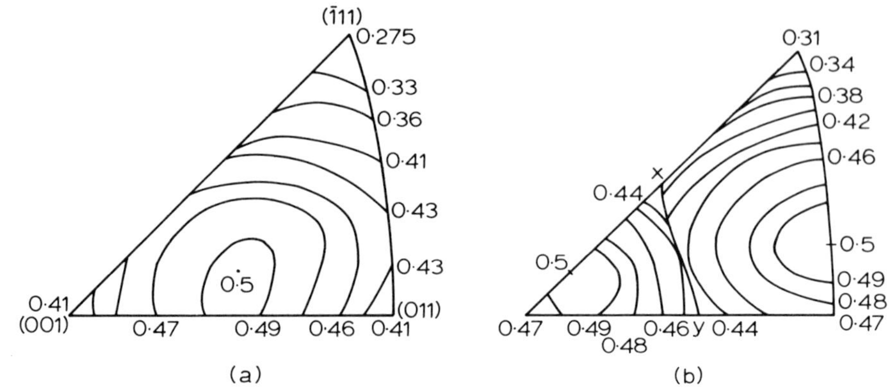


**Figure 5.13 Schmid factor contours for slip on (a) {111} ⟨1̄01⟩ and (b) {111} ⟨112⟩**

high symmetry and have constructed the guide shown in Fig. 5.15 relating crystal orientation to creep rupture life.

The implications of these considerations to the selection of the single crystal orientations that maximize the stress rupture life are obvious. However, the anisotropy of yielding also has a considerable effect on the mechanical behaviour of polycrystalline aggregates. In the case of the columnar grains of ⟨100⟩ oriented crystals produced by directional solidification, the stress is fairly uniformly distributed between the various crystals which are likely to yield at the same strain. Consequently, little internal stressing should develop during the deformation of directionally solidified material and this may account for the high ductilities that are obtained. This contrasts with the low ductilities that are obtained in conventionally cast superalloys.

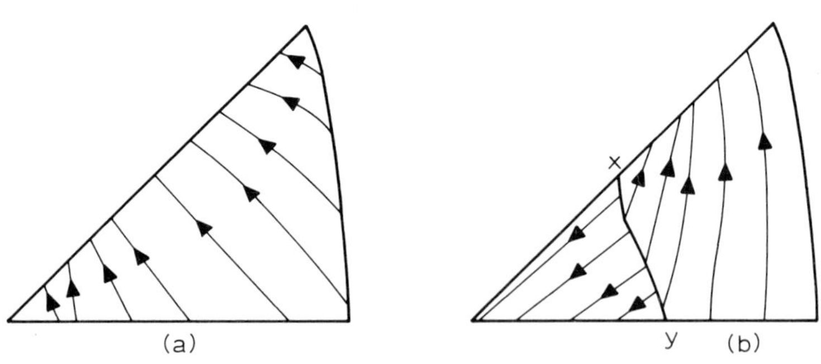


**Figure 5.14 Lattice rotations associated with single slip on (a) {111} ⟨1̄01⟩ and (b) {111} ⟨112⟩**

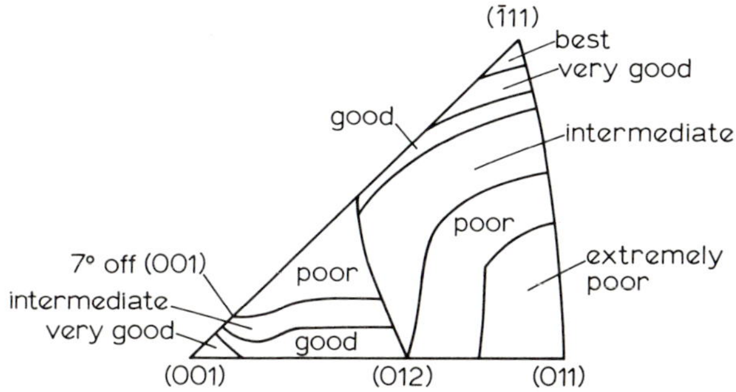


**Figure 5.15 Variation in stress rupture properties of single crystals of MarM247 at 760°C as a function of crystallographic orientation [after McKay *et al.*[14]]**

Why, then, do polycrystalline superalloys exhibit work hardening and multiple slip while most of the constituent crystals, if tested separately, would deform on a single slip system without work hardening? Figure 5.16 shows schematically the deformation of a cube shaped crystal that is in a plastically 'soft' orientation (i.e. with a high Schmidt coefficient) surrounded by grains in 'hard' orientations. Yield of the 'soft' crystal will occur at a relatively low stress while the other crystals are deforming elastically. If we consider separately the ideal plastic deformation of the crystal and of the elastic surrounding complete with the hole that should be occupied by the crystal it is readily

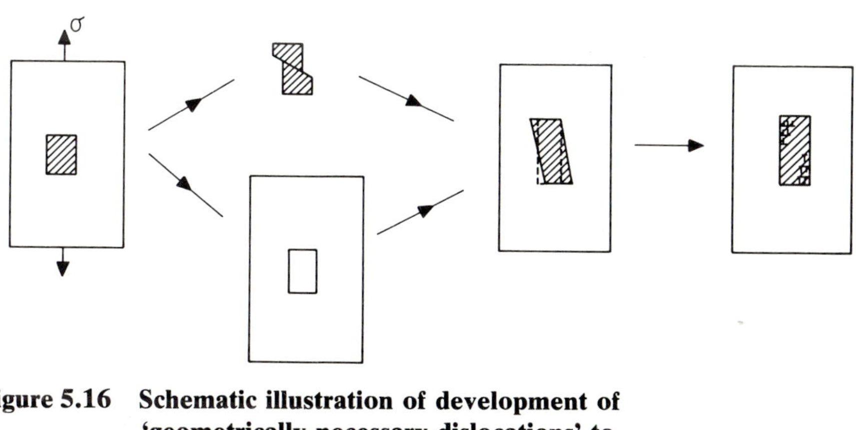


**Figure 5.16 Schematic illustration of development of 'geometrically necessary dislocations' to allow compatibility between a plastically deforming grain surrounded by elastically deforming grains**

seen that the elastically strained hole and the plastically strained crystal will generally be

(a) different sizes due to volume being conserved during plastic flow but not during elastic deformation
(b) different shapes because of the asymmetric plastic and symmetric elastic deformation.

However, the crystal and the hole it occupies cannot adopt different shapes. Consequently compatibility stresses will develop at the grain boundaries. Ashby[16] has proposed that these can be relieved by 'geometrically necessary dislocations' as shown in Fig. 5.16 which may propagate on some of the secondary slip systems and therefore lead to work hardening. Since the transverse strain of the soft grain $\varepsilon_{ij}$ will exceed that of the surrounding material it follows that, in an applied stress of uniaxial tension, there will be local normal tensile stresses across grain boundaries that are parallel to the applied stress and compressive stresses across boundaries that are transverse to the applied stress. Dyson *et al.*[17] have shown by direct observation using high voltage electron microscopy that single slip dominates during the deformation of polycrystalline Nimonic 80A which is a wrought nickel-base superalloy. Moreover, where the incompatibility stresses are tensile microcracks often form where the slip bands intersect grain boundaries and these can provide nuclei for the growth of creep cavities. An example is shown in Fig. 5.17. We shall return to this topic when comparing the rupture lives of superalloys in the cast and directionally solidified forms.

## 5.4 CREEP BEHAVIOUR

The main thrust of much of the early work on the directional solidification of superalloys appears to have been directed to the control of grain structure which was successfully achieved using the power-down and high rate solidification processes. This led to the large increases in creep elongation and variable improvements in rupture lives described in § 5.1 and illustrated in Fig. 5.2. However, we have shown in Chapter 2 that directional solidification affects aspects of the microstructure other than the grain boundary morphology and that the solidification conditions, particularly the temperature gradient $G_L$ and rate of solidification $V_I$, have a large influence on the form and dimensions of microstructural features such as dendrites, $\gamma'$ precipitation, carbide particles and microporosity but appear to have relatively little effect on the shape and density of the grain boundaries. If grain boundary effects are dominant, then the mechanical behaviour should be relatively insensitive to the solidification conditions; if the other microstructural features are important then there should be considerable scope for property optimization through process control. Since it is well established that the temperature and stress dependences of creep of conventionally cast superalloys are dependent on

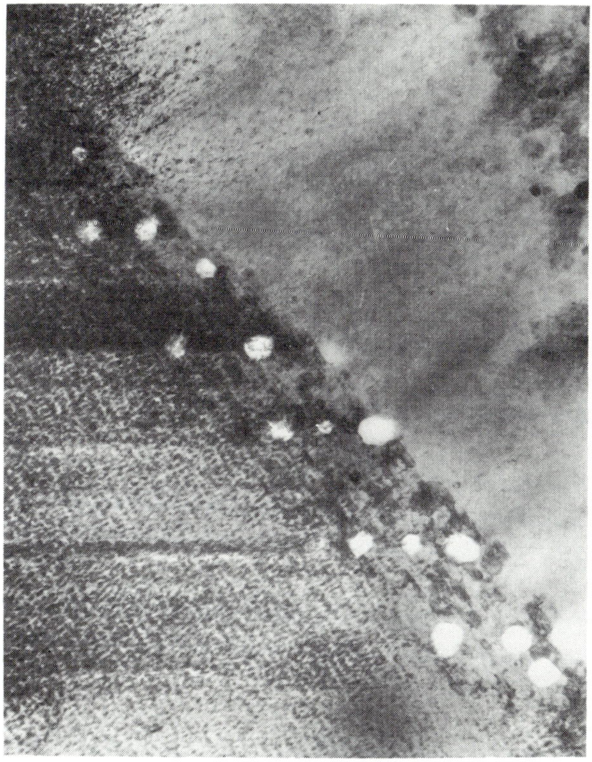

**Figure 5.17 Electron micrograph showing grain boundary cavities associated with slip bands in the wrought superalloy Nimonic 80A. The cavities were produced by prestraining before creep testing [after Dyson *et al.*[17]]**

the shape and distribution of the $\gamma'$ precipitate, on the nature and morphology of the phases precipitated within the interdendritic region and at grain boundaries and on the structure and size of grain boundaries it is unlikely that the creep deformation and fracture of directionally solidified material can be fully understood without reference to the detailed microstructure. The paucity of data relating mechanical behaviour to directional solidification conditions is surprising and is more likely to be due to protection of commercial interests than complacency with the property advantages already obtained.

Northwood and Homewood[4] showed that the stress rupture life of IN713, directionally solidified by the high rate solidification process, varied with the rate of solidification at the two test conditions indicated in Fig. 5.18. There

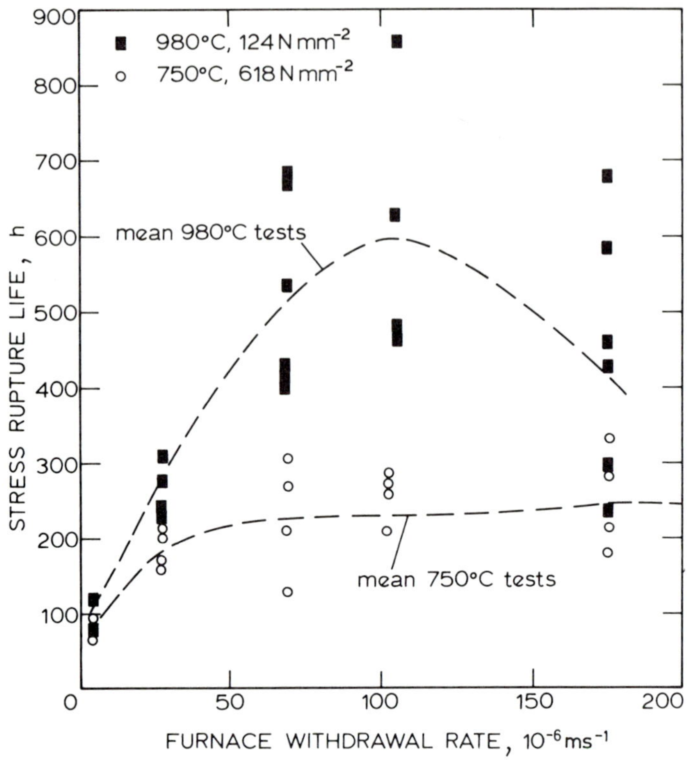


**Figure 5.18 Variations in stress rupture lives at 750 and 980°C of directionally solidified IN713LC as a function of solidification rate. The directional solidification was carried out in a low temperature gradient typical of commercial rigs [after Northwood and Homewood[4]]**

was an improvement with increasing growth rate up to about 380 mm h$^{-1}$ (10$^{-4}$ m s$^{-1}$) after which the lives deteriorated or a wide scatter in properties was obtained. These findings were correlated with the formation of spurious grains, increased spread of orientations and finer dendrites at the higher solidification rates. A more detailed study of various aspects of creep behaviour in alloys directionally solidified in different temperature gradients and at various rates has been carried out at NPL.[18,19] Figure 5.19 compares the creep curves of the low chromium alloy MarM246 that has been prepared at the same rate of 300 mm h$^{-1}$ (0·083 mm s$^{-1}$) but in different temperature gradients obtained in a commercial high rate solidification rig ($G \sim 4$ K mm$^{-1}$)

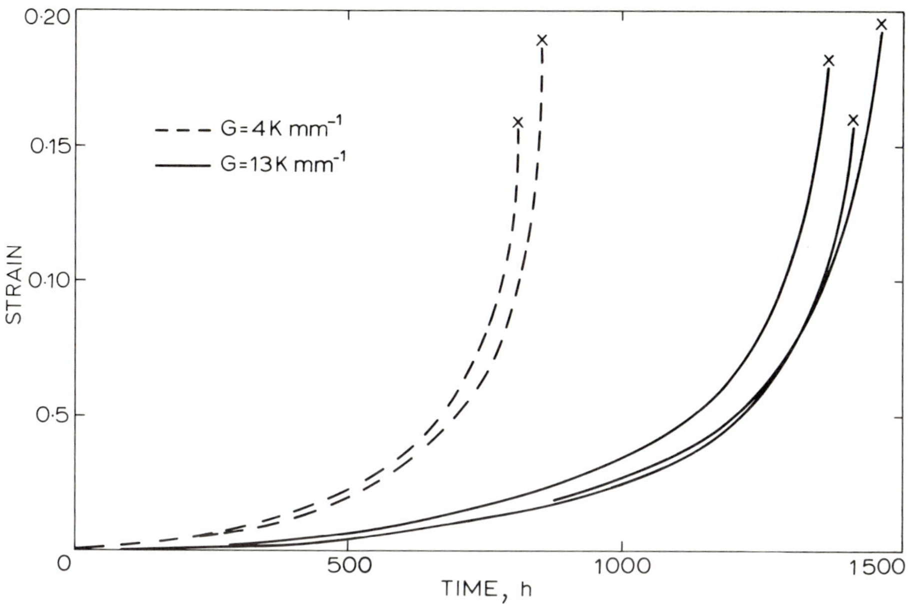


**Figure 5.19 Comparison of creep curves for MarM246 directionally solidified at 0·083 mm s$^{-1}$ (300 mm h$^{-1}$) in low and high temperature gradients ($G$ = 4, 15 K mm$^{-1}$ respectively). The tests were performed in air at 1123 K and 250 MPa**

and in laboratory conditions using liquid metal cooling ($G \sim 13$ K mm$^{-1}$). Both materials display much higher creep elongations (15 to 20%) than would be obtained with the conventionally cast alloy (~3%), but there are very significant differences in both the minimum creep rate and rupture lives. This trend is confirmed in the fuller analysis of minimum creep rate, rupture life and elongation of the alloy directionally solidified at various rates and related in Fig. 5.20 to the cooling rate during directional solidification $G_L V_I$ which is an important parameter in determining the forms and sizes of both dendrites and $\gamma'$ precipitates. There is a general increase in rupture life and decrease in minimum creep rate as the cooling rate increases (i.e. as the microstructure is refined) reaching an optimum creep behaviour at a growth rate of 300 mm h$^{-1}$. Similar trends have been noted in other nickel-base superalloys. This gives a clue to why some alloys are *reputed* to show no property advantage on directional solidification; for example, Woodford and Frawley[5] showed that when the high chromium alloy IN738 was directionally solidified using

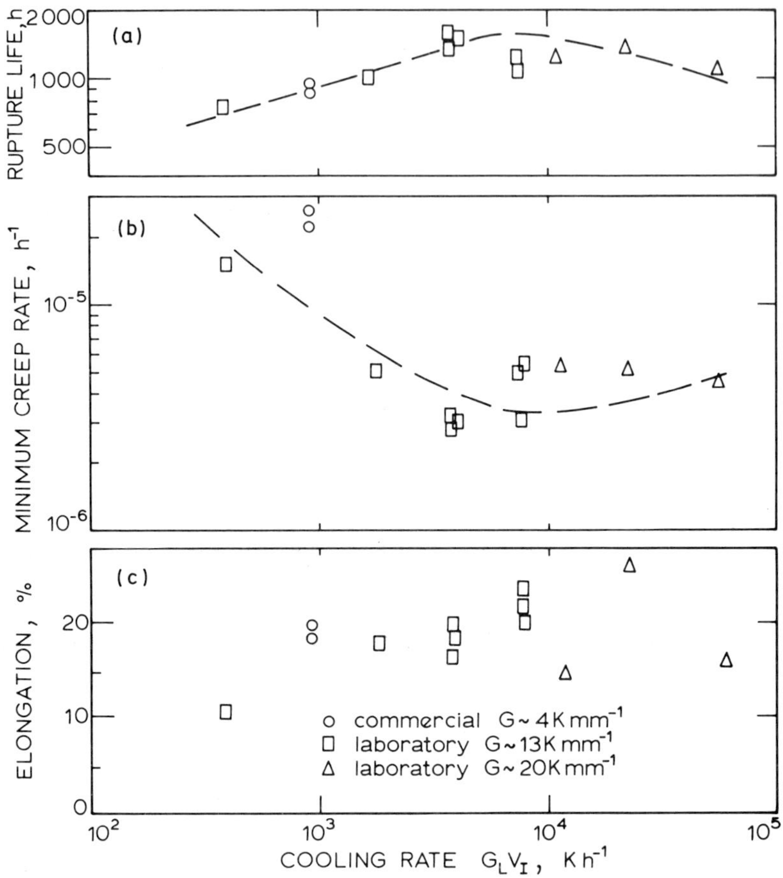


**Figure 5.20 Summary of creep data for MarM246 directionally solidified at different rates and in different temperature gradients plotted as functions of the cooling rate $G_L V_I$: (a) rupture life, (b) minimum creep rate and (c) extension at fracture [350 MPa, 1123 K]**

the power-down process, which gives a low cooling rate, there was no improvement in either rupture life or creep strength over the conventionally cast alloy although there was an increasing ductility. However, when processed in high temperature gradients and/or at high rates even this alloy gave longer creep lives than in the conventionally cast form, suggesting that the microstructures leading to increased rupture lives were not achieved in such alloys using the early directional solidification procedures.

Most of the extra rupture life resulting from directional solidification is a consequence of the extension of the tertiary creep regime rather than of increased creep strength (i.e. reduced creep rate for a given stress). Indeed, in most cases the creep rates of directionally solidified alloys are greater than those of the conventionally cast and heat treated material. Since for many applications the effective failure criterion is related to the time to reach a specified elongation limit (e.g. 1%), much of the extended rupture life may not be useful. Figure 5.21 shows that when directionally solidified IN738LC is given a two stage heat treatment designed first to partially dissolve and then to age the $\gamma'$ precipitate to a standard morphology, namely 2 h at 1120°C + 16 h at 845°C with air cooling after each stage, then (a) most of the variations in creep properties with solidification conditions are removed, and (b) the creep strength is increased to the level of the *best* conventionally cast alloy. This points to the rate of creep deformation being largely controlled by the $\gamma'$ characteristics rather than by the solidification morphology. (We shall discuss the importance of the solidification features on the fracture behaviour in § 5.7.) It might be argued that one potential benefit of directional solidification in high gradients and at fast rates would be to control the heat treatment that occurs on cooling in order to obtain the desired microstructure.

Because of the segregation that occurs during dendritic solidification the $\gamma'$ solution temperatures in the dendrites and in the interdendritic regions are generally different. As superalloys have been progressively strengthened the solvus temperature has tended to increase and in many advanced alloys complete $\gamma'$ solution does not occur before the onset of incipient melting. It is therefore often not possible to completely dissolve the $\gamma'$ precipitates throughout the material and then reprecipitate the $\gamma'$ in the optimum form. For example the standard heat treatment for IN738LC only dissolves the precipitate in the dendrite cores, leading to rapid growth of the interdendritic precipitate. It has now been recognized that alloys that have been developed to improve the performance of conventional castings are not necessarily appropriate for the directional solidification of single crystal forms and new alloy compositions specific to these processing routes are being developed to permit effective heat treatment control.

Advanced alloys for conventional castings include elements designed to strengthen the grain boundaries (e.g. C, B, Zr). Further additions of hafnium were found to modify the grain boundary behaviour of columnar grained d.s. alloys inhibiting grain boundary cracking during directional solidification and leading to improved transverse ductility. However, all of these additives designed to modify the grain boundaries extend the melting ranges of the alloys, restricting the range of temperature over which heat treatment is possible (if at all). Such grain boundary strengtheners are redundant in single crystals and new alloys in which these elements have been eliminated are currently being developed. The composition of the Pratt and Whitney

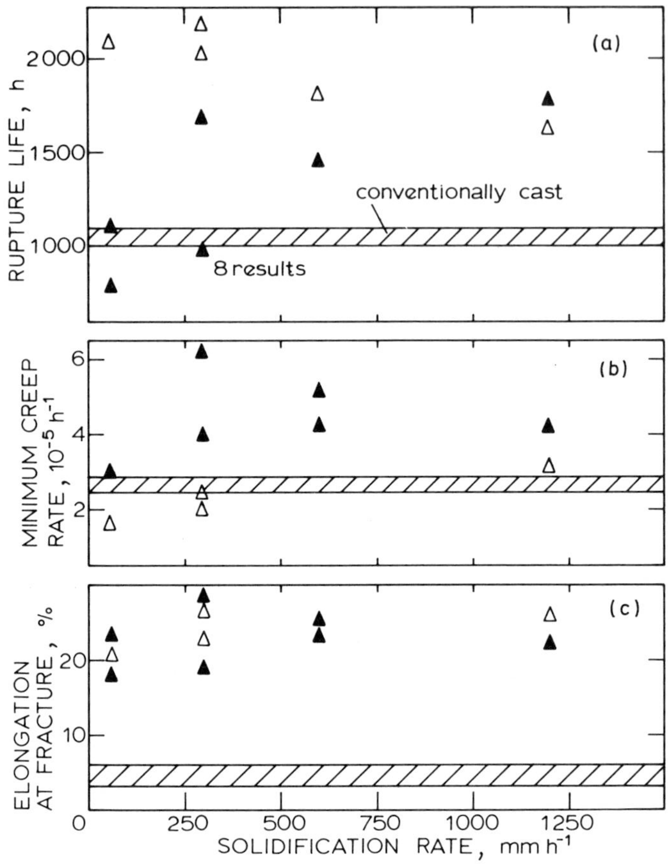


**Figure 5.21 Creep data for IN738LC directionally solidified in a high temperature gradient ($G \sim 13$ K $mm^{-1}$) at various rates both in the directionally solidified condition (solid symbols) and with additional heat treatment (open symbols): (a) rupture life, (b) minimum creep rate and (c) extension at fracture [250 MPa, 1123 K]**

developmental single crystal alloy 454 is compared in Table 5.2 with that of MarM200 containing hafnium. With this composition the incipient melting point has been increased sufficiently to allow heat treatment at 1288°C where complete homogenization and $\gamma'$ solution is claimed after 4 h. The creep rupture properties, shown in Fig. 5.22, offer between 25 and 50°C advantages over columnar grained MarM200. However, this has been achieved at the

**Table 5.2 Comparison of compositions of commercial alloy MarM200 and developmental single crystal alloy 454 (wt%)**

| | Ni | Cr | Al | Co | W | Ti | Nb | Zr | Ta | C | B | Hf |
|---|---|---|---|---|---|---|---|---|---|---|---|---|
| Alloy 454 | 62·5 | 10 | 5 | 5 | 4 | 1·5 | — | — | 12 | — | — | — |
| MarM200 | 59·0 | 9 | 5 | 10 | 12 | 2 | 1 | 0·08 | — | 0·14 | 0·02 | 1·8 |

expense of higher density by virtue of the 12% Ta addition that is made to improve the casting characteristics of the alloy. Perhaps the greatest concern over the use of such specific single crystal alloys is for the danger of grain boundaries being accidentally introduced into a component either through faulty processing and inspection or by recrystallization during service; clearly the total absence of grain boundary strengtheners would make such defects catastrophically weak. Northwood *et al.*[53] have shown that foreign damage to blades is unlikely to affect creep properties unless the blade is subsequently heated to very high temperatures (≳1160°C); however, accidental damage prior to solution heat treatment could be deleterious. It is anticipated that there will be continuing effort both on alloy and process development and on quality assurance before these materials are introduced into service.

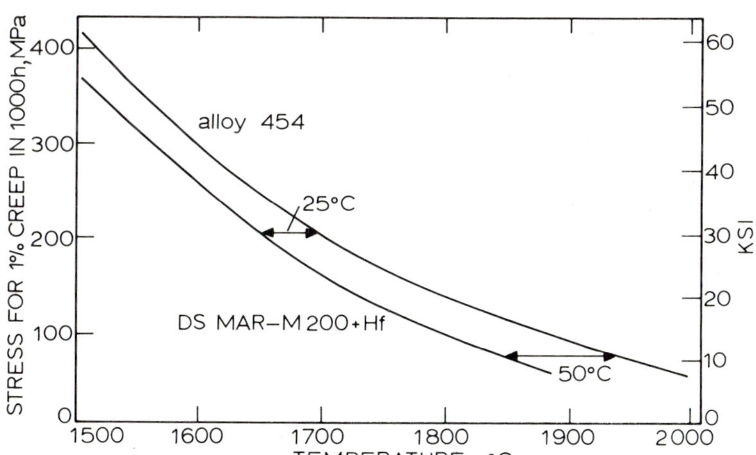


**Figure 5.22 Comparison of the creep strengths of the Pratt and Whitney single crystal alloy 454 and directionally solidified MarM200 + Hf**

## 5.5 CREEP MECHANISMS

The relationships between the steady state creep rate $\dot{\varepsilon}$ and the applied stress $\sigma$ for a wide range of polycrystalline materials can be represented by the common curve shown in Fig. 5.23. Sherby and Burke[21] observe that three distinct regimes can be identified where quite different deformation mechanisms are operative:

*Regime* 1  At low stresses $\dot{\varepsilon} = K\sigma$ where $K$ is a constant. This applies when creep occurs by a diffusional flux of vacancies from transverse to longitudinal grain boundaries either by volume diffusion through the grain (Herring–Nabarro creep) or by grain boundary diffusion around the grains (Coble creep). Most evidence of this mechanism derives from work on pure single phase alloys and there is some evidence that the mechanism may be inhibited by the presence of impurities or particles that may influence the efficiency of grain boundaries as sources or sinks of vacancies. The grain size $d$ is an

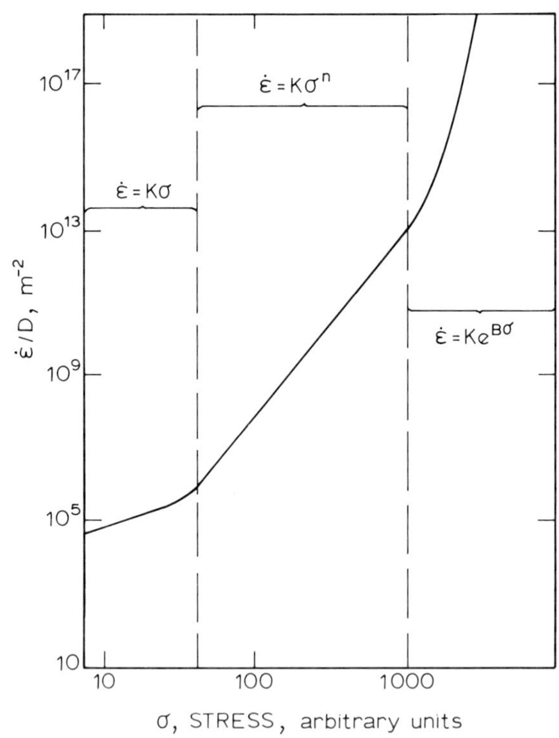


**Figure 5.23 Influence of stress on the diffusion compensated steady state creep rate of metals [after Sherby and Burke[21]]**

important parameter in determining the magnitude of diffusion creep which scales as $1/d^2$ for Herring–Nabarro and as $1/d^3$ for Coble creep.

*Regime* 2 Over a wide intermediate stress range at sufficiently high temperatures the steady state creep rate can be represented by a power law function of stress, $\dot{\varepsilon} = K'\sigma^n$ (where $K'$ and $n$ are constants). Under these conditions deformation occurs by a combination of glide and climb of dislocations. Sub-categories can be identified depending on whether dislocation climb occurs by lattice diffusion (at high temperatures) or by dislocation core diffusion (at low temperatures). Dislocation creep will be discussed more fully below.

*Regime* 3 At sufficiently high stresses ($\sim 10^{-3}$ $\mu$) the power law representation of creep breaks down and $\dot{\varepsilon}$ varies exponentially with applied stress, $\dot{\varepsilon} = K'' \exp(B\sigma)$ where $K''$ and $B$ are constants. There is no generally accepted model of this behaviour, but it appears to occur when the rate of viscous glide is rate controlling. The important difference between this mechanism and low temperature yield by glide is that a steady state dislocation structure can be maintained by rapid recovery processes such as dynamic recrystallization. Regimes 2 and 3 may be regarded as extremes of a single mechanism whereby recovery processes create the conditions that allow the activation of glide events; in Regime 2 the recovery stage is slow and intermittent glide bursts are rapid whereas in Regime 3 recovery is rapid relative to the viscous glide events and indeed many more dislocation segments are in motion.

Ashby[22] has developed the approach of delineating conditions under which various mechanisms are operative by means of deformation mechanism maps. Figure 5.24 shows two such maps for the conventionally cast nickel-base superalloy MarM200 for two different grain sizes (0·1 and 10 mm); the range of operation of a typical turbine blade is indicated. Thus for the smaller grain size diffusion creep is predicted whereas power law creep is preferred for the coarser grained alloy. In the directionally solidified condition where very large elongated grains are produced, the dramatic reduction in the density of transverse grain boundaries will proportionately reduce the diffusion creep rate and, of course, this will further restrict the extent of the diffusion creep regime to even lower stresses and temperatures. Consequently, in further discussion of creep deformation of directionally solidified superalloys, only the dislocation creep mechanism will be considered.

It has become customary to represent creep data for most materials over a wide stress range by the power law relationship attributed to Bailey and Norton

$$\dot{\varepsilon} = A\sigma^n \exp\left(-\frac{Q}{RT}\right) \qquad (5.11)$$

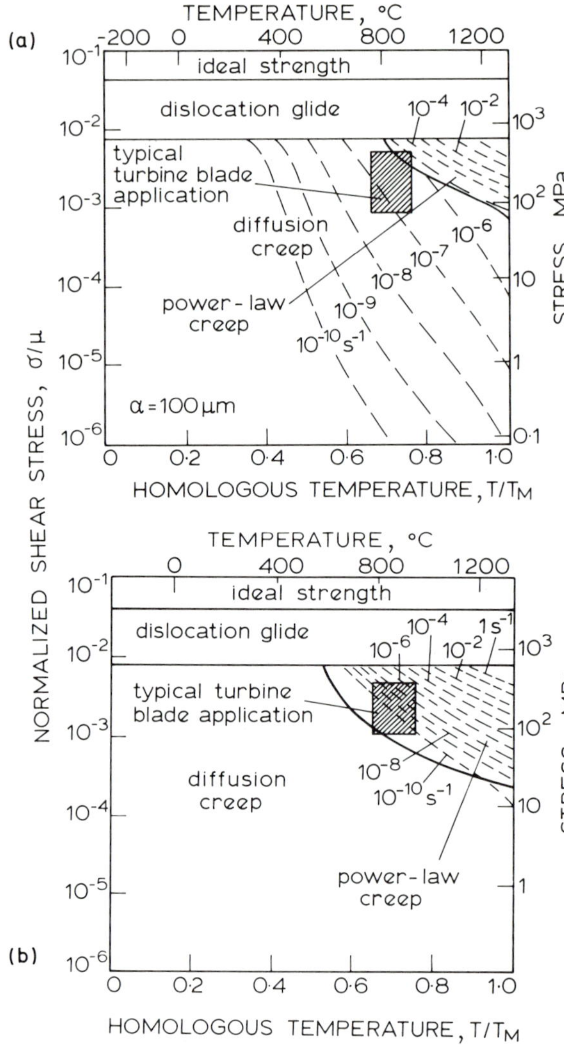


**Figure 5.24 Deformation maps for equiaxed MarM200 with two different grain sizes: (a) 100 µm or (b) 10 mm [Ref. 22]**

where $\dot{\varepsilon}$ is the minimum creep rate, $\sigma$ is the applied stress, $T$ is the temperature in K, $R$ is the gas constant and $A$, $n$ and $Q$ are constants to be empirically determined. This has been generalized by Barrett and Sherby[23] to account for differences between materials by normalizing the creep rate and stress with respect to the volume diffusion coefficients $D$ and the Young's moduli

$E$ of the alloys respectively.

$$\frac{\dot{\varepsilon}}{D} = A'\left(\frac{\sigma}{E}\right)^{n} \tag{5.12}$$

The results for pure metals and microstructurally simple alloys yield values of $n \sim 3$ to 4 and $Q \sim Q_{SD}$, the activation energy for self-diffusion, in the regime where dislocation rather than diffusion creep operates; this is in accord with the theoretical expectation from models[24,25] of recovery controlled creep which envisages steady state creep as being a dynamic balance between (i) the destruction of dislocations by recovery, or coarsening, of the dislocation network, motivated by the reduction in dislocation line energy, and (ii) the creation of new dislocations by interaction of gliding dislocation links and by the activation of dislocation sources (i.e. work hardening). Orowan[26] expressed this balance

$$\begin{aligned} d\sigma &= \frac{d\sigma}{dt}\,dt + \frac{d\sigma}{d\varepsilon}\,d\varepsilon \\ &= -r\,dt + h\,d\varepsilon \end{aligned}$$

where $r = -d\sigma/dt$, $h = d\sigma/d\varepsilon$, and, at steady state, $d\sigma/dt = 0$. Then

$$\dot{\varepsilon} = \frac{d\varepsilon}{dt} = \frac{r}{h} \tag{5.13}$$

D. McLean[24] made use of Freidel's[27] theory of dislocation recovery which predicts the parabolic growth with time $t$ of the average dislocation link length $\bar{\lambda}$, i.e.

$$\bar{\lambda}^2 = \bar{\lambda}_0^2 + Kt \tag{5.14}$$

where $\bar{\lambda}_0$ is the value of $\bar{\lambda}$ at $t = 0$ and $K$ a mobility that is proportional to the self-diffusivity of the alloy. Then

$$K = K_0 \exp\left(-\frac{Q_{SD}}{RT}\right) \tag{5.15}$$

A strength $\bar{\sigma}$ may be attributed to the dislocation network through the Orowan yielding condition

$$\sigma = \frac{\mu \mathbf{b}}{\bar{\lambda}} \tag{5.16}$$

where $\mu$ = shear modulus and $\mathbf{b}$ = Burgers vector.

Differentiation with respect to time gives

$$r = -\frac{d\sigma}{dt} = -\frac{\mu\mathbf{b}}{\bar{\lambda}^2}\frac{d\bar{\lambda}}{dt} = -\frac{\mu\mathbf{b}K}{2\bar{\lambda}^3}$$
$$= -\frac{K}{2\mu^2\mathbf{b}^2}\sigma^3 \tag{5.17}$$

It is more difficult to relate the hardening coefficient $h$ to the applied stress. However, there are measurements[28] that indicate that the dislocation density $\rho$ increases linearly with strain, and since $\bar{\lambda} \propto \rho^{-1/2}$ we may write

$$\sigma = \mu\mathbf{b}\rho^{1/2} = \alpha\mu\mathbf{b}\varepsilon^{1/2} \tag{5.18}$$

Then

$$h = \frac{d\sigma}{d\varepsilon} = \tfrac{1}{2}\alpha\mu\mathbf{b}\varepsilon^{-1/2} = \tfrac{1}{2}(\alpha\mu\mathbf{b})^2\frac{1}{\sigma} \tag{5.19}$$

where $\alpha$ is a constant. Combination of eqns (5.13), (5.15), (5.17) and (5.19) gives

$$\dot{\varepsilon} = A\sigma^4 \exp\left(-\frac{Q_{SD}}{RT}\right) \tag{5.20}$$

where $A = K_0/[\alpha^2(\mu\mathbf{b})^2]$. Equations (5.11) and (5.20) are clearly identical in form, giving considerable credence to the model.

When creep data for microstructurally complex materials such as superalloys are analysed by eqn (5.11) they yield values of $n \gg 4$ (up to $\sim$40 have been reported) and $Q \gg Q_{SD}$. It is still valid to use eqn (5.11) as an empirical representation of the data, but clearly the values of the parameters derived have little physical significance and there must be dangers in extrapolating beyond the range of the data. Various authors[29,30] have proposed that eqn (5.11) be framed in terms of an effective, rather than applied stress.

$$\dot{\varepsilon} = A'(\sigma - \sigma_0)^m \exp\left(-\frac{Q'}{RT}\right) \tag{5.21}$$

where $\sigma_0$ is a measure of the inherent resistance of the material to dislocation motion and has been variously described as an internal-, friction-, back- or resisting-stress. For appropriate values of $\sigma_0$ the same data that yield physically unacceptable values of $n$ and $Q$ from eqn (5.11) give values of $m \sim 4$ and $Q' \sim Q_{SD}$ from eqn (5.21). A typical set of data for directionally solidified IN738LC is shown in Fig. 5.25. Whereas the correlation with applied stress gives values of $n \sim 14$ to 20 and $Q = 900$ kJ mol$^{-1} \gg Q_{SD}$, when the $\sigma_0$ correlation is made $m \sim 4$ and $Q \sim 300$ kJ mol$^{-1}$ (cf. $Q_{SD} = 275$ kJ mol$^{-1}$).

Before $\sigma_0$ can be used as an adjustable parameter to constrain data to fit eqn (5.21) it is essential to fully test the validity of the approach. This

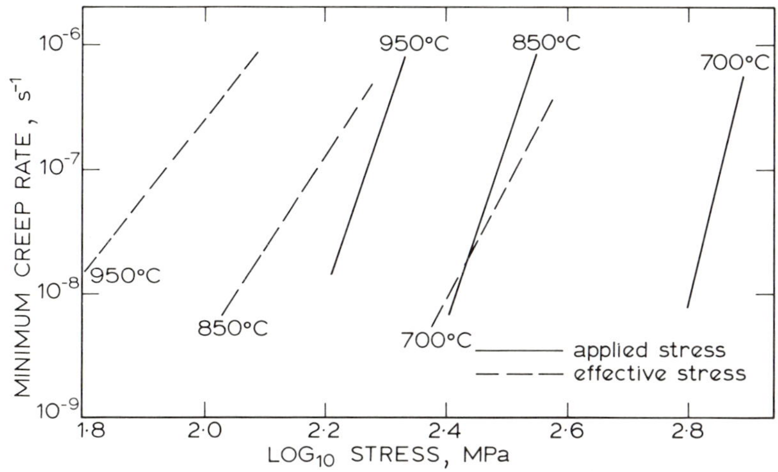


**Figure 5.25 Creep rate data for directionally solidified IN738LC expressed as functions of applied and of effective stress**

requires the *independent* measurement of $\sigma_0$. Several approaches have been proposed, but that used most extensively is the stress dip method pioneered by Wilshire and co-workers.[30] In this technique steady state creep is allowed to be established before a stress reduction $\Delta\sigma_1$ is made suddenly; the transient creep behaviour follows the schematic pattern shown in Fig. 5.26 consisting

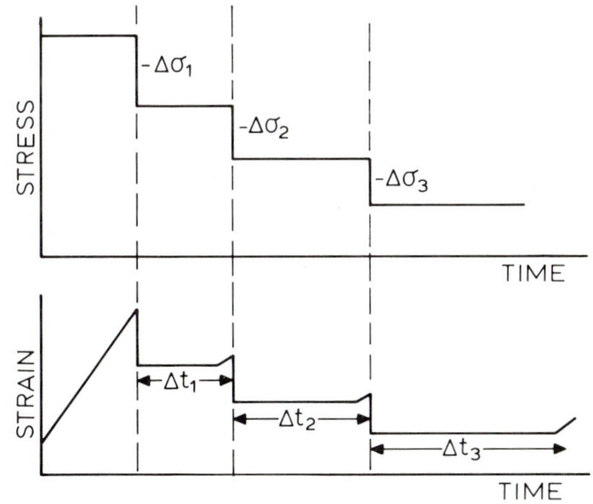


**Figure 5.26 Schematic illustration of the test sequence and transient strain associated with the stress-dip creep test procedure**

of (i) an immediate contraction $\Delta\varepsilon$ that exceeds the expected elastic strain, (ii) a delay period $\Delta t_1$ of zero creep and (iii) the resumption of creep at a lower rate than the steady state $\dot{\varepsilon}$ that would be associated with the reduced stress $(\sigma - \Delta\sigma_1)$. After forward creep has recommenced further progressively smaller stress reductions $\Delta\sigma_i$ are made and the appropriate incubation periods $\Delta t_i$ noted. In plots of the total stress reduction $\sum_i \Delta\sigma_i$ against the cumulative incubation period $\sum_i \Delta t_i$ (Fig. 5.27), $\sum_i \Delta\sigma_i$ appears to approach asymptotic values as $\sum_i \Delta t_i \rightarrow \infty$ and the stress remaining on the specimen $\sigma - (\sum_i \Delta\sigma_i)_{max}$ is taken to be $\sigma_0$. This approach has gone a long way towards rationalizing the high values of $n$ and $Q$ obtained from the Bailey–Norton representation. However, it can be criticized for the lack of an objective criterion for determining $\sigma_0$ from the experimental data on at least two grounds.

(a) It is difficult to establish that curves such as those in Fig. 5.27*a* do indeed approach an asymptotic value rather than tending to infinity as $\sum_i \Delta t_i \rightarrow \infty$ and, even if a limiting value is approached, there will be large errors in determining this limit particularly if the analytical form of the curve is not known.

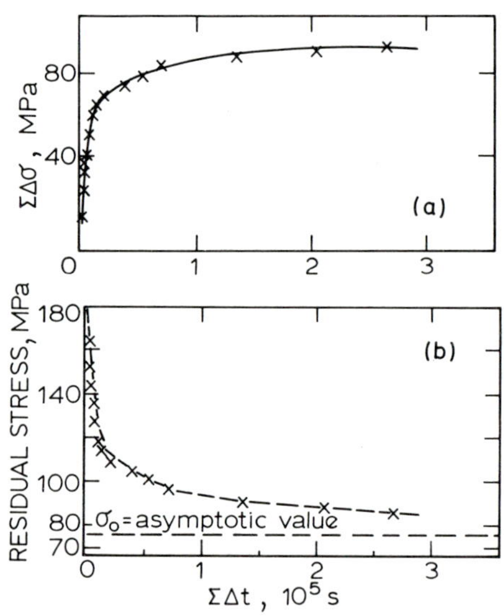


**Figure 5.27 Typical data set obtained by stress-dip testing for directionally solidified IN738LC**
**(a) $\sum_i \Delta\sigma_i$ versus $\sum_i \Delta t_i$ [after Wilshire and co-workers[32,33]]**
**(b) $\sigma_R$ versus $\sum_i \Delta t_i$ with best fit of eqn (5.24) to the data [after McLean[31]]**

(b) When the residual stress after several stress decrements is reduced to a value that gives minimum creep rates that are lower than those resolvable by the measurement techniques available then $\sum_i \Delta t_i$ will *appear* to be infinite even when creep is in fact occurring below the detection rate.

In an attempt to provide an objective basis for data analysis in stress dip creep testing, McLean[31] has extended the concepts of recovery controlled creep to model the transient creep behaviour following stress reductions. Following previous workers (e.g. D. McLean,[24] Lagneborg[25]) it is assumed that in steady-state creep there is a distribution of dislocation link lengths $\lambda$ with an upper cut-off defined by the Orowan yield criterion

$$\lambda = \frac{\mu \mathbf{b}}{\sigma - \sigma_0} \tag{5.22}$$

Such a distribution is shown schematically in Fig. 5.28. As the dislocation network coarsens (recovers) some links grow to satisfy eqn (5.22) and immediately yield, interacting to form smaller links (i.e. harden). If the stress is reduced from $\sigma$ to $\sigma - \Delta\sigma$, the new yield criterion occurs at longer link lengths $\lambda + \Delta\lambda$,

$$\lambda + \Delta\lambda = \frac{\mu \mathbf{b}}{\sigma - \Delta\sigma - \sigma_0} \tag{5.23}$$

However, since the original dislocation distribution is inherited, no creep can occur until sufficient recovery has occurred to allow the *longest* links to grow

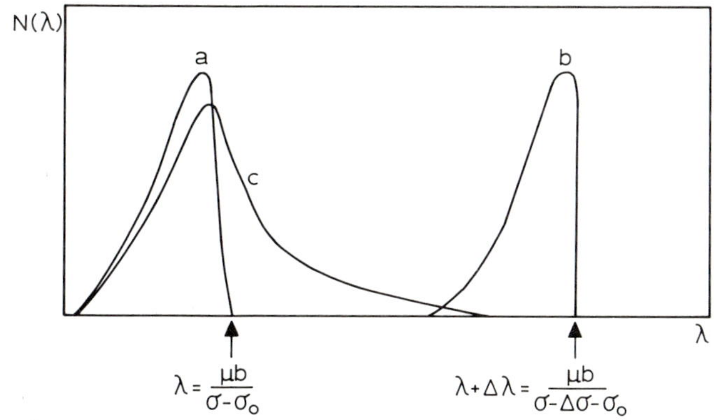


**Figure 5.28 Schematic illustration of the distributions of dislocation link lengths: (a) steady state for stress $\sigma$; (b) steady state for stress $\sigma - \Delta\sigma$; (c) transient for stress $\sigma - \Delta\sigma$**

to satisfy eqn (5.23). McLean[31] has extended Friedel's[27] model of dislocation recovery to account for the behaviour of a distribution of dislocation link lengths showing that the *longest* links grow with parabolic kinetics at a faster rate than the shorter links following the form of eqn (5.14) but with an increased kinetic constant. Consequently, combining this with eqns (5.22) and (5.23) it can easily be shown that the delay period $\Delta t$ before creep recommences is given by the expression

$$\Delta t = \frac{\mu^2 \mathbf{b}^2}{K}[(\sigma_R - \sigma_0)^{-2} - (\sigma_I - \sigma_0)^{-2}] \tag{5.24}$$

where $\sigma_R$ and $\sigma_I$ are the remaining and initial stresses and $K$ is a kinetic constant that is proportional to the diffusion coefficient. Noting that a plot of $\sigma_R$ versus $\sum_i \Delta t_i$ is completely equivalent to that of $\sum_i \Delta\sigma_i$ vs. $\sum_i \Delta t_i$ with a simple change of axes ($\sigma_R = \sigma_I - \sum_i \Delta\sigma_i$), Fig. 5.27*b* shows a least squares fit of eqn (5.24) to a typical set of data. The agreement between the theoretical curve and the experimental data is very good. However, extrapolation to the asymptotic value of $\sigma_R$ occurs over a very long time scale, indicating significantly lower values of $\sigma_0$ than have been obtained by visual inspection of the data. Moreover, the detailed forms of such curves depend on temperature and initial stress in addition to $\sigma_0$. This is illustrated in Figs. 5.29*a–c* where the consequences of varying $\sigma_0$, $(\mu^2\mathbf{b}^2)/K$ or $\sigma_I$ while keeping the others constant on the shapes of the $\sigma_R$ versus $\sum_i \Delta t_i$ curves are demonstrated. Clearly if all such variations were attributed to differences in $\sigma_0$ quite misleading results would be obtained. Consequently, unless a reliable method of data analysis is employed, apparent rather than real variations of $\sigma_0$ with testing conditions may be erroneously identified. In addition, this approach provides a quantitative measure of the recovery kinetics that has always been contained in such data but never previously extracted, and that may be of value in interpreting the cumulative strain during cyclic creep.

The influences of microstructural parameters and testing conditions have been investigated by several authors; Wilshire and co-workers[32,33] and Evans and Harrison[34] used the earlier approach to data analysis by inspection and McLean[35] and Henderson and McLean[36] used the numerical approach described above. The results on directionally solidified IN738LC by the latter method suggest that $\sigma_0$ is relatively independent of $\sigma_I$ and is less dependent on temperature and microstructural parameters than earlier work has indicated. Although it seems clear that $\sigma_0$ is largely governed by the volume fraction, size and morphology of the $\gamma'$ precipitate in nickel-base alloys, the creep rate is also particularly dependent on the recovery kinetics that may also be affected by these same parameters.

It can be very misleading to attribute changes in creep rate only to $\sigma_0$ without regard to the kinetic factors. For example, it has been suggested[32] that the characteristic shapes of the creep curves of superalloys, and indeed of many complex engineering alloys, which take the form of continuously increasing creep rates rather than of well-defined regimes of steady state creep (Fig. 5.30*a*) are due to $\sigma_0$ decreasing as the microstructures of the alloys coarsen during the creep tests. This interpretation qualitatively predicts the correct forms of the creep curves; inspection of eqn (5.21) indicates that, for a constant applied stress $\sigma$ and parameters $A$ and $m$, a steadily decreasing $\sigma_0$ will give increasing $\dot{\varepsilon}$. However, Dyson and McLean[37] have considered the full quantitative implications of such a model. Combining eqn (5.21) with expressions describing (a) the coarsening of the $\gamma'$ precipitate particles

$$r^3 = r_0^3 + Bt \tag{5.25}$$

where $r$ is the particle radius, $B$ is the coarsening rate constant and $t$ is time (particle coarsening will be discussed more fully in Chapter 7) and (b) the reduction in $\sigma_0$ with increasing $r$

$$\sigma_0 = a + br^{-q} \tag{5.26}$$

where $a$, $b$ and $q$ are constants, it is readily shown that

$$\left(\frac{\dot{\varepsilon}}{\dot{\varepsilon}_{\min}}\right)^{1/4} = 1 + \alpha t \tag{5.27}$$

where

$$\alpha = \frac{1}{3}\,\frac{bqr_0^{-(q+3)}B}{\{\sigma - (a + br_0^{-q})\}}$$

and $\dot{\varepsilon}_{\min}$ is the minimum creep rate measured. This indicates that replotting creep curves as $(\dot{\varepsilon}/\dot{\varepsilon}_{\min})^{1/4}$ versus $t$ should give straight lines with slopes that are either independent of $\sigma$ if $\sigma_0 \propto \sigma$ or have increasing slopes with decreasing $\sigma$ if $\sigma_0$ is invariant. Results on several nickel base superalloys show that although acceptable straight lines are obtained over much of the creep data the highest slopes are obtained for tests at high $\sigma$ contrary to the theoretical expectation. Typical data for MarM246 are shown in Fig. 5.31*a*. Dyson and McLean[37] show that a better, though empirical, correlation can be made with the accumulated strain rather than time, ratioing $\dot{\varepsilon}$ to $\dot{\varepsilon}_{0\cdot 01}$ the creep rate at 1% strain (Fig. 5.31*b*), and this has been interpreted as indicating a change in recovery kinetics as strain accumulates. Measurements of $\sigma_0$ and $(\mu^2\mathbf{b}^2)/K$ at different stages during creep tests indicate that the variation in $\sigma_0$ is much less than that required to account for the observed increase in creep rate but that $(\mu^2\mathbf{b}^2)/K$ progressively decreases (or recovery increases)—see Fig. 5.32. If recovery is not taken into account during its measurement then $\sigma_0$ would *appear* to decrease as recovery kinetics increase since the asymptotic value is more rapidly approached.

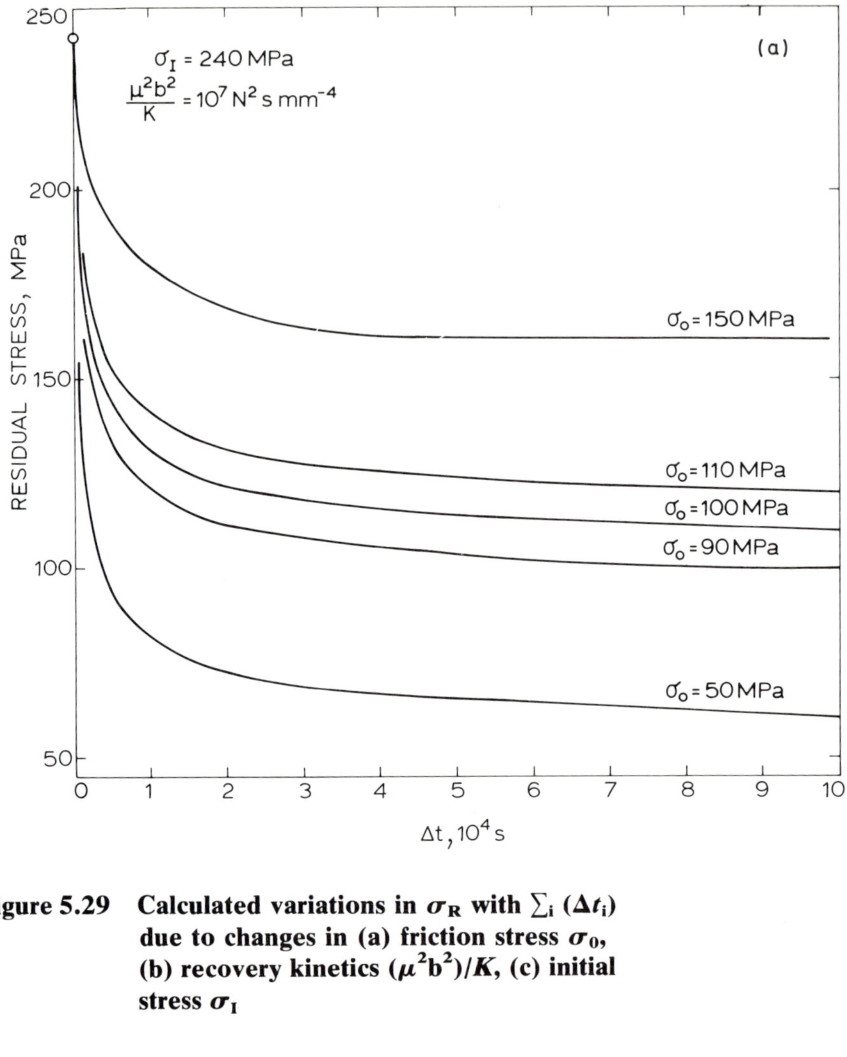


**Figure 5.29 Calculated variations in $\sigma_R$ with $\sum_i (\Delta t_i)$ due to changes in (a) friction stress $\sigma_0$, (b) recovery kinetics $(\mu^2 b^2)/K$, (c) initial stress $\sigma_I$**

When the creep data for many engineering alloys, not only superalloys, are expressed in terms of eqn (5.21) using the values of $\sigma_0$ determined by, for example, the stress decrement technique, values of $n$ and $Q$ that are consistent with theoretical expectation are obtained. Evans and Harrison[38] have analysed data on a wide range of alloys and have shown that much of the scatter in properties between different alloys can be removed and a 'universal' description obtained. If validated this should have considerable advantages in eventually reducing the quantity of experimental testing required for engineering design purposes.

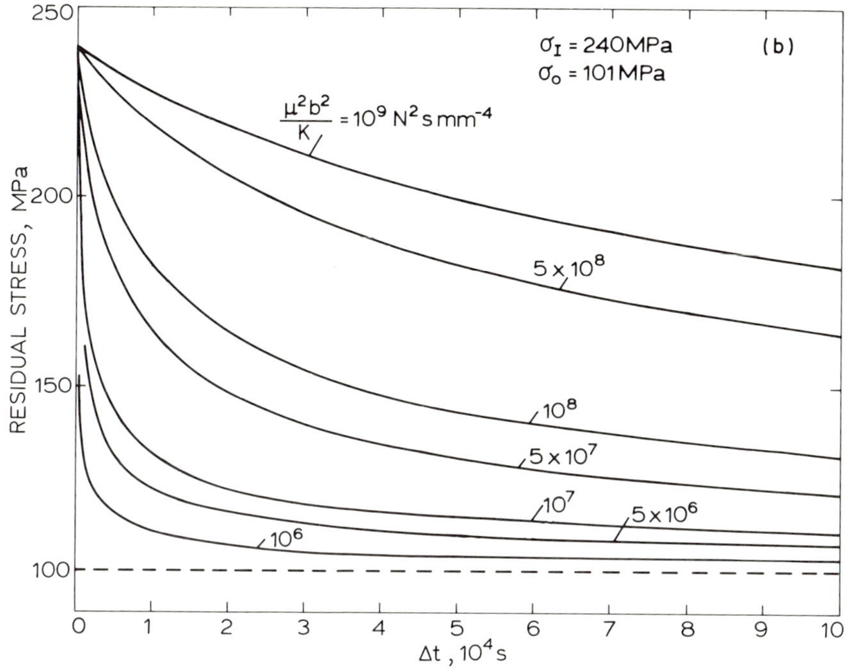

$\sigma_I$ = 240MPa
$\sigma_o$ = 101MPa
(b)
$\frac{\mu^2 b^2}{K}$ = $10^9$ N$^2$ s mm$^{-4}$
5 x $10^8$
$10^8$
5 x $10^7$
$10^7$
5 x $10^6$
$10^6$
RESIDUAL STRESS, MPa
Δt, $10^4$ s


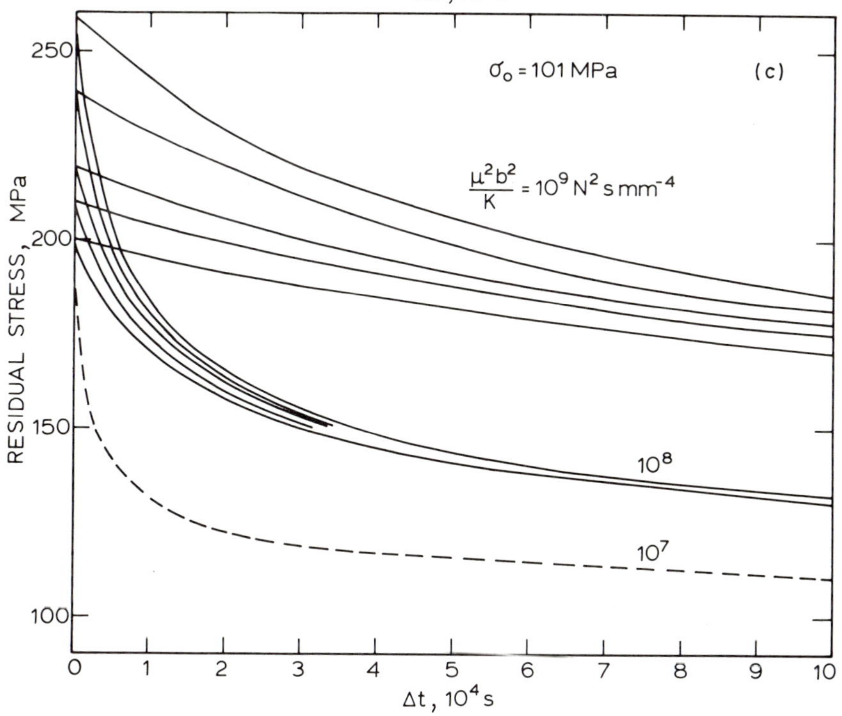

$\sigma_o$ = 101 MPa
(c)
$\frac{\mu^2 b^2}{K}$ = $10^9$ N$^2$ s mm$^{-4}$
$10^8$
$10^7$
RESIDUAL STRESS, MPa
Δt, $10^4$ s

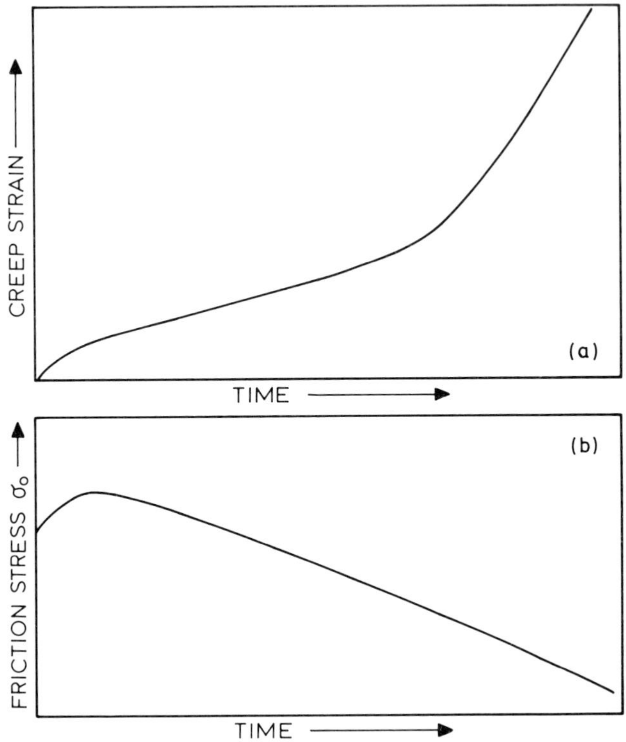


**Figure 5.30 Schematic illustration of the proposed relationship between (a) progressively increasing creep rate during a creep test, and (b) decreasing $\sigma_0$**

## 5.6 STRENGTHENING MECHANISMS

The high temperature strengths of superalloys result from a range of phenomena:

(i) solid solution hardening where large substitutional atoms such as tungsten, molybdenum, niobium and rhenium locally compress the lattice, causing an increase in the Peierl's stress for dislocation motion

(ii) grain boundary strengthening by various carbides (e.g. MC, $M_6C$ and $M_{23}C_6$) which predominantly inhibit grain boundary sliding. The important carbide forming elements are Ti, Ta, Hf and V

(iii) particle strengthening by high volume fractions (0·3–0·6) of ordered intermetallic $\gamma'$ $Ni_3Al$ particles that are coherently orientated with respect to the fcc matrix and which are precipitated from the solid state. The $\gamma'$ forming elements are Al and Ti.

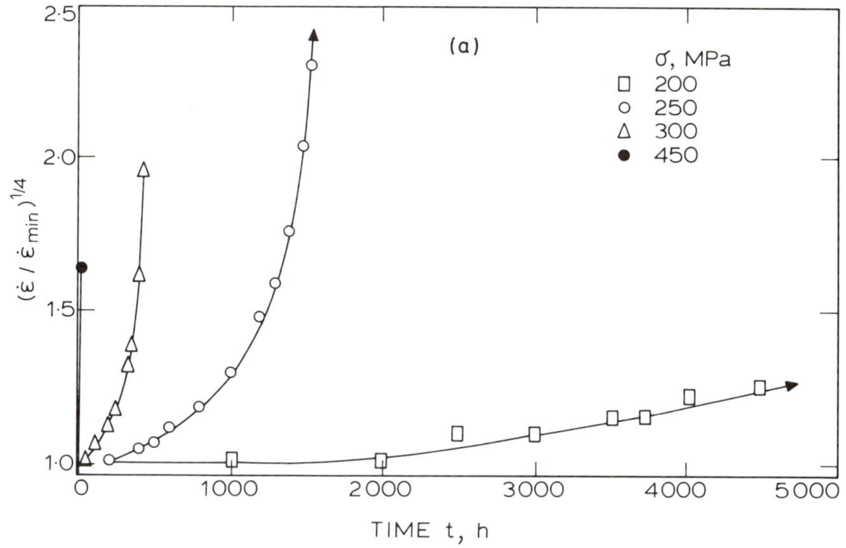


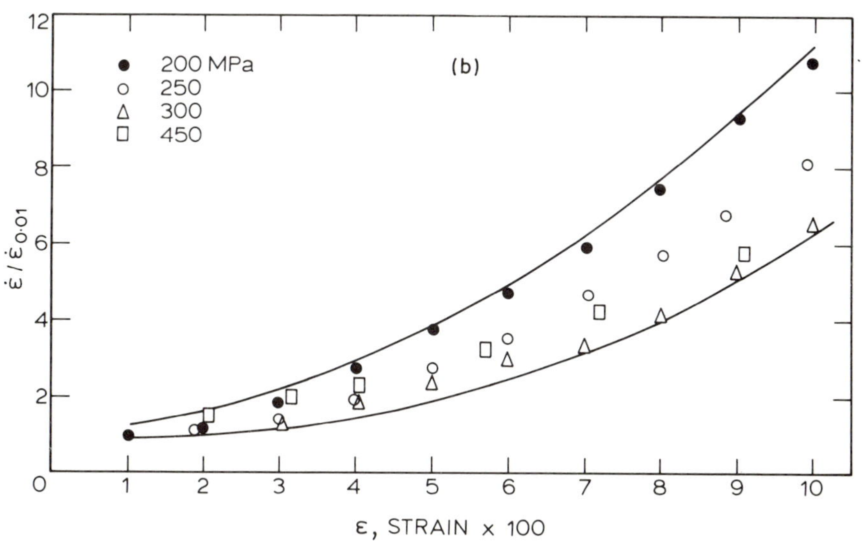


**Figure 5.31 Creep data for directionally solidified MarM246 tested at 1123 K with various stresses and replotted as (a) $(\dot{\varepsilon}/\dot{\varepsilon}_{min})^{1/4}$ versus $t$ and (b) $\dot{\varepsilon}/\dot{\varepsilon}_{0\cdot01}$ versus $\varepsilon$**

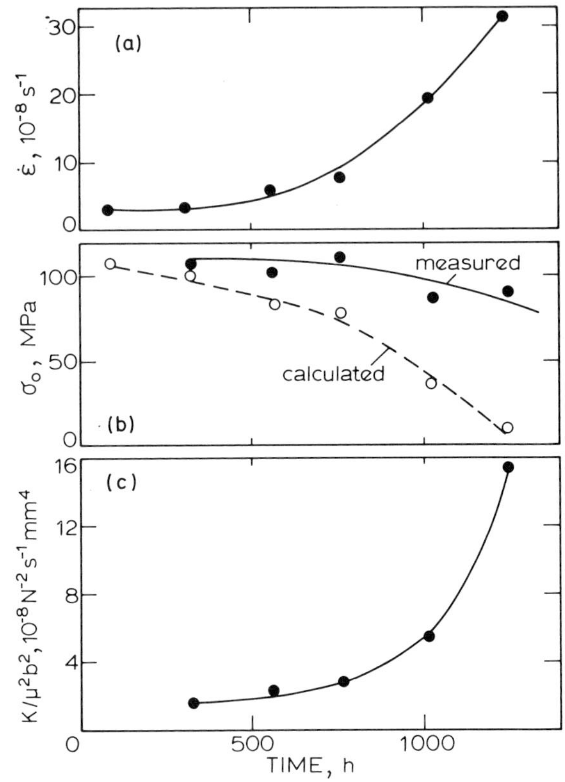


**Figure 5.32** **(a) Variation in creep rate during creep test on directionally solidified IN738LC at 850°C and a stress of 250 MPa. The test was interrupted periodically by stress dip procedures to measure friction stress and recovery kinetics.**
**(b) Variation of $\sigma_0$ during the creep test and the calculated variation of $\sigma_0$ required to account for increasing creep rate in terms of eqn (5.21).**
**(c) Variation in recovery kinetics $K/(\mu^2 b^2)$ during creep test**

There have been several very full reviews of strengthening mechanisms in superalloys and the reader is referred to these (e.g. Stoloff[39]) for a complete survey of the various contributing factors. Here we shall be concerned only in examining the likely origins of the friction stress introduced in the previous section.

Although mechanisms (i) and (ii) are important to all superalloys and, indeed, are the principal factors in strengthening cobalt- and iron-base superalloys it is now generally accepted that $\gamma'$ strengthening is the most important feature of advanced nickel-base superalloys. The particles act as barriers to the motion of dislocations through the matrix and flow cannot occur until a critical, or threshold, stress that enables the barrier to be surmounted has been exceeded. There are four mechanisms by which dislocations can overcome these obstacles and a separate threshold stress can be associated with each of them. Only those processes having thresholds less than the applied stress can be activated and the fastest of these will be the dominant deformation mode. The four mechanisms that have been identified are

1. bowing between particles
2. cutting of particles
3. dislocation climb over particles
4. dragging of particles by dislocations.

The last of these may occur in a few special circumstances but appears to be of little relevance to superalloys. Consequently it will not be discussed further here.

It has been customary to discuss the various particle-bypass mechanisms as acting independently with no interaction between them and to assume that only one of the mechanisms can be operative. In the case of a single crystal with a plane front solidification morphology this may be true, but is less likely to be so when there are microstructural and orientation variations associated with dendritic structures and polycrystals.

Figure 5.33 shows the dislocation configuration schematically when a tensile stress $\sigma$ is applied at an angle $\phi$ to the surface normal of the active glide plane in a material with cuboid particles. Then the shear stress $\tau$ leading to glide in and the normal stress $\sigma_n$ giving climb out of that plane are

$$\tau = \sigma \sin \phi \cos \phi \tag{5.28}$$

$$\sigma_n = \sigma \sin^2 \phi \tag{5.29}$$

The dislocation bows between the particles with a radius

$$r = \frac{\Gamma}{\tau \mathbf{b}} \tag{5.30}$$

and makes an angle $\theta$ with the obstacles where

$$\cos \theta = \frac{\tau \mathbf{b} L}{2\Gamma} \tag{5.31}$$

($\Gamma$ is the dislocation line energy, $\mathbf{b}$ is the Burgers vector and $L$ is the interparticle spacing.) When the resolved force components of the dislocations

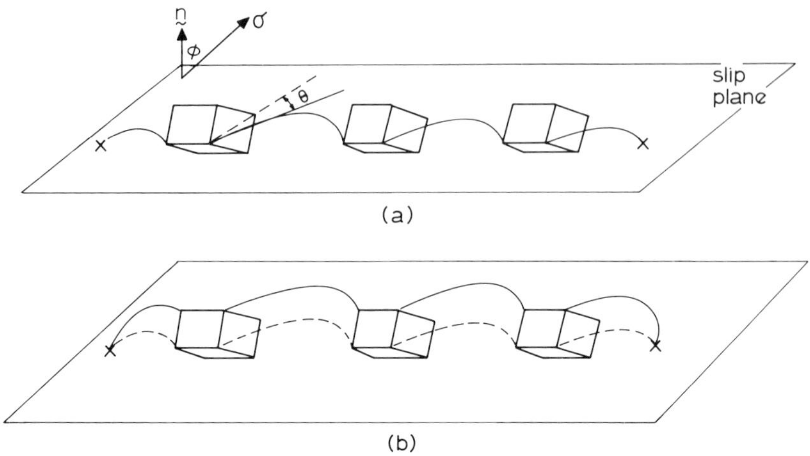


**Figure 5.33 Schematic illustration of a partially bowed dislocation in a particle strengthened alloy (a) without and (b) with a climb component**

create a shear stress on the particle exceeding the critical shearing stress of the particle $\tau_c$ then particle cutting will occur, i.e.

$$\tau_c < \frac{2\Gamma \cos\theta}{\sqrt{2}\, a^2} = \frac{\tau \mathbf{b} L}{\sqrt{2} a^2} \tag{5.32}$$

or

$$\tau > \frac{\sqrt{2} a^2}{\mathbf{b} L} \tau_c = \frac{\sqrt{2} a f}{\mathbf{b}(1-f)} \tau_c \tag{5.33}$$

If the dislocation completely bows ($\theta = 0$) before the critical cutting stress is reached then it becomes unstable and escapes from the obstacle leaving a loop around the particle. Both of these mechanisms are essentially time independent, occurring very rapidly after the threshold condition has been satisfied.

At lower stresses the particles can be avoided by the dislocations extending by climb as shown schematically in Fig. 5.33*b*; however, since this involves diffusion it will occur much more slowly than the others. Consequently dislocation climb will only be important when the others cannot occur. However, the transitions between the various mechanisms are extremely sensitive to the particle characteristics (e.g. volume fraction, spacing) and the importance of climb relative to the others depends on the distribution of stress between shear and normal stress components (orientation). It is therefore quite feasible that deformation can occur by different mechanisms in different grains of a polycrystal or in different regions of a dendritic structure.

The detailed treatments of the athermal processes, particle cutting and bowing as described by Brown and Ham[40] are fairly generally accepted and give critical shear stresses for these processes of

(a) *Orowan bowing*

$$\tau_c = A \frac{\mu \mathbf{b}}{2\pi\lambda} \ln\left(\frac{\lambda}{2\mathbf{b}}\right) \tag{5.34}$$

where $A = \frac{1}{2}((2-\nu)/(1-\nu))$, $\lambda$ is the particle spacing and $\nu$ is Poisson's ratio.

(b) *Particle cutting*
As the matrix dislocation passes through an ordered precipitate, like $\gamma'$, it creates disorder which increases the energy of the system by the antiphase-boundary energy $\gamma$. Brown and Ham[40] show that for passage of a single dislocation

$$\tau_c = \left(\frac{\gamma^{3/2}}{\mathbf{b}}\right)\left[\frac{2af}{\pi\Gamma}\right]^{1/2} \tag{5.35}$$

where $f$ is the precipitate volume fraction, $a$ is the particle diameter and $\Gamma$ is the dislocation line tension. However, pairs of dislocations can pass through particles with a lower stress where the second annihilates the antiphase boundary created by the first. Three cases have been identified by Brown and Ham:

(i) When the dislocation pairs are straight, i.e. $a < \pi\Gamma f/2\gamma$, then the particles have little influence on the flow stress.
(ii) When the first dislocation bows but the second remains straight, i.e. $\pi\Gamma f/2\gamma < a < 2\Gamma/\gamma$, then

$$\tau_c = \frac{\gamma}{2\mathbf{b}}\left[\left(\frac{2af\gamma}{\pi\Gamma}\right)^{1/2} - f\right] \tag{5.36}$$

(iii) When both dislocations are contoured ($a > 2\Gamma/\gamma$) then

$$\tau_c = \frac{\gamma}{2\mathbf{b}}\left[\left(\frac{4f}{\pi}\right)^{1/2} - f\right] \tag{5.37}$$

Although there have been several treatments of these mechanisms they generally predict the same functional dependences on the microstructural parameters and independence of applied stress. However, there is much less unanimity on the particle climb mechanism.

(c) *Climb around particles*
Brown and Ham describe a treament that gives a simple analytical expression for the threshold stress for climb. This is obtained by performing an energy

balance to determine the stress required to create sufficient new dislocation to contour the particles and takes the form of a condition like that for Orowan bowing expressed in terms of the particle size rather than spacing. Various modified treatments[41,42] have been proposed that come to contradictory conclusions. McLean[43] has critically reviewed these and argues that a physically realistic treatment must (i) allow co-operative climb around groups of particles, (ii) account for partial bowing in the slip plane and (iii) allow relaxation of the dislocation between particles. The relevant equations must be solved numerically and do not give an explicit expression for the threshold stress. However, the calculations show that the threshold stress for climb around particles is

(i) sensitive to the applied stress
(ii) weakly dependent on the microstructural parameters
(iii) influenced by $a$ and $\lambda$ which define the transitions to bowing and shearing mechanisms.

The results of calculations for a single crystal superalloy stressed parallel to $\langle 100\rangle$ and yielding on $\{111\}$ are shown in Fig. 5.34. The values of the parameters taken ($\mu \mathbf{b} = 10\ \mathrm{N\ m^{-1}}$, $f = 0{\cdot}4$) are relevant to the superalloy IN738LC tested at 850°C for which experimental values of $\sigma_0$ are available. The figure clearly shows the testing conditions and microstructural dimension leading to the activation of the various mechanisms. Moreover, within the climb regime, the calculated values of $\sigma_0$ for $200 < \sigma_a < 300$ MPa are close to the experimental values of 80 to 100 MPa.

An interesting proposal to improve the creep strength of some nickel-base superalloys through control of the morphology rather than size of $\gamma'$ particles has been made by Pearson *et al.*[44] The mechanism by which $\gamma'$ in some alloys changes in shape when heat treated under stress will be described more fully in Chapter 7. However, Pearson *et al.* show that this effect can considerably reduce the rate of deformation of the experimental Ni–13Al–9Mo–2Ta (atomic %) alloy. Rafts of $\gamma'$ oriented normal to the tensile axis develop at 1311 K with an applied stress of 207 MPa; when subsequently creep tested to failure at a lower temperature the stress-annealed material shows a significantly lower minimum creep rate and longer life than the heat treated virgin alloy as shown in Fig. 5.35. The interpretation of this strengthening is that lateral extension of the $\gamma'$ particles increases the distances over which climb must occur whereas particle cutting would not be affected.

## 5.7 CYCLIC DEFORMATION

Components in service normally experience varying loads and the design engineer must generally anticipate the response of the component to cyclic stressing through the application of constant load creep data or from attempts to reproduce specific component stress/strain histories. Several procedures

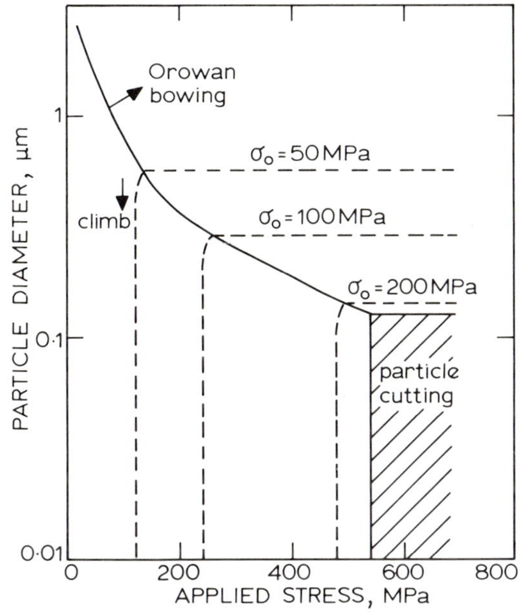


**Figure 5.34 Diagram indicating the particle by-pass mechanisms that operate for different values of applied stress and particle size. The calculations are for a single crystal superalloy stressed parallel to ⟨100⟩ that contains 40 vol.-% $\gamma'$**

have been proposed to account for the accumulation of strain and/or damage in materials during both stress- and strain-controlled cycles; however, these have tended to be based on empirical or continuum engineering concepts and have not considered microstructural features. Factors affecting cracking of directionally solidified alloys will be discussed in § 5.8. However, the high intrinsic ductilities of these materials would suggest that a much higher proportion of the fatigue life of directionally solidified than of conventionally cast superalloys is spent prior to the formation of crack nuclei which is governed by plastic or creep strain accumulation.

The low elastic modulus $E$ of the natural ⟨100⟩ growth axis in directionally solidified superalloys ensures that under strain controlled conditions the stress range $\Delta\sigma$ is lower than in the equivalent commercially cast material

$$\Delta\sigma = E\Delta\varepsilon \tag{5.38}$$

This appears to be particularly important during temperature cycling where differential cooling and heating rates of different parts of a component cause

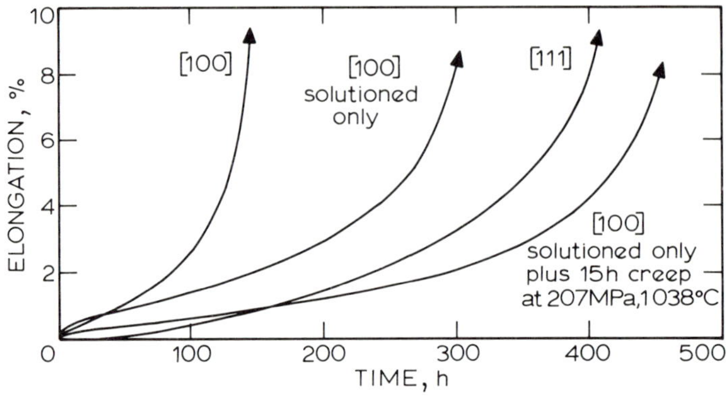


**Figure 5.35 Creep curves for ⟨100⟩ oriented single crystals of Ni–13Al–9Mo–2Ta after different heat treatments including one that was subjected to a stress anneal of 15 h at 1311 K/207 MPa; testing conditions 1172 K at 413 MPa**

strain controlled thermal fatigue that is determined by differential thermal expansion. The combination of minimum $E$ and good stress rupture properties of ⟨100⟩ orientated crystals ensures improved performance over conventional casting in both steady load and thermal fatigue. However, although ⟨111⟩ oriented crystals are reported to have better stress rupture properties than ⟨100⟩, the maximum modulus for ⟨111⟩ probably militates against this orientation being used in service conditions because of the likelihood of reduced performance in cyclic temperatures.

Although the same considerations would anticipate inferior performance of ⟨100⟩ orientations relative to conventional castings under stress controlled high cycle fatigue, the improved ductility of directionally solidified material overcomes the disadvantage of the lower modulus to give significant improvements in fatigue endurance. Thus directionally solidified MarM200 when stressed longitudinally has about a 10% improvement in the $10^7$ cycle endurance limit over the conventionally cast alloy; transverse stressing gives similar results to the conventional casting.

Although creep and fatigue behaviour can be separately characterized with some precision there remains considerable difficulty in predicting either the strain accumulation or life of material that is exposed to a combination of steady and periodically fluctuating stresses; this, of course, is the situation for components in service. These remarks are valid for all engineering materials; however, the relative structural simplicity of directionally solidified materials may make them rather more amenable to study than polycrystalline alloys. Several life prediction methods accounting for creep/fatigue interactions have been proposed with varying success.

### (a) Life fraction rule

This approach, also known as the Robinson Rule,[45] assumes linear damage in both creep and fatigue so that failure occurs when the sum of the fractional lives by each mechanism considered separately is unity.

$$\frac{N}{N_f} + \frac{\sum t_h}{t_f} = 1 \tag{5.39}$$

where $N$ is the elapsed number of cycles, $N_f$ is the fatigue life to failure in continuous cycling, $\sum t_h$ is the accumulated life in the creep load and $t_f$ is steady state creep rupture life. No interaction between creep and fatigue is assumed. The method is most appropriate to applications where one mechanism is dominant.

### (b) Strain range partitioning

A rather complex theoretical approach has been developed by Manson and co-workers[46,47] which assumes that accumulated creep strain causes the reduced fatigue lives that are often observed by the insertion of hold times at high temperatures. The effect of creep on the fatigue properties, however, depends on where it occurs in the cycle. Four independent types of inelastic strain range are identified:

$\Delta\varepsilon_{pp}$ plastic strain reversed by plastic strain
$\Delta\varepsilon_{cc}$ creep strain reversed by creep strain
$\Delta\varepsilon_{pc}$ tensile plastic strain reversed by compressive creep strain
$\Delta\varepsilon_{cp}$ tensile creep strain reversed by compressive plastic strain.

The fatigue lives for each type of cycle can be determined experimentally. An arbitrary closed loop fatigue cycle can be considered as the sum of up to three of the basic strain ranges and the cycle life can be estimated by summing the fractional lives for the basic strain ranges, again assuming a linear damage rule.

Some modifications to these techniques have been proposed to improve the estimates of creep strain by, for example, integrating over all stresses rather than only considering the contribution during hold periods. However, most of these treatments have been of an empirical nature, whereas interactions are most likely to be accounted for by a better understanding of the physical processes occurring during stress cycling at elevated temperatures. The transient creep behaviours following both reductions and increments in stress that have been characterized to elucidate basic creep mechanisms should also be relevant to high temperature fatigue studies. If it were possible to predict the strain–time sequence following an arbitrary stress cycle, this information could be included in summing the total creep strain during the entire cycle. It would be particularly useful to have explicit analytical descriptions of the strain contributions from various factors such as anelasticity and recovery. Henderson and McLean[48] have made a detailed study of unloading/reloading strain transients of a typical *in situ* composite and directionally solidified superalloy; full details are given in Chapter 6 since the analysis appeals to the theory of composite materials. It utilizes the expression for the incubation period $\Delta t$ before creep recommences following a stress reduction (eqn (5.24)) to quantitatively characterize creep stagnation periods in both the unloading and loading transients in order to explain the greater severity of slow relative to fast stress cycling. It further suggests that much of the time dependent transient strain, namely anelasticity, can be accounted for by stress redistributions between 'hard' and 'soft' elements of the material, just as stress is off-loaded from fibre and matrix in a composite material, rather than by dislocation interactions. On this basis single crystals and directionally solidified alloys should exhibit much less anelastic strain than equiaxed material consisting of randomly oriented grains with different elastic and plastic properties.

## 5.8 FRACTURE

Creep fracture of superalloys with equiaxed grain structures is generally associated with grain boundaries. There remains some debate over the precise mechanism but creep cracks usually propagate along the grain boundaries that lie transverse to the direction of applied stress. The fracture surface of the conventionally cast superalloy IN738LC, shown in Fig. 5.36*a*, is quite typical of this failure mode. Two general causes of grain boundary fracture during creep that have been proposed are briefly outlined below.

### (a) Creep cavitation

During or prior to creep cavitation voids can nucleate at grain boundaries in materials due to differential yielding of neighbouring grains as discussed in § 5.3. If these voids exceed a critical size $r \gtrsim 2\gamma/\sigma$ (where $\gamma$ is the surface energy and $\sigma$ is the applied stress) then the cavity can continue to grow by diffusional processes. Many detailed theories of the development of cavitation have been proposed, and these have recently been reviewed by Svensson and Dunlop.[49] All predict the same gradual degradation of microstructure until the final failure occurs by linkage of the cavities. This process has been confirmed many times in studies of pure metals and is certainly relevant to the simpler wrought superalloys such as Nimonic 80A[17] and probably also to cast superalloys at low stresses.

### (b) Edge crack propagation

Several studies of the more advanced wrought and cast superalloys at higher stress levels have suggested that creep failure occurs by the initiation of

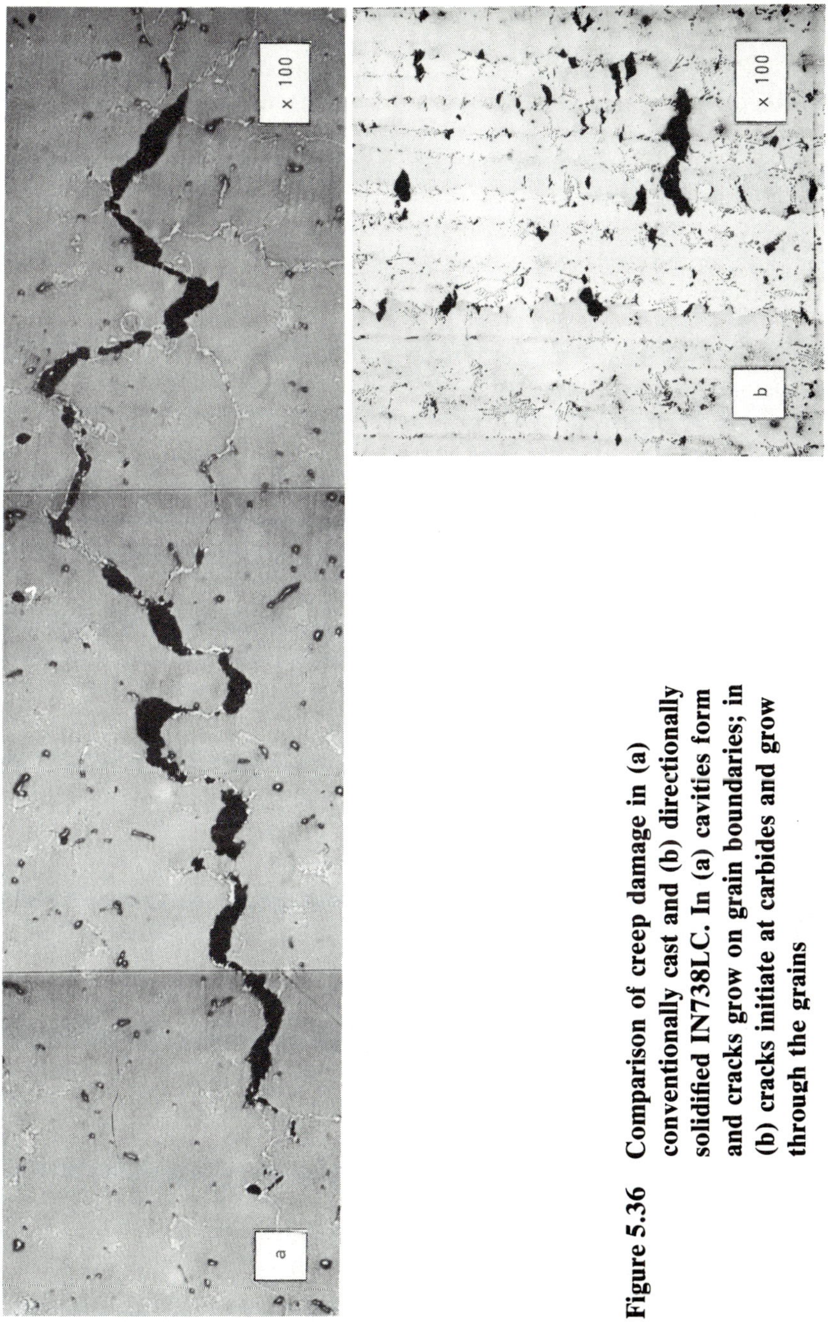


**Figure 5.36** Comparison of creep damage in (a) conventionally cast and (b) directionally solidified IN738LC. In (a) cavities form and cracks grow on grain boundaries; in (b) cracks initiate at carbides and grow through the grains

cracking at the edge of the specimen or component and its rapid propagation through the section along transverse grain boundaries. Thus damage is isolated to the vicinity of the final failure and, perhaps, to a few subcritical cracks. This mechanism is most likely to occur in alloys such as MarM200 that show little ductility and virtually no tertiary creep regime. The initial nucleation of the edge crack may occur by the ingress of oxidation or corrosion products along the grain boundary.

Since the transverse grain boundaries are essentially eliminated in directionally solidified structures neither of these mechanisms can occur and cracks are constrained to propagate through, rather than between, grains as illustrated in Fig. 5.36*b* which shows a creep failure in directionally solidified IN738LC. Moreover, any transverse crack must extend through the aligned dendritic structures that are produced by directional solidification. The crack nuclei appear to form in a variety of ways depending on the detailed microstructure. Two such examples are indicated in Fig. 5.37 where (a) small crack segments are associated with transverse components of longitudinal grain boundaries (either at the surface or within the specimen) and (b) wedge cracks starting at interdendritic carbides extend into the bulk alloy. Irrespective of how they nucleate it seems clear that failure is determined by the rate of propagation of these cracks. Consequently damage is not as highly localized as for conventionally cast superalloys and this may lead to the higher ductilities associated with directional solidification.

Since the density of potential crack nuclei (e.g. interdendritic carbides) and the mean free path of crack segments (i.e. the interdendritic spacing) are known to be controlled by the solidification processing conditions, it might

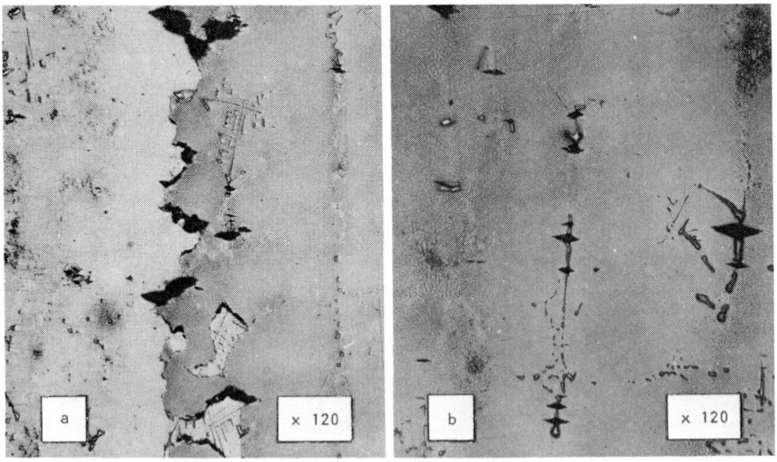


**Figure 5.37 Fracture initiation in IN738LC: (a) transverse segments of grain boundary; (b) interdendritic carbides**

be expected that the fracture behaviour of directionally solidified superalloys would be sensitive to the solidification structure. This author knows of no such detailed study. However, preliminary experiments on IN738LC directionally solidified at a wide range of rates and in different temperature gradients have failed to show any significant differences in rupture ductility. This, perhaps, suggests that there are insufficient potential cracks to allow interaction by linkage and that failure must occur by a single crack slowly propagating across the specimen section.

In view of the importance of turbine blade cooling and the increasing use of thin section hollow blades, it is important to retain the beneficial mechanical properties of superalloys in thin sections. The rupture lives of conventionally cast superalloys decrease with decreasing section thickness as the probability increases of a grain boundary traversing the specimen thickness. Thus, 1 mm thick specimens can have as little as 20% of the rupture

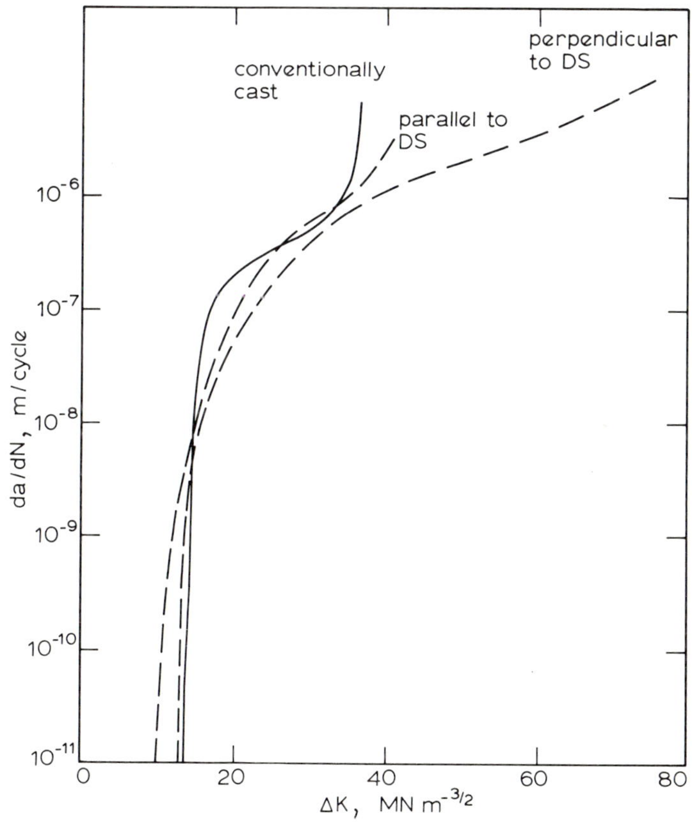


**Figure 5.38 Growth rates of fatigue cracks in conventionally cast and directionally solidified IN738LC**

lives of bulk alloys. The likelihood of such an event is substantially lower in the directionally solidified material and its sensitivity to specimen thickness is consequently reduced. Winstone and Northwood[50] have shown on five nickel-base superalloys that there is no significant reduction in rupture properties of directionally solidified thin sections if they are prepared in moulds that have a fine surface finish. Excessive surface roughness of moulds can prevent casting contraction and lead to cracking of longitudinal grain boundaries.

The crack propagation behaviour, under both fatigue and steady loading conditions, of directionally solidified IN738LC has been considered in some detail and compared with that of the conventionally cast alloy by Scarlin and Speidel.[51,52] During creep or in fatigue at high temperatures crack growth rates perpendicular to the solidification direction are considerably reduced due to branching parallel to the interdendritic boundaries. On the other hand, fatigue cracking resistance in this direction is reduced at low temperatures because of the easy crystallographic fracture on the {100} planes that are preferentially orientated normal to the solidification direction. Their experimental results at 850°C are shown in Fig. 5.38. The results for fatigue crack growth parallel to the direction of solidification and for conventional castings are indistinguishable whereas those perpendicular to the solidification direction have about twice the fracture toughness values (although the threshold stress for cracking is similar). This is the temperature regime where thermal fatigue damage is greatest and indicates why thermal cracking does not occur as readily in directionally solidified alloys as in conventionally cast material.

## REFERENCES CHAPTER 5


1 F. L. VERSNYDER AND R. W. GUARD: *Trans. ASM*, 1960, **52**, 485

2 B. H. KEAR AND B. J. PIEARCEY: *Trans. Metall. Soc. AIME*, 1967, **239**, 1209

3 B. J. PIEARCEY AND B. E. TERKELSON: *Trans. Metall. Soc. AIME*, 1967, **239**, 1143

4 J. E. NORTHWOOD AND T. HOMEWOOD: *Metallurgia and Metal Forming*, 1974, **41**, 254

5 D. A. WOODFORD AND J. J. FRAWLEY: *Metall. Trans.*, 1974, **5**, 2005

6 P. R. SAHM AND F. SCHUBERT: 'Solidification and casting of metals', 389, 1979, London, Metals Society

7 J. S. ERIKSON, C. P. SULLIVAN AND F. L. VERSNYDER in 'High temperature materials for gas turbines', eds. P. R. Sahm and M. O. Speidel, 1974, Amsterdam, Elsevier

8 J. M. DRAPIER, P. VIATOUR, D. COUTSOURADIS AND L. HABRAKEN: 1st Progress Report on COST 50 Programme B9/1, July, 1975
9 J. P. HIRTH AND J. LOTHE: 'Theory of dislocations', 1968, New York, McGraw-Hill
10 C. H. WELLS: *Trans. ASM*, 1967, **60**, 270
11 'Handbook of mathematical functions', eds. M. Abramowitz and I. A. Stegun, 1965, New York, Dover
12 D. LEFEVRE: thesis, University of Liège, Belgium, 1977
13 J. F. NYE: 'Physical properties of crystals', 1957, Oxford University Press
14 R. A. McKAY, R. L. DRESHFIELD AND R. D. MAIER: Proc. 4th Int. Symp. Superalloys, eds. J. K. Tien *et al.*, 1980, Metals Park, Ohio, ASM
15 G. R. LEVERANT AND B. H. KEAR: *Metall. Trans.*, 1970, **1**, 491
16 M. F. ASHBY in 'Strengthening methods in crystals', eds. A. Kelly and R. B. Nicholson, 1971, London, Applied Science
17 B. F. DYSON, M. F. LOVEDAY AND M. J. RODGERS: *Proc. R. Soc.*, 1976, **349A**, 245
18 P. N. QUESTED AND M. McLEAN: 'Solidification technology in the foundary and casthouse', 586, 1983, London, Metals Society
19 M. McLEAN AND P. N. QUESTED: 'Creep and fracture of engineering materials and structures', eds. B. Wilshire and H. Evans, 1981, Swansea, Pineridge Publishing
20 M. GELL, A. F. GIAMEI AND D. N. DUHL: Proc. 4th Int. Symp. Superalloys, eds. J. K. Tien *et al.*, 1980, Metals Park, Ohio, ASM
21 O. D. SHERBY AND P. M. BURKE: *Prog. Mater. Sci.*, 1967, **13**, 325
22 M. F. ASHBY: Proc. ICSMA 3, sponsored by Inst. of Metals and Iron and Steel Inst., 1973, **2**, 8
23 C. R. BARRETT AND O. D. SHERBY: *Trans. Metall. Soc. AIME*, 1965, **233**, 1116
24 D. McLEAN: *Rep. Prog. Phys.*, 1966, **29**, 1
25 R. LAGNEBORG: *J. Mater. Sci.*, 1968, **3**, 596
26 E. OROWAN: *J. West of Scotland ISI*, 1946–7, **54**, 45
27 J. FRIEDEL: 'Dislocations', 1964, Oxford, Pergamon Press

28 T. B. GIBBONS AND M. HENDERSON-BROWN: *Scr. Metall.*, 1975, **9**, 15

29 C. N. AHLQUIST AND W. D. NIX: *Acta Metall.*, 1971, **19**, 373

30 P. L. THREADGILL AND B. WILSHIRE: 'Creep strength of steels', 8, 1974, London, Metals Society

31 M. McLEAN: *Proc. R. Soc.*, 1980, **A371**, 279

32 H. BURT, J. P. DENNISON AND B. WILSHIRE: *Met. Sci.*, 1979, **13**, 295

33 K. R. WILLIAMS AND B. WILSHIRE: *Met. Sci. J.*, 1973, **1**, 176

34 W. J. EVANS AND G. F. HARRISON: *Met. Sci.*, 1976, **10**, 307

35 M. McLEAN: *Proc. R. Soc.*, 1980, **A373**, 93

36 P. J. HENDERSON AND M. McLEAN: to be published

37 B. F. DYSON AND M. McLEAN: *Acta Metall.*, 1983, **31**, 17

38 W. J. EVANS AND G. F. HARRISON: *Met. Sci.*, 1979, **13**, 346

39 N. S. STOLOFF in 'The superalloys', eds. C. T. Sims and W. C. Hagel, 1979, New York, Interscience

40 L. M. BROWN AND R. K. HAM in 'Strengthening methods in crystals', eds. A. Kelly and R. B. Nicholson, 1971, London, Applied Science

41 R. LAGNEBORG: *Scr. Metall.*, 1973, **7**, 605

42 G. KNOWLES AND H. E. EVANS: *Met. Sci.*, 1980, **14**, 262

43 M. McLEAN: to be published

44 D. D. PEARSON, F. D. LEMKEY AND B. H. KEAR: Proc. 4th Int. Symp. Superalloys, Seven Springs, Pa., September 1980, eds. J. K. Tien *et al.*, p. 513, Metals Park, Ohio, ASM

45 E. L. ROBINSON: *Trans. ASME*, 1952, **74**, 777

46 S. S. MANSON: 'Design for elevated temperature environment', p. 12, 1971, ASME

47 S. S. MANSON: ASTM STP 520, 1973, p. 744

48 P. J. HENDERSON AND M. McLEAN: *Acta Metall.*, 1982, **30**, 1121

49 L. E. SVENSSON AND G. DUNLOP: *Int. Met. Rev.*, 1981, **26** (2), 109

50 M. R. WINSTONE AND J. E. NORTHWOOD: 'Solidification technology in the foundry and cast-house', 298, 1983, London, Metals Society

51 R. B. SCARLIN AND M. O. SPEIDEL: Final report on COST 50 Programme CH3/4, July 1976
52 R. B. SCARLIN: *Metall. Trans.*, 1976, **7A**, 1535
53 J. E. NORTHWOOD, M. R. WINSTONE AND J. C. GREENBANK: 19th Conference of European Investment Casters Federation, Palma, Majorca, October 1982, to be published

# 6
# Mechanical behaviour: *in situ* composites

## 6.1 INTRODUCTION

The main role of directional solidification when applied to superalloys, described in Chapter 5, has been to produce a grain morphology that eliminates certain microstructural features (namely, transverse grain boundaries) where fracture of conventional castings normally originates. Although some modification of the creep strengths of superalloys results from the preferred crystallographic textures produced by directional solidification and from the improved heat treatment procedures that can be applied to some advanced single crystal superalloys, the temperature capabilities of these materials are ultimately determined by the temperatures at which the strengthening $\gamma'$ particles dissolve in their nickel matrix. Consequently the main approach to extending the useful temperature range of conventional superalloys is to increase the $\gamma'$ solvus temperature, as indeed has been done over the past forty years or so; there remains relatively little scope for substantial further improvements in this way. Any dramatic increase in the temperature capability of nickel-base alloys must depend on the introduction of new strengthening mechanisms that operate beyond the $\gamma'$ solution temperature. The development of *in situ* composites by directional solidification of eutectic alloys introduces reinforcement of superalloy-like matrices (i.e. $\gamma$-nickel plus equiaxed $\gamma'$ particles) by refractory fibres or lamellae that continues to be effective beyond the $\gamma'$ solvus. Other parallel developments leading to dispersion strengthened and more conventional wire reinforced superalloys are beyond the scope of this book; however, the principal characteristics of these rival materials will be summarized in Chapter 8.

*In situ* composites, which form a new type of highly anisotropic material, derive their improved high temperature mechanical properties from a combination of strengthening mechanisms; most retain the $\gamma'$ hardening of conventional superalloys in addition to fibre reinforcement. It is important to derive physically realistic descriptions of both the deformation and fracture of these materials in order to provide a basis for both (a) the optimization of alloys by, for example, compositional and microstructural control and (b) the engineering design and life prediction of components that are likely to be subject to complex service histories.

In the more conventional composite materials, reinforced by relatively large wires or fibres (~100 μm diameter, say), much of the mechanical behaviour can be modelled quite simply by considering the load to be partitioned between the matrix and the reinforcing phases in proportion to their volume fractions. Although this *direct* strengthening continues to be important, the very fine microstructures (~1 μm diameter fibres) associated with *in situ* composites can, in addition, lead to substantial *indirect* strengthening. Consequently it will be necessary to examine the available theories of composite strengthening critically before applying them to describe available experimental data with particular regard to both the *quality* of phase alignment and the scale of the microstructure.

This chapter will attempt to describe the evolution in strengthening of high temperature *in situ* composites within the framework of micromechanical models, often developed retrospectively, that provide a basis for predicting likely service behaviour. It will deal with the relationship of the microstructures produced by different solidification procedures to a wide range of mechanical properties, due both to steady state and cyclic loading and to the effects of material anisotropy and heterogeneity that are associated with such composite materials. The primary aims will be to establish whether current extrapolation procedures developed for more conventional alloys, and used extensively in design, have any relevance to this new family of alloys and to identify problems specific to them that make them vulnerable to new rapid deformation and fracture modes.

## 6.2 PRINCIPLES OF REINFORCEMENT

There are many materials that are inherently stronger than current metallic superalloys, both in tensile and creep tests, but for a variety of reasons are not suitable to be used directly in engineering components. Thus many ceramics, intermetallic compounds and carbides are very brittle, being particularly vulnerable to the catastrophic growth of isolated cracks which lead to a wide scatter in properties; on the other hand, refractory metals such as tungsten and molybdenum are chemically unstable in likely service environments. In principle, these shortcomings can be ameliorated by forming a composite material with the 'strong' phase dispersed as fine fibres (or lamellae) in a weaker but tough and corrosion resistant matrix. Physical separation of the fibres prevents the macroscopic growth of a crack nucleus that would occur if the brittle phase were continuous and this leads to improvements in both the absolute levels and reproducibility of strength. The role of the matrix is much more than to provide a neutral medium separating the fibres; it allows load transfer between the fibres, can protect the fibre from environmental interactions and, since the properties of the composite are a compromise of those of its constituent phases, a strong matrix can make a significant contribution to the strength of the composite. In addition, the fine reinforcing phases in *in situ* composites can be in the form of 'whiskers' which have few defects and strengths that approach the theoretical values.

### 6.2.1 Elasticity

The factors appropriate to averaging the elastic constants of the constituent phases to describe the behaviour of the composite are similar to those described in § 5.2 for elongated crystals. The principal differences are:

(a) The composite consists of two distinct populations of elastic elements (fibre and matrix) with well-defined characteristics whereas the superalloy was made up of grains with a spectrum of elastic properties.
(b) The stiff fibres are surrounded by a matrix which constrains the transverse dilatation of the fibres whereas the superalloy grains were assumed to be able to deform independently of their neighbours.

If the fibre/matrix interaction is ignored, then the Young's modulus $E_c$ for the composite stressed parallel to the fibres, given by the Voigt average which assumes stresses are partitioned to allow each element to have the same strain, is described by the Rule of Mixtures.[1]

$$E_c = V_f E_f + V_m E_m \quad \text{(parallel to fibres)} \tag{6.1}$$

The subscripts *c*, *f* and *m* refer to the composite, fibre and matrix respectively and $V_i$ is the volume fraction of the *i*-th phase. In a two-phase composite volume conservation requires that

$$V_f + V_m = 1 \tag{6.2}$$

Similarly, for deformation in a plane transverse to the fibres application of the Reuss average, which partitions the strain to allow each element to carry the same stress, leads to the expression

$$\frac{1}{E_c} = \frac{V_f}{E_f} + \frac{V_m}{E_m} \quad \text{(normal to fibres)} \tag{6.3}$$

Kelly[1] has reviewed the theoretical treatments of the elastic deformation of fibre reinforced composites that accounts for the differential transverse contractions of fibre and matrix. The expression for Young's modulus of the composite in a direction parallel to the fibres extends eqn (6.1) with an additional term that is dominated by the difference in Poisson's ratio for fibre $\nu_f$ and matrix $\nu_m$.

$$E_c = E_f V_f + E_m V_m + 4 V_f V_m (\nu_m - \nu_f)^2 \left( \frac{V_m}{K_{pf}} + \frac{V_f}{K_{pm}} + \frac{1}{\mu_m} \right)^{-1} \tag{6.4}$$

where $K_{pi}$ is the plane strain bulk modulus and $\mu_i$ is the shear modulus relating to strain in a plane parallel to the fibres. Thus $E_c$ can never be less than the Rule of Mixtures estimate; the deviation from the value given by eqn (6.1) is very small (being less than 2% for parameters relevant to *in situ* composites) and indeed (6.1) is rigorously valid when $\nu_m = \nu_f$. The analysis of the transverse Young's modulus is considerably more complicated, but again for the purposes of this text there is negligible error in adopting eqn (6.3) for the volume fractions and elastic constants of phases relevant to *in situ* composites. Figure 6.1 compares the variations in the longitudinal and transverse Young's moduli as functions of $V_f$ for an imaginary composite where $E_f = 100\ E_m$. Thus, while the properties of the fibres largely control the value of the longitudinal Young's modulus, other elastic characteristics of the composite, such as bulk and shear moduli (both parallel and transverse to the fibre axes), are more dependent on the matrix behaviour.

### 6.2.2 Yield behaviour (time-independent)

The various stages during longitudinal tensile testing of a composite consisting of strong fibres in a relatively weak but ductile matrix are illustrated schematically in Fig. 6.2. The stresses carried by matrix and fibre are indicated and it is assumed that the stress on the composite is given by the Rule of Mixtures which sums the stress components in the ratio of the volume fractions of the phases. Four possible regimes depending on the mode of deformation of the

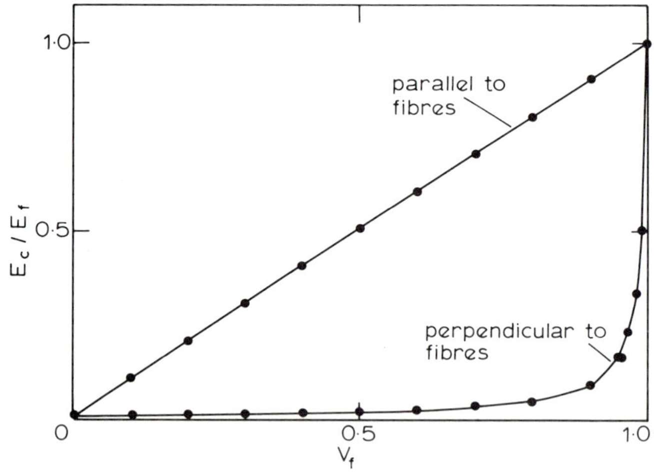


**Figure 6.1 Variations in Young's modulus of longitudinally and transversely loaded composites with volume fraction of the reinforcing phase for $E_f = 100\ E_m$**

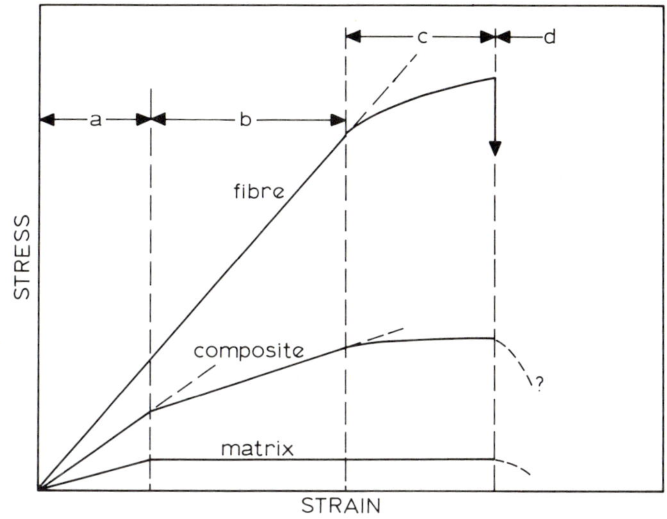


**Figure 6.2 Schematic illustration of the different regimes of the stress/strain curve of an ideal composite indicating the equivalent behaviours of the constituent matrix and reinforcing phases**

constituent phases can be identified but all need not occur for every composite. In the following discussion we assume that the reinforcing phase is continuous unless otherwise stated; this is a realistic assumption for *in situ* composites where the fibres (or lamellae) grow with very high aspect ratios ($\geqslant 100$). A consequence of this geometry is that the longitudinal strains $\varepsilon$ of the two phases must be identical in order to maintain material continuity. Differential longitudinal strains can only occur, without causing cavitation of one phase, for equiaxed-particle or short-fibre reinforced materials where matrix flow can occur between the reinforcing elements.

*Regime* **a**: *elastic fibres, elastic matrix*
This stage has been described in detail above and the composite stress is given by a simple extension of eqn (6.1)

$$\sigma_c = E_f V_f \varepsilon + E_m V_m \varepsilon \tag{6.5}$$

*Regime* **b**: *elastic fibres, plastic matrix*
Beyond a critical strain the matrix yields while the fibres continue to deform elastically. The details of this stage depend on whether or not the matrix work

hardens. The composite stress is given by the expression

$$\sigma_c = E_f V_f \varepsilon + \frac{d\sigma_m}{d\varepsilon} V_m \varepsilon \tag{6.6}$$

where $d\sigma_m/d\varepsilon$ is the slope of the *matrix* tensile curve evaluated at the current strain $\varepsilon$. The behaviour of the composite in this regime is often regarded as being pseudo-elastic with a Young's modulus given by the following expression

$$E'_c = E_f V_f + \frac{d\sigma_m}{d\varepsilon} V_m \tag{6.7}$$

*Regime* **c**: *plastic fibres, plastic matrix*
If the fibres can deform plastically the appropriate equation for the composite stress is simply given by

$$\sigma_c = \frac{d\sigma_f}{d\varepsilon} V_f \varepsilon + \frac{d\sigma_m}{d\varepsilon} V_m \varepsilon \tag{6.8}$$

This regime often does not occur in *in situ* composites since the reinforcing phase often fails in a brittle fashion. In principle yield of fibres entrained in a plastic matrix should occur at a smaller strain than when the fibres are tested in isolation. However, this effect is too small to be of practical significance.

*Regime* **d**: *fracture*
The nature of failure of the composite depends very much on both the tensile strength and failure strain of each phase. In the unlikely event that both matrix and reinforcement should fail at the same strain, then the ultimate tensile stress of the composite is rigorously given by the Rule of Mixtures average of those of the individual phases

$$\sigma_{uc} = \sigma_{uf} V_f + \sigma_{um} V_m \tag{6.9}$$

where $\sigma_{ui}$ is the UTS of the relevant phase. This variation in $\sigma_{uc}$, represented diagrammatically by the line $AB$ in Fig. 6.3, is the theoretical upper limit to the tensile strength of the composite. If the failure strain of the fibre is less than that of the matrix two types of behaviour are possible.

(a) If $\sigma_{um} V_m > \sigma_{uf} V_f$, failure of a fibre does not influence neighbouring fibres since the matrix, if it work hardens, carrys most of the stress. Rather, as the test continues segmentation of the fibres occurs randomly throughout the material until the criterion for *matrix* failure is satisfied

$$\sigma_{uc} = \sigma_{um} V_m \tag{6.10}$$

This is represented by line $AC$ in Fig. 6.3. The microstructural manifestation of this mode of failure is the extensive fracture of fibres far from the fracture surface.

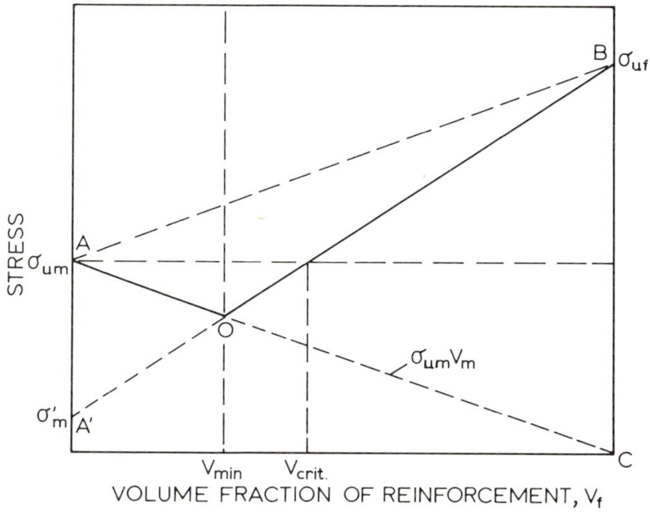


**Figure 6.3 Schematic illustrations of the variation in the fracture stress as a function of the fibre volume fraction for a composite in which the fibre and matrix have different fracture strains**

(b) If $\sigma_{um} V_m < \sigma_{uf} V_f$, when a fibre breaks the crack rapidly propagates through the matrix to fracture neighbouring fibres. Consequently fibre damage is restricted to the neighbourhood of the final failure surface. In this case, the contribution of the matrix to the Rule of Mixtures average is $\sigma'_m$, the matrix stress at the fibre failure strain.

$$\sigma_{uc} = \sigma_{uf} V_f + \sigma'_m V_m \tag{6.11}$$

This is represented by the line $A'B$ in Fig. 6.3.

The failure mechanism that operates for a given $V_f$ will be that giving the *maximum* value of $\sigma_{uc}$ (from eqn (6.10) or (6.11)) which is described by $AOB$ in Fig. 6.3. Fibres can only carry stress for values of $V_f > V_{min}$ where

$$V_{min} = \frac{\sigma_{um} - \sigma'_m}{\sigma_{uf} + (\sigma_{um} - \sigma'_m)} \tag{6.12}$$

and at this condition $\sigma_{uc} < \sigma_{um}$. There can only be an effective strengthening of the composite for values of $V_f > V_{crit}$ where

$$V_{crit} = \frac{\sigma_{um} - \sigma'_m}{\sigma_{uf} - \sigma'_m} \tag{6.13}$$

In the multiple fracture regime (i.e. for values of $V < V_{min}$) the composite can extend beyond the failure strains of the fibres by progressive segmentation and matrix flow between the broken fibre ends. Kelly[1] has shown that there is a critical length of fibre $l_c$ that must be exceeded in order to transfer stress to the fibres. Plastic flow of the matrix causes a shear stress $\tau$ at the fibre/matrix interface that transfers load to the fibre. The effective tensile stress on the fibre $\sigma_f$ due to $\tau$ can be estimated; for a fibre of radius $r$ and length $l$ the variations in $\sigma_f$ and $\tau$ take the form shown in Fig. 6.4 and are related by the force balance equation

$$2\pi r\tau \, dx = \pi r^2 \, d\sigma_f \qquad (6.14)$$

i.e.

$$\frac{d\sigma_f}{dx} = \frac{2\tau}{r} \qquad (6.15)$$

Integration of (6.15) gives the condition that the failure stress will be achieved at the mid-point of a fibre of length $l_{crit}$ when

$$l_{crit} = \frac{\sigma_{uf} r}{\tau} \qquad (6.16)$$

$l_{crit}$ being the *shortest* length of fibre that can be broken by flow of the matrix; for fibres shorter than this deformation can occur by matrix flow around the

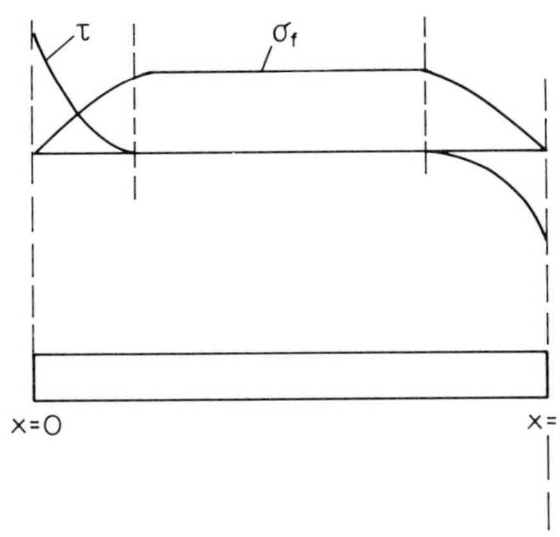


**Figure 6.4 Schematic illustrations of the variation in tensile stress carried by a fibre and shear stress in the adjacent matrix for a composite containing fibres of finite length**

fibres. If $\sigma_f$ increases linearly with distance from the end of the fibre, then the mean stress carried by the fibres is simply given by the expression $\bar{\sigma}_f = \sigma_f[1-(l_{crit}/2l)]$ where $\sigma_f$ is the stress carried by an infinitely long fibre.

There have been many observations of changes in stress–strain curves for different composite materials that have been interpreted as indicating transitions between the various deformation modes listed above. Not all of the alternatives occur for different materials. In particular some materials have completely brittle reinforcement whereas in others the fibres, or lamellae, exhibit significant ductility. In the following discussion we will only be concerned with composites having a ductile matrix although the reinforcement may be either ductile or brittle.

Bibring and co-workers[2] and Buchanan and Tarshis[3,4] have made particularly detailed studies of the deformation of ductile/brittle *in situ* composites in which nickel- or cobalt-base matrices of varying degrees of complexity are reinforced by monocarbide fibres (i.e. the NITAC and COTAC alloys). These materials are characterized by a very regular distribution of apparently unfaulted fibres; consequently there is a very narrow distribution in the strengths of the fibres and the composites exhibit near ideal behaviour. Thus the stress–strain curves of the simple Ni–NbC alloy, reproduced in Fig. 6.5, show the first two stages of deformation (i.e. elastic/elastic and elastic/plastic) ending in a sharply defined yield point. Metallographic examination of tests interrupted before yield shows no evidence of fibre fracture; however, yield is correlated with the development of deformation bands in

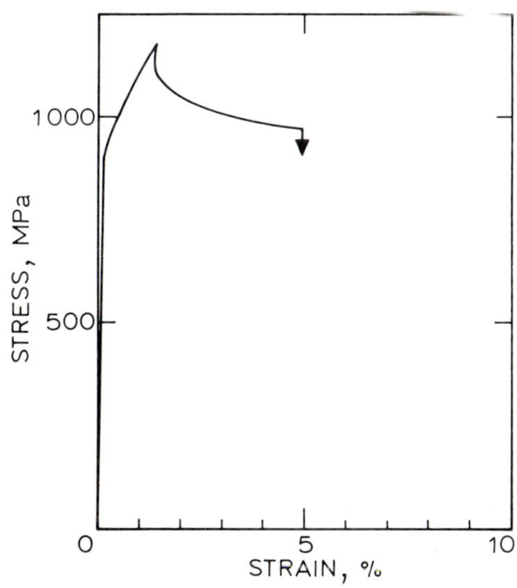


**Figure 6.5 Stress/strain curve for the simple Ni–NbC *in situ* composite at 20°C [Ref. 2]**

which there is extensive fibre fragmentation. Thus there is no ductile/ductile regime but brittle failure of the fibres. The deformation band (Fig. 6.6) that forms at the yield point does not lead to immediate failure of the composite. Rather, there is a sudden drop in stress and a plateau of this reduced stress is maintained while the volume of fibre fragmentation extends. Failure occurs at 5–7% strain before this zone extends over the entire gauge length.

The evolution of the NITAC and COTAC series of alloys has incorporated precipitation and solution strengthening to optimize matrix work hardening and ensure that full advantage is taken of fibre strengthening. Figure 6.7 compares the stress–strain curves at two temperatures, 20 and 927°C, of a Co–15Cr–13TaC *in situ* composite and of a single phase Co–17·3Cr alloy estimated to have the matrix composition. In this case there is no general stress reduction on yield of the composite at low temperature; on the contrary after a small dip the stress continues to rise until final failure occurs at >20% strain when fibre segmentation has occurred over the entire gauge length. At the higher temperature where work hardening is less effective the behaviour reverts to that of the simple alloy showing a decrease in

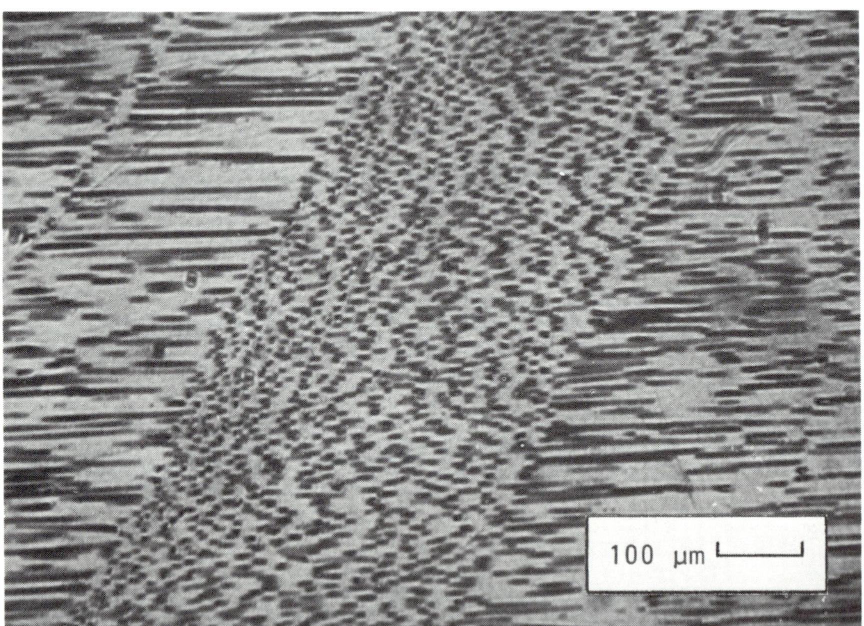


**Figure 6.6 Deformation band containing fractured carbides that develops during tensile testing of the Ni–NbC *in situ* composite at 20°C [Ref. 2]**

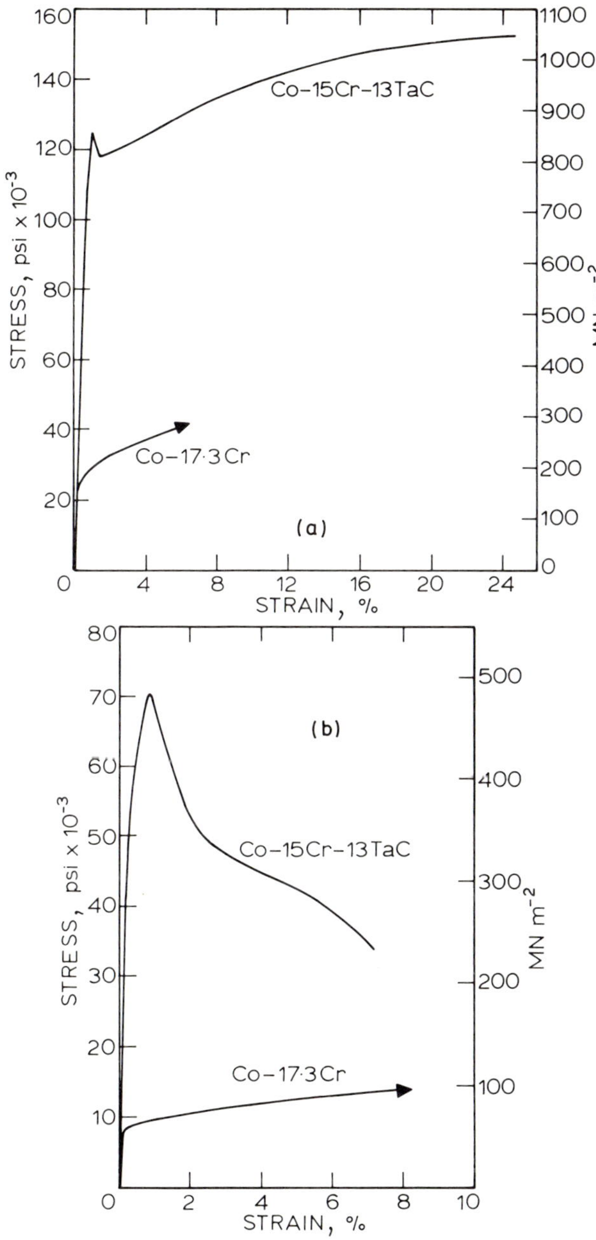


**Figure 6.7 Stress/strain curves of the highly alloyed *in situ* composite, Co–15Cr–13TaC, and of a directionally solidified alloy estimated to have the matrix composition, Co–17·3Cr, at (a) 20°C and (b) 927°C [Ref. 3]**

post-yield stress with increasing strain and a much lower fracture elongation (~7%) that is associated with more localized fibre fracture.

The results cannot be rationalized simply in terms of Fig. 6.3 which correlates multiple fibre fracture with the absence of fibre strengthening. Clearly the fibres contribute very substantially to the tensile strength of the composite (see Fig. 6.7) yet there is extensive fibre segmentation. The simple theory outlined above assumes that the properties attributed to the individual phases do not change by interaction when they coexist in a composite. This assumption may be valid for composites constructed with relatively coarse wires, but for the sub-micron scale reinforcement of *in situ* composites there may be substantial indirect strengthening of the matrix by the fibres by a dispersion strengthening mechanism. Thus in Fig. 6.3, $\sigma_{um}$ and $\sigma'_m$ may be substantially greater than the equivalent values for the matrix in the absence of the fibres. Thus while it may be true that the volume fraction of fibres is below the critical value for *direct* fibre strengthening, the indirect effect can lead to quite significant strengthening.

The microstructures of most *in situ* composites are much less regular than those of the NITAC and COTAC alloys, and the deformation behaviours can be rationalized in terms of a wide range of breaking strengths of the reinforcing elements. The stress–strain curves of the Ni–$Ni_3Nb$ lamellar[5] and of the $\gamma$–$\gamma'$–$Cr_3C_2$ fibrous[6] alloys, shown in Fig. 6.8, are somewhat similar in form although the first is a ductile/ductile and the second a ductile/brittle system. Rather than a sharp yield criterion there is a gradual transition into a region of pseudo-plastic deformation. In the case of Ni–$N_3Nb$ this may be attributed to some plasticity of the reinforcement due to mechanical twinning of the $\delta$–$Ni_3Nb$. However, in alloys such as $\gamma$–$\gamma'$–$Cr_3C_2$ and Co,Cr–(Co,Cr)$_7C_3$ where the reinforcing phase is brittle, the results are more likely to indicate the progressive failure of carbides of increasing strength and an associated increase in stress carried by the stronger fibres.

### 6.2.3 Deviations from the Rule of Mixtures

The simple Rule of Mixtures takes no account of the differing quality and scale of microstructures that can be produced by different processing conditions; it simply partitions the stress in the ratio of the volume fractions of the constituent phases. However, as discussed above there is considerable evidence that the yield and fracture behaviours of *in situ* composites can be very sensitive to the fibre or lamella dimensions. Figure 6.9*a* shows the variations in yield and fracture stresses for the $Ni_3Al$–$Ni_3Nb$ lamellar composite at two temperatures as functions of the $Ni_3Al$ lamellar thickness.[7] Similarly the ultimate tensile strength of the fibrous $\gamma$–$\gamma'$–$Cr_3C_2$ alloy at 20°C is shown as a function of fibre spacing.[6] In each case there is a substantial element of strengthening that is attributable to the microstructural dimensions.

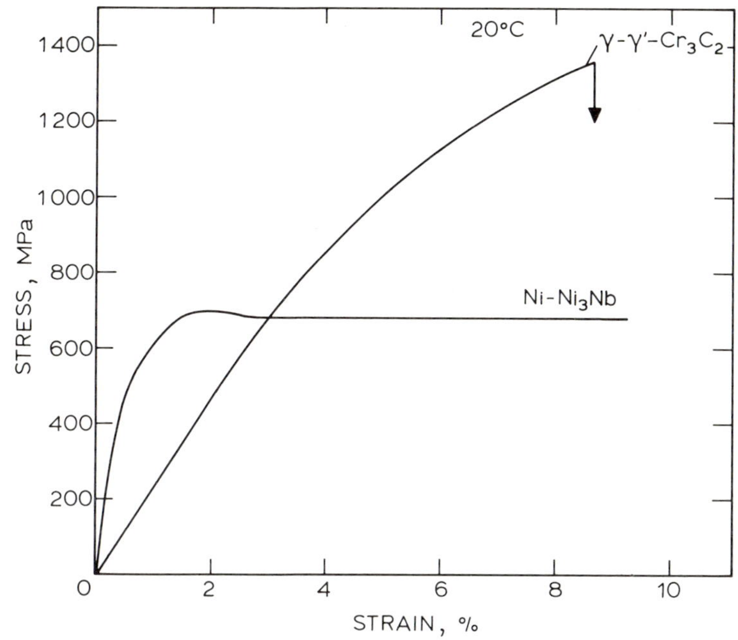


**Figure 6.8 Stress/strain curves for (a) the Ni–Ni$_3$Nb lamellar composite and (b) $\gamma$–$\gamma'$–Cr$_3$C$_2$ fibrous composite at 20°C**

Such data are often presented in the form of a Hall–Petch plot of stress $\sigma$ versus $\lambda^{-1/2}$ (where $\lambda$ is the fibre or lamella spacing) as indeed has been done in Fig. 6.9. However, this practice has been questioned by Chadwick[8] who shows that equally acceptable empirical relationships of $\sigma$ versus $\lambda$ or $\lambda^{-1}$ can be obtained. The main argument for the Hall–Petch formalism depends on the development of dislocation pile-ups due to slip being blocked by the reinforcing fibres or lamellae. Armstrong *et al.*[9] considered the quantitative implications of this model and demonstrated that the basic parameters required to fit the theory to *in situ* composites are within the range of those required to describe grain boundary strengthening of the matrix-type alloys. This must lend some credence to the Hall–Petch approach. However, there have been very few studies of dislocation structure during deformation; dislocation pile-ups have certainly been observed (e.g. Fig. 6.10) but further quantitative analysis is required to confirm or reject the hypothesis. Nevertheless, it is clear that a fine composite microstructure leads to higher tensile strength. While bearing in mind the uncertainties in applying Hall–Petch analyses to these materials Fig. 6.9 indicates that we may nevertheless use it empirically and modify the Rule of Mixtures expression to take into account

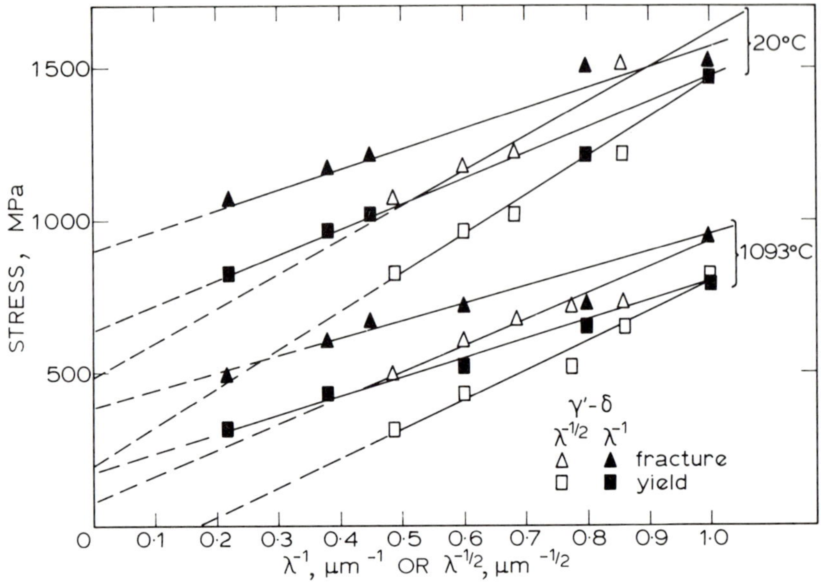


(a)

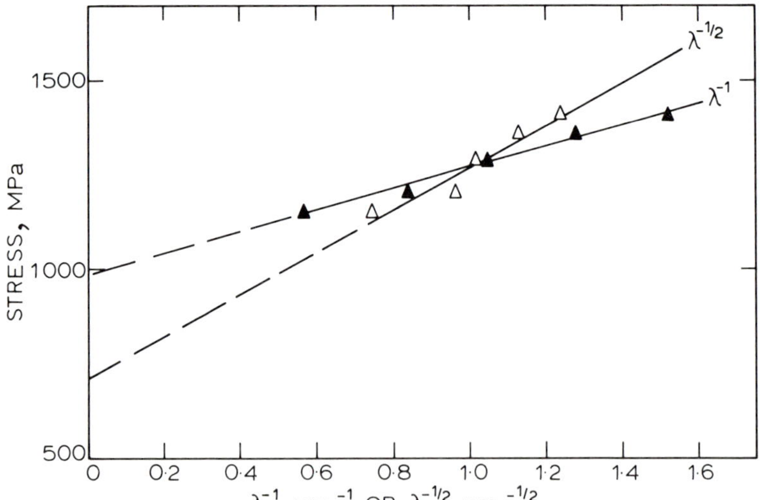


(b)

**Figure 6.9 Effect of fibre or lamella spacings on: (a) yield and ultimate tensile stress of $Ni_3Al$–$Ni_3Nb$ at 20°C [Ref. 7]; (b) ultimate tensile stress of $\gamma$–$\gamma'$–$Cr_3C_2$ at 20°C [Ref. 6]**

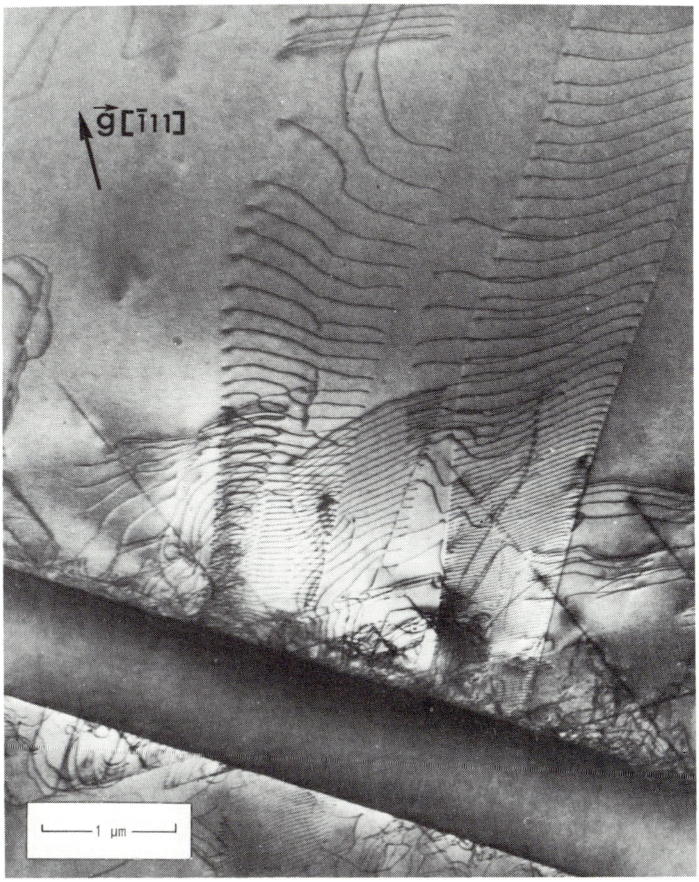


**Figure 6.10 Transmission electron micrograph of the deformed COTAC alloy showing a dislocation pile-up in the matrix impinging on a carbide fibre [Ref. 34]**

the microstructural strengthening of the matrix

$$\sigma_c = \sigma_f V_f + (\sigma_m + \alpha\lambda^{-1/2}) V_m \quad (6.17)$$

where $\lambda$ is the fibre or lamella spacing and $\alpha$ is a constant. Since $\lambda$ is related to the solidification rate $V_I$ by a simple $\lambda^2 V_I$ = constant relationship, eqn (6.17) can be expressed in the alternative form

$$\sigma_c = \sigma_f V_f + (\sigma_m + \alpha' V_I^{1/4}) V_m \quad (6.17a)$$

The solidification conditions leading to the formation of fine fibres or lamellae, namely high rates of solidification and high temperature gradients, may lead to less obvious effects that influence strength.

(a) High cooling rates may cause residual stresses in the composite due to differential thermal contraction of fibre and matrix. This effect will be greatest in the conditions leading to the finest fibres.
(b) The strengths of the finer, more regular fibres may be greater than those produced in coarse microstructures as they approach whisker-like characteristics. Alternately, for a constant defect concentration within the fibres, crack initiation and propagation in fine fibres should be statistically less likely to occur than in the coarse microstructures. In either case there should be a progressive increase in strength with decreasing microstructural scale.

### 6.2.4 Non-axial loading

All of the earlier discussion in this chapter has considered the effects of simple tensile stresses applied parallel to the aligned reinforcement. For such axial loading and high aspect ratio fibres or lamellae, the matrix and fibres (or lamellae) must strain by the same amount to ensure material continuity; otherwise cavities must form in one phase. If the tensile stress is applied at an angle $\phi$ to the direction of reinforcement then alternative deformation and fracture modes can become operative. Kelly[1] has described three possible modes of deformation that result from different components of applied stress. It is convenient to consider the components of the applied stress referred to Cartesian coordinates based on the symmetry of the microstructure with the $z$-axis parallel to the fibres and the $xz$ plane containing the applied stress. Then the only non-zero components of stress are:

$$\left.\begin{aligned}\sigma_{zz} &= \sigma \cos^2 \phi \\ \sigma_{xx} &= \sigma \sin^2 \phi \\ \sigma_{xz} &= \sigma \sin \phi \cos \phi.\end{aligned}\right\} \qquad (6.18)$$

Each component contributes to a different deformation mode and the relative importance of each depends on $\phi$.

(i) $\sigma_{zz}$ leads to the coupled deformation of fibre and matrix that has been the subject of previous discussion in this chapter. Failure of the composite by fibre fracture requires that $\sigma_{zz}$ equals the longitudinal composite failure stress, $\sigma_{uc}$ say. Thus the applied stress $\sigma$ to cause failure is given by the expression

$$\sigma = \sigma_{uc} \sec^2 \phi \qquad (6.19)$$

This fracture mode becomes increasingly unlikely as $\phi$ increases.

(ii) $\sigma_{xz}$ causes shear of the matrix between the fibres and this becomes more important as $\phi$ approaches 45°. If $\tau_{um}$ is the failure stress of the matrix in shear, then this fracture mode occurs when

$$\sigma = \frac{\tau_{um}}{\sin\phi\cos\phi} \tag{6.20}$$

(iii) $\sigma_{xx}$ can lead to flow of the matrix between the fibres without any extension of the fibres. If $\sigma_{um}$ is the tensile failure stress of the matrix then the condition for composite failure by this mode is:

$$\sigma = \frac{\sigma_{um}}{\sin^2\phi} \tag{6.21}$$

The variations of these failure stresses with $\phi$ are indicated in Fig. 6.11 as broken lines and the minimum envelope is shown as a solid curve. The curves have been calculated for the Al–$CuAl_2$ alloy[10] using experimental values of $\sigma_{uc}$ and estimated values of $\sigma_{um}$ and $\tau_{um}$. The agreement with the experimental data shown is quite satisfactory. It indicates that there is little variation in UTS as $\phi$ deviates from 0 to 10° where the fibres contribute significantly to strengthening, but at larger misorientations where matrix yield dominates there are very large reductions in strength.

When the directional solidification conditions lead to a cellular rather than a well-aligned composite microstructure then a range of fibre (or lamellar) orientations are obtained that may deviate by up to 10° from the solidification direction. Integration between $\phi = 0$ and $\phi = 10°$ suggests that there should be little difference between the tensile strengths of well-aligned and cellular microstructures and this is borne out by experiment. However, we shall show later that cellular microstructures have inferior creep strengths and stress rupture characteristics.

## 6.3 TIME-DEPENDENT DEFORMATION

For the case of very long fibres where material continuity requires that each phase must strain by the same amount in the direction parallel to the fibres in order to prevent the formation of voids or cracks in one phase, then the stress must be partitioned to allow each phase to deform at the same rate. Consequently the stress $\sigma_c$ required to cause the composite to have a creep rate $\dot{\varepsilon}$ is simply the volume fraction weighted average of the stresses $\sigma_m$, $\sigma_f$ to cause the same $\dot{\varepsilon}$ in the matrix and fibres respectively (provided that the creep strengths of the individual phases are identical in the composite and bulk form).

$$\sigma_c = \sigma_m V_m + \sigma_f V_f \tag{6.22}$$

Thus there is, in principle, no critical volume fraction for fibre reinforcement during creep deformation as there was for tensile strength (Fig. 6.3). Rather

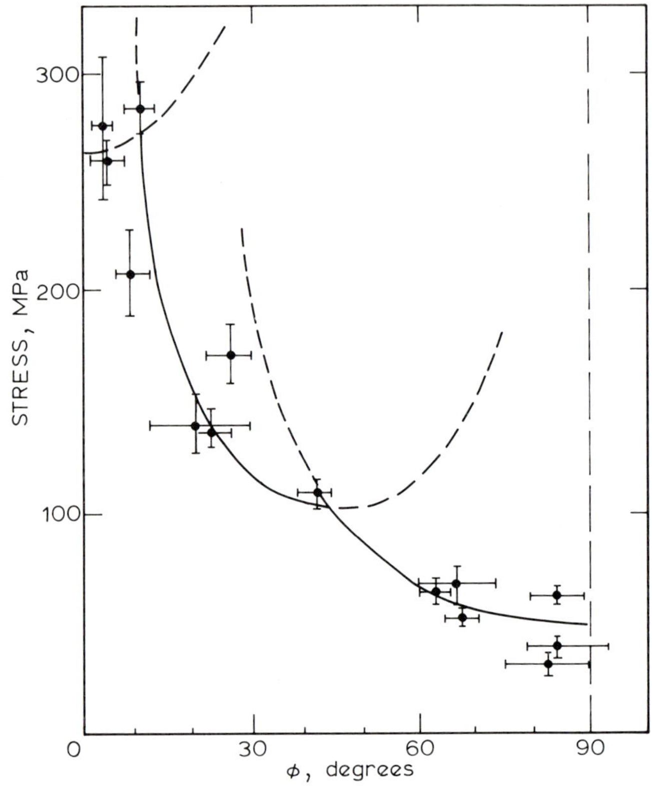


**Figure 6.11 Variations in the stresses leading to failure by each of three possible mechanisms as a function of loading angle $\phi$ and the calculated net failure stress of the composite. The calculations are for Al–CuAl$_2$ at 20°C and are compared with experimental data [Ref. 10]**

the composite stress to give an arbitrary $\dot{\varepsilon}$ as a function of $V_f$ should fall on a straight line such as *AB* and there should be no transition from a multiple to localized fracture mode, the latter dominating for all $V_f$. This tends to be confirmed by experiment. For example, fracture of $\gamma$–$\gamma'$–Cr$_3$C$_2$ is characterized by extensive fibre segmentation throughout the specimen gauge length during tensile tests, but in creep fibre damage is localized to a zone around the creep fracture surfaces.[6] The high strength of the advanced NITAC alloys,[11] in spite of a very low volume fraction of fibres (~2%) which is below that required for time-independent strengthening, confirms the absence of such limitations for creep strengthening.

Two modes of time-dependent deformation that are of practical significance to advanced *in situ* composites may be distinguished on the basis of the manner of deformation of the reinforcing and matrix phases.

**(a) Creeping matrix/elastic fibres**

If the reinforcing phase can only deform elastically, then the proportion of the load carried by the fibres progressively increases with increasing strain and the load on the matrix is progressively reduced. Consequently there is no steady state rate of deformation; rather, in principle, equilibrium can be achieved at a definite strain $\varepsilon_c$ where all of the applied load is carried by the fibres, i.e.

$$\varepsilon_c = \frac{\sigma_c}{E_f V_f} \tag{6.23}$$

The rate at which this equilibrium configuration is approached is determined entirely by the creep kinetics of the matrix. Figure 6.12 shows schematically both the initial stress distribution on loading and the form of the subsequent redistribution under steady load $\sigma_c$.

The stress on the matrix, by manipulation of eqn (6.1), is

$$\sigma_m = \frac{\sigma_c - \varepsilon E_f V_f}{1 - V_f} \tag{6.24}$$

Since both phases are constrained to deform at the same rate $\dot{\varepsilon}$, then

$$\begin{aligned} \dot{\varepsilon} &= \frac{1}{E_f} \cdot \frac{d\sigma_f}{dt} \qquad \text{for the fibre} \\ &= \frac{1}{E_m} \frac{d\sigma_m}{dt} + B_m \sigma_m^{n_m} \qquad \text{for the matrix} \end{aligned} \tag{6.25}$$

where we assume that matrix creep is described by a power law. Combination of eqns (6.24) and (6.25) gives the following differential equation describing the transient creep of the composite

$$\begin{aligned} \dot{\varepsilon} &= B_m \sigma_m^{n_m} \left(1 + \frac{E_f V_f}{E_m (1 - V_f)}\right)^{-1} \\ &= \alpha B_m \sigma_c^{n_m} \left(1 - \frac{E_f V_f}{\sigma_c} \varepsilon\right)^{n_m} \\ &= \alpha B_m \sigma_c^{n_m} \left(1 - \frac{\varepsilon}{\varepsilon_c}\right)^{n_m} \end{aligned} \tag{6.26}$$

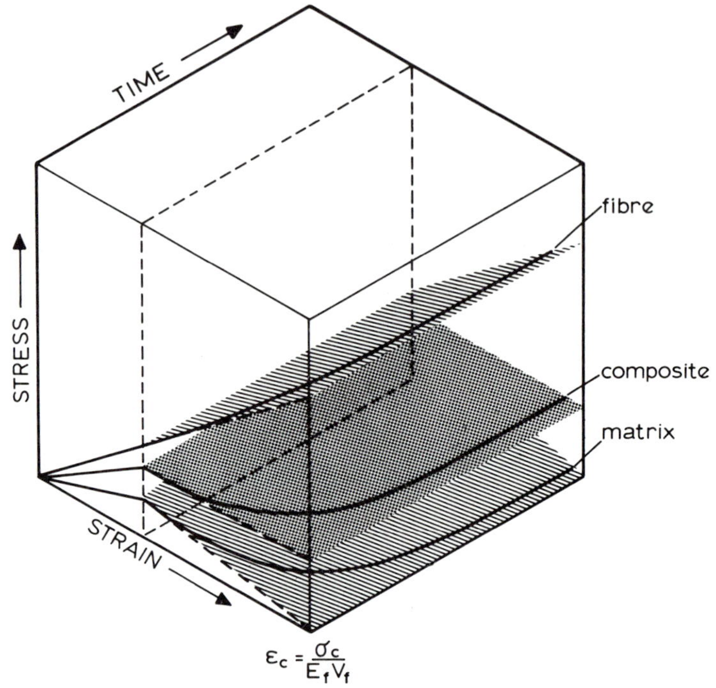


**Figure 6.12 Schematic plot of the changing distribution of stresses between fibres and matrix during loading and subsequent creep under a steady load of a composite consisting of elastic fibres and creeping matrix**

where

$$\alpha^{-1} = \left[1 + \frac{E_f V_f}{E_m(1-V_f)}\right](1-V_f)^{n_m}$$

The instantaneous strain on loading, $\varepsilon_0 = \sigma_c/(E_m V_m + E_f V_f)$ (i.e. at $t = 0$), provides the boundary condition for solution of eqn (6.26). When $n_m > 1$,

$$\varepsilon = \varepsilon_c - (\varepsilon_c - \varepsilon_0)\left[1 + \left(1 - \frac{\varepsilon_0}{\varepsilon_c}\right)^{n_m - 1} \frac{(n_m - 1)\alpha B_m \sigma_c^{n_m} t}{\varepsilon_c}\right]^{-1/(n_m - 1)} \tag{6.26a}$$

When $n_m = 1$,

$$\varepsilon = \varepsilon_c\left[1 - \left(1 - \frac{\varepsilon_0}{\varepsilon_c}\right)\exp\left(-\frac{\alpha B_m \sigma_c t}{\varepsilon_c}\right)\right] \tag{6.26b}$$

Equation (6.26*a*) is probably most appropriate in describing creep of *in situ* composites containing single crystal or large grain metallic matrices which deform by power law creep. However, eqn (6.26*b*) will represent the data for composites with linearly viscous matrices such as polymers or metals with fine grain sizes exhibiting diffusion creep.

Figure 6.13 plots eqn (6.26*a*) evaluated for parameters appropriate to $\gamma$–$\gamma'$–$Cr_3C_2$; the creep rate progressively decreases with increasing strain as the creep curve approaches an asymptote of $\varepsilon_c = \sigma_c / E_f V_f$. The tertiary creep regime associated with *in situ* composites could be interpreted as being due to increasing fibre fracture with increasing strain which reduces the volume fraction of reinforcement.

Since there is no true steady state, or even minimum creep rate, on this model it is convenient to consider the times to reach and the creep rates achieved at a given strain. The onset of tertiary creep for many *in situ* composites appears to occur at ~1% strain. Figure 6.14*a* and *b* shows those quantities calculated using parameters relevant to $\gamma$–$\gamma'$–$Cr_3C_2$ at 825°C and available experimental data for the composite and matrix are included for comparison. Although the composite creep behaviour is completely determined by the matrix, the rapid decrease in $\dot{\varepsilon}$ as $\sigma_c$ approaches $\varepsilon_c V_f E_f$ would indicate a changing stress exponent if the data were (erroneously) interpreted in terms of the Bailey–Norton power law formalism. Thus at high stresses the composite creep behaviour would indicate values of stress exponent and activation energy for creep that would be similar to the matrix behaviour.

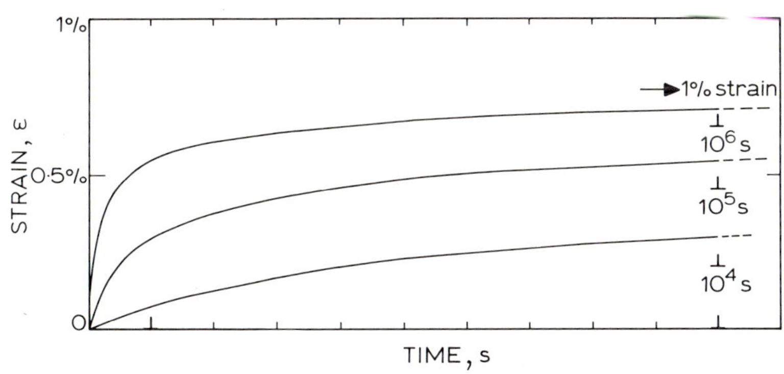


**Figure 6.13 Creep curve plotted to different maximum times calculated from a model of elastic fibres in a creeping matrix assuming the following values of parameters: $V_f = 0{\cdot}11$, $n = 6$, $E_f = 10^5$ MPa, $B_m = 10^{-21}$ MPa$^{-6}$ s$^{-1}$, $\sigma_c = 100$ MPa**

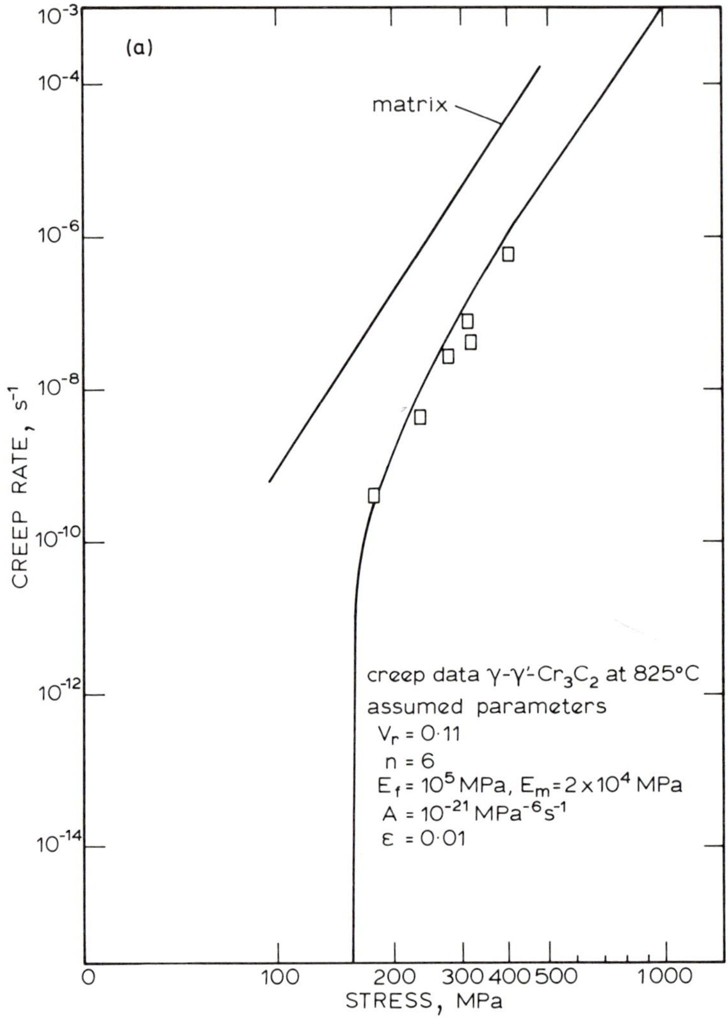


**Figure 6.14** **(a) calculated creep rate at 1% strain for a composite consisting of elastic fibres and creeping matrix as a function of applied stress for the parameters given in Fig. 6.13. The data are for $\gamma$–$\gamma'$–$Cr_3C_2$**

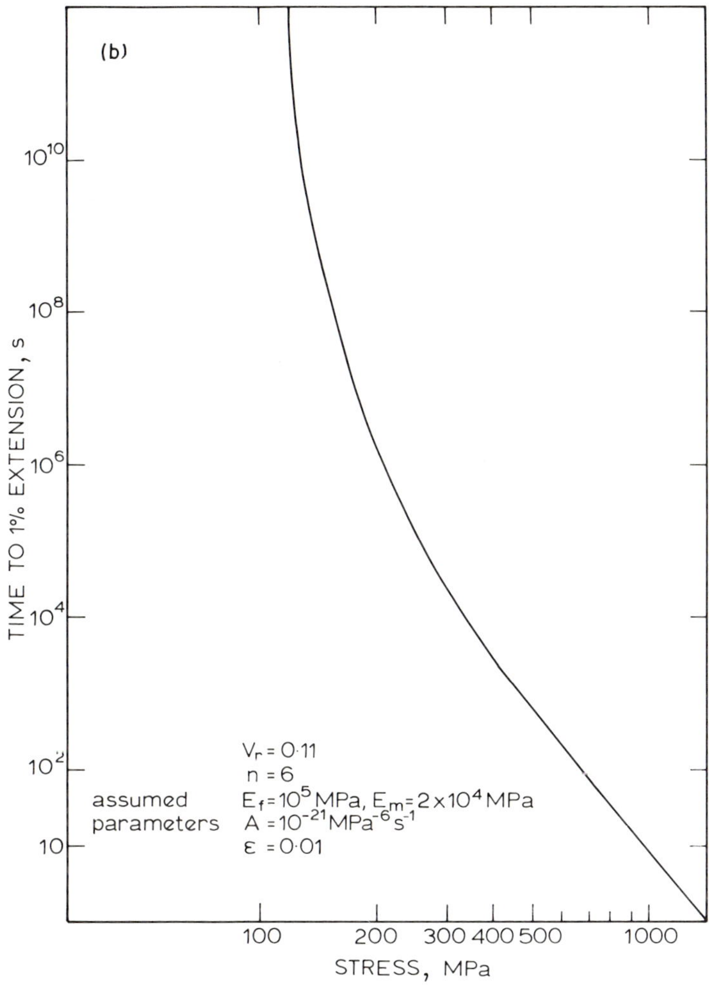


**Figure 6.14 (b) calculated time to reach 1% strain**

However, at low stresses there would be an apparent increase in stress exponent while the temperature dependence of deformation would reflect that of $E_f$. Similarly, the calculated time to achieve 1% extension increases very rapidly as the equilibrium strain $\varepsilon_c = \sigma_c / E_f V_f$ is approached, suggesting that conventional approaches to data extrapolation might underestimate the long-term lives of those materials; this is in accord with several observations.

The elastic strain in the fibres should be recoverable by heat treatment of the composite after creep testing. The results of Khan *et al.*[12] for COTAC 74, shown in Table 6.1, confirm that almost all of the creep strain is removed

**Table 6.1 Stress rupture life improvement of COTAC 744 after repetitive creep-heat treatment experiments**

| Creep test conditions | | Gauge length, mm | | | | | Time to rupture, h | |
|---|---|---|---|---|---|---|---|---|
| Temp., °C | Stress, MPa | Initial | 1st creep | Heat treat. | 2nd creep | Heat treat. | Cumulative | Uninterrupted |
| 850 | 400 | 30·02 | 30·40 | 30·06 | 30·42 | 30·10 | 1040 | 700* |
| 900 | 310 | 29·98 | 30·36 | 30·08 | 30·35 | 30·08 | 1303 | 700* |
| 1000 | 220 | 30·00 | 30·36 | 30·06 | — | — | 302 | 200* |

* Average of three tests

by high temperature heat treatment where the matrix creep resistance is sufficiently low to allow complete relaxation of the elastically strained fibres. This phenomenon, known as the length memory effect, offers the potential of almost perpetual property regeneration. However, not all *in situ* composites exhibit this behaviour. The post-creep relaxation of $\gamma$–$\gamma'$–$Cr_3C_2$ shown in Fig. 6.15 achieves only partial strain recovery and the extent of the contraction is very sensitive to the test conditions, being greatest when following the lowest temperature creep tests. This clearly indicates an increasing plastic (creep) contribution to deformation of the fibres with increasing test temperature.

### (b) Creeping matrix, creeping fibres

The situation when both phases are creeping at rates described by different forms of the Bailey–Norton equation has been considered by McDanels *et al.*[13] with particular relevance to microstructurally coarse wire reinforced composites. For tensile stresses applied parallel to infinitely long fibres the common strain rate of fibres and matrix can be described at constant temperature by the following equations:

$$\dot{\varepsilon} = A_f \sigma_f^{n_f} \exp\left(-\frac{Q_f}{RT}\right) = B_f \sigma_f^{n_f}$$
$$\dot{\varepsilon} = A_m \sigma_m^{n_m} \exp\left(-\frac{Q_m}{RT}\right) = B_m \sigma_m^{n_m} \qquad (6.27)$$

where the subscripts *m*, *f* refer to the matrix and fibres respectively. The Rule of Mixtures average of the stresses $\sigma_m$ and $\sigma_f$ gives, to a first approximation,

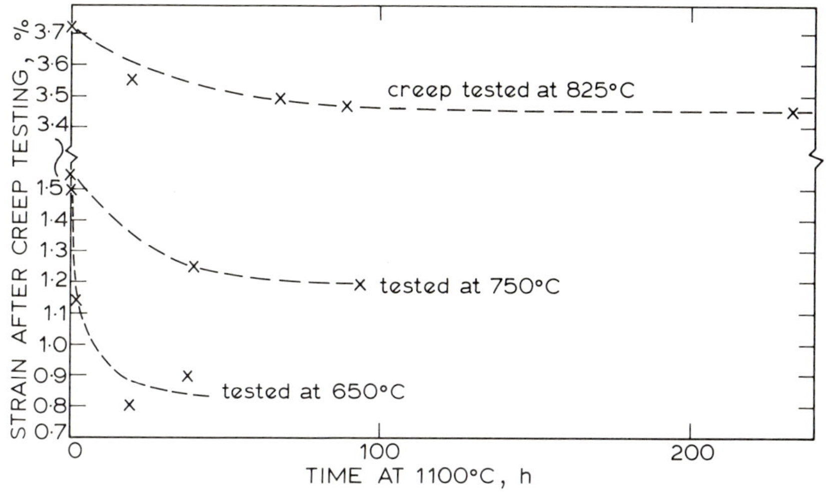


**Figure 6.15 Change in length with time of creep specimens of $\gamma$–$\gamma'$–$Cr_3C_2$ after different prior test conditions during heat treatment at 1100°C**

the stress required to cause the composite to creep at a rate $\dot{\varepsilon}$

$$\sigma_c - \sigma_f V_f + \sigma_m V_m = \left(\frac{\dot{\varepsilon}}{B_f}\right)^{1/n_f} V_f + \left(\frac{\dot{\varepsilon}}{B_m}\right)^{1/n_m} (1 - V_f) \tag{6.28}$$

It is experimentally convenient to continue to describe the creep behaviour of the *in situ* composites using the familiar formalism, such as the Bailey–Norton equation, developed for more simple materials. However, eqn (6.28) does not have the necessary analytical form to give a straight line from a plot of log $\dot{\varepsilon}$ versus log $\sigma$ and considerable care must be taken in interpreting such data. Figure 6.16 shows the creep behaviour predicted by eqn (6.28) for a composite consisting of 10 vol.-% of fibres where the bulk fibre and matrix phases creep in the form indicated. The dashed lines show components of $\sigma_c$ carried by fibre and matrix (i.e. $\sigma_f V_f$ and $\sigma_m V_m$). Where most of the stress is carried by one phase then the composite creep characteristics (i.e. stress exponent and activation energy) will parallel those of the dominant phase. However, there is a wide transitional zone where this does not occur. Early attempts to 'explain' the high values of stress exponent and activation energy as indicating that fibre creep is rate-determining are not valid. Indeed, we saw in Chapter 5 that the creep of superalloys, which constitute the matrices of most advanced *in situ* composites, also required high values of $n$ and $Q$

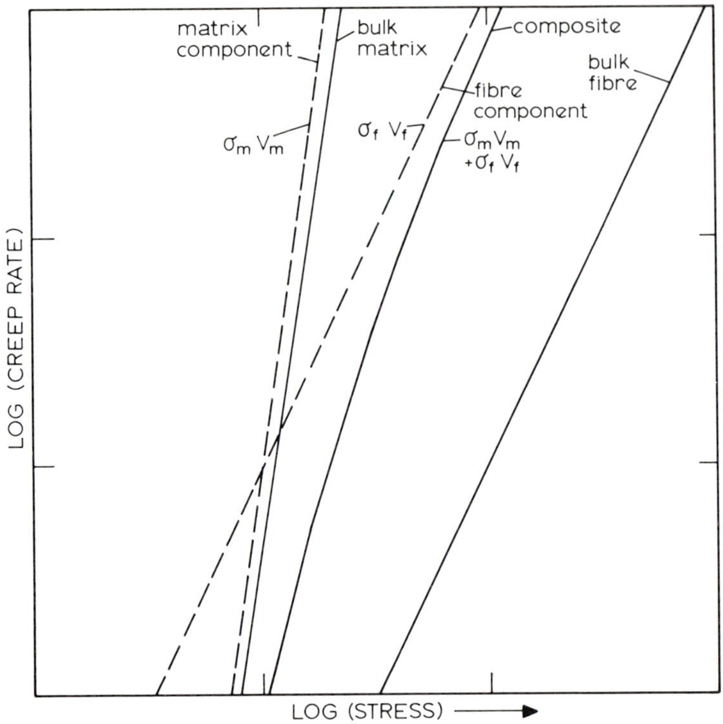


**Figure 6.16 Calculated variation in log $\dot{\varepsilon}$ versus log $\sigma$ assuming Rule of Mixtures for a composite consisting of fibre and matrix in which the creep behaviours are described by the Bailey–Norton equation ($V_f = 0{\cdot}1$)**

when fitted to the simple Bailey–Norton equation. The possibility of transitions between matrix and fibre limited creep must have important implications for the reliability of life prediction procedures.

Equation (6.28) contains no parameter to account for microstructural dimensions; it assumes that the stresses are homogeneous within each phase and that there is a discontinuous change in stress at the fibre/matrix interface as shown in Fig. 6.17*a*. Any obstacle to slip on an active matrix plane, such as an incoherent fibre which cannot be penetrated or an image force at the interface with a coherent fibre, will inhibit free deformation and generate incompatibility stresses since the two phases will usually tend to deform by different amounts. In structural terms, dislocations will be generated near the particle/matrix interface, in much the same manner as indicated in Fig. 5.16,

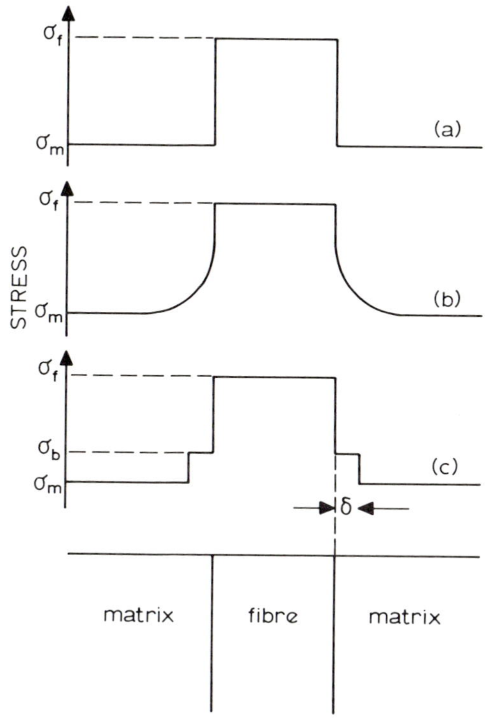


**Figure 6.17 Schematic illustrations of the relationships between the stresses carried by matrix and fibre with the underlying microstructure: (a) no fibre/matrix interaction (Rule of Mixtures); (b) matrix hardening; (c) assumed distribution**

leading to a local hardening of the matrix. Consequently the stress distribution to produce homogeneous straining of the matrix will take the schematic form indicated in Fig. 6.17*b*.

The effects of matrix hardening by the fibres can be accounted for, to a first approximation, by considering the composite to consist of three rather than two phases—matrix, fibre and interface zone of thickness $\delta$ as shown in Fig. 5.16*c*. The interface zone has no chemical reality but is a mathematical convenience to represent the higher stresses carried in the matrix adjacent to the interface to ensure uniform straining. Since the reinforcing phase is usually considerably harder than the matrix it is reasonable to take the fibre volume fraction $V_f$ to be constant and assume that the 'interface zone' forms part of the original matrix. Then, for the case of fibres of radius $\lambda$,

$$\left.\begin{aligned} V_f &= \text{constant} \\ V'_m &= 1 - V_f - \frac{2V_f\delta}{\lambda} \\ V_b &= \frac{2V_f\delta}{\lambda} \end{aligned}\right\} \qquad (6.29)$$

where $V'_m$ and $V_b$ are the volume fractions of 'unaffected' matrix and 'interface zone' respectively. Then the stress distribution to cause an arbitrary creep rate is:

$$\sigma_c = \sigma_f V_f + \sigma_m\left(1 - V_f - \frac{2V_f\delta}{\lambda}\right) + \sigma_b \frac{2V_f\delta}{\lambda} \qquad (6.30)$$

If the creep of each phase, including the notional interface zone, can be described by a Bailey–Norton type of expression $\dot{\varepsilon} = B_i\sigma^{n_i}$ where $i = m, f$ or $b$ denotes matrix, fibre or boundary zone, then the analogue to eqn (6.28) is simply

$$\sigma_c = \left(\frac{\dot{\varepsilon}}{B_f}\right)^{1/n_f} V_f + \left(\frac{\dot{\varepsilon}}{B_m}\right)^{1/n_m}\left(1 - V_f - \frac{2V_f\delta}{\lambda}\right) + \left(\frac{\dot{\varepsilon}}{B_b}\right)^{1/n_b}\frac{2V_f\delta}{\lambda} \qquad (6.31)$$

$$= \sigma_c^{\infty} + \frac{2V_f\delta}{\lambda}\left[\left(\frac{\dot{\varepsilon}}{B_b}\right)^{1/n_b} - \left(\frac{\dot{\varepsilon}}{B_m}\right)^{1/n_m}\right] \qquad (6.32)$$

where

$$\sigma_c^{\infty} = \left(\frac{\dot{\varepsilon}}{B_f}\right)^{1/n_f} V_f + \left(\frac{\dot{\varepsilon}}{B_m}\right)^{1/n_m}(1 - V_f) \qquad (6.33)$$

Equations (6.28) and (6.33) are identical, so $\sigma_c^{\infty}$ is the stress carried by the composite as $\lambda \to \infty$.

How then does the creep rate $\dot{\varepsilon}$ vary for composites having the same volume fractions but different diameters of fibres? If the stress exponents of the creep equations were equal ($n_f = n_m = n_b = n$), then eqn (6.31) could be re-expressed

$$\dot{\varepsilon}^{-1/n} = \dot{\varepsilon}_{\infty}^{-1/n} + \frac{1}{\sigma_c}\cdot\frac{2V_f\delta}{\lambda}\left(\frac{1}{B_b^{1/n}} - \frac{1}{B_m^{1/n}}\right) \qquad (6.34)$$

where

$$\dot{\varepsilon}_{\infty}^{-1/n} = \frac{1}{\sigma_c}\left[\frac{V_f}{B_f^{1/n}} + \frac{(1 - V_f)}{B_m^{1/n}}\right]$$

$\dot{\varepsilon}_{\infty}$ is the creep rate given by eqn (6.28) when $n_f = n_m$ and is appropriate to composites as $\lambda \to \infty$. Algebraic manipulation of eqn (6.34) gives the expression

$$\frac{\dot{\varepsilon}}{\dot{\varepsilon}_\infty} = \left(1+\frac{\beta}{\lambda}\right)^{-n} \tag{6.35}$$

where $\beta = 2V_f\delta\left(\frac{B_f}{B_b}\right)^{1/n} \cdot \frac{(B_m^{1/n} - B_b^{1/n})}{(V_f B_m^{1/n} + V_m B_f^{1/n})}$

Taking logarithms

$$\ln \dot{\varepsilon} - \ln \dot{\varepsilon}_\infty = -n \ln\left(1+\frac{\beta}{\lambda}\right) \sim -\frac{n\beta}{\lambda} \quad \text{if} \quad \frac{\beta}{\lambda} \ll 1$$

The model predicts that if composites of the same $V_f$ but different $\lambda$ are loaded with the same stress, then $\log \dot{\varepsilon} \propto 1/\lambda$. Bullock *et al.*[14] have shown by a graphical argument that this conclusion remains valid when the stress exponents $n_f$, $n_m$ and $n_b$ are different.

Because of the stringent solidification conditions required to process most *in situ* composites, few data are available to test the model. Figure 6.18 shows results for $\gamma$–$\gamma'$–$Cr_3C_2$ which are quite consistent with the form of the predictions and which extrapolate to a lower $\dot{\varepsilon}_\infty$ (i.e. $\lambda \to \infty$) than the matrix creep rate at the same stress. Although the data can be fitted to a Hall–Petch relationship, $\log \dot{\varepsilon} \propto \lambda^{-1/2}$, it makes the unreasonable prediction that as $1/\lambda \to 0$ the matrix creep strength is greater than that of the composite. Irrespective of the validity of the detailed mechanism, it is clear that a number of *in situ*

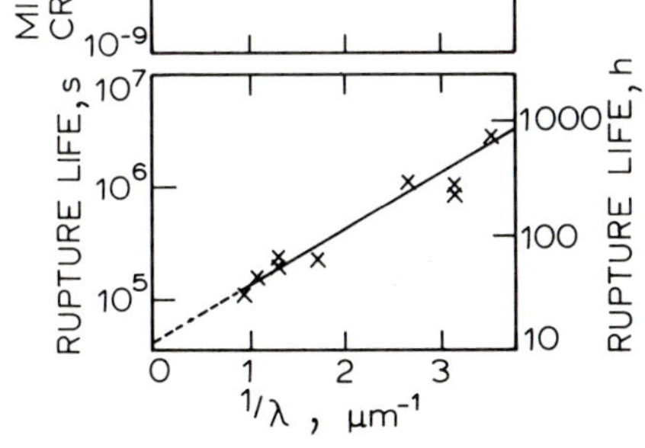


**Figure 6.18 Effect of fibre diameter $\lambda$ on minimum creep rate and rupture life for $\gamma$–$\gamma'$–$Cr_3C_2$ tested at 980°C and a stress of 122 MPa [Ref. 14]**

composites can be strengthened considerably by refining the microstructure (e.g. $\gamma$–$\gamma'$–$\delta$,[15] NITAC[16]). The few alloys that show the opposite trend of decreasing creep strength with decreasing $\lambda$ are microstructurally unstable, being vulnerable to matrix allotropic transformation and/or degradation of the reinforcing phase.[17] For example, in the alloys (Co,Cr)–$Cr_7C_3$ and (Fe,Cr)–$Cr_7C_3$ the carbide transforms during testing to $M_{23}C_6$.

In the foregoing discussion it has been assumed that a well-aligned composite microstructure has been achieved. However, in Chapter 3 it was shown that transitions between cellular and plane front solidification modes made large differences to the composite microstructure. Figure 6.19 shows for the $Ni_3Al$–$Ni_3Nb$ eutectic the variation in both minimum creep rate and elongation for various solidification rates that encompass the conditions for transition between plane front and cellular solidification. Within the plane front regime there is a clear trend of improving creep behaviour with increasing solidification rate $V_I$ (i.e. decreasing $\lambda$). However, when plane front solidification breaks down at a critical $V_I$ there is a sudden degradation in creep behaviour. This points to two conditions to achieve optimum creep resistance:

(a) ensure plane front solidification $(G_L/V_I > (G_L/V_I)_c)$
(b) solidify at the fastest rate compatible with (a) to achieve a fine microstructure.

In alloys where it is difficult to achieve plane front solidification because of either high intrinsic values of $(G_L/V_I)_c$ or of gravitational segregation of phases that cause deviations from the eutectic composition uncontrolled transitions to cellular microstructures can lead to unexpected scatter in properties. This point is illustrated by the creep curves for $\gamma'$–$\delta$ shown in Fig. 6.20; those for material produced in a low temperature gradient show a considerable range of creep rates, elongations and rupture lives that can be related *a posteriori* to the quality of the microstructure. However, the use of a high temperature gradient to ensure that plane front conditions are obtained reproducibly leads both to an absolute improvement in creep performance and, more importantly, to considerable reduction in the scatter of results.

As in the case of tensile testing, off-axis loading is expected to modify the creep deformation and fracture behaviour of composite materials considerably. For a tensile stress $\sigma$ applied at an angle $\phi$ to the fibre axes, the three non-zero components of stress referred to the symmetry of the microstructure lead to three possible time-dependent deformation modes:

(1) $\sigma_{zz} = \sigma \cos^2 \phi$ is in the direction of the fibres and will lead to the coupled extension of both fibres and matrix. If this composite creep behaviour can be represented by a Bailey–Norton equation (which we have shown is only strictly valid when most of the stress is carried by one phase) then

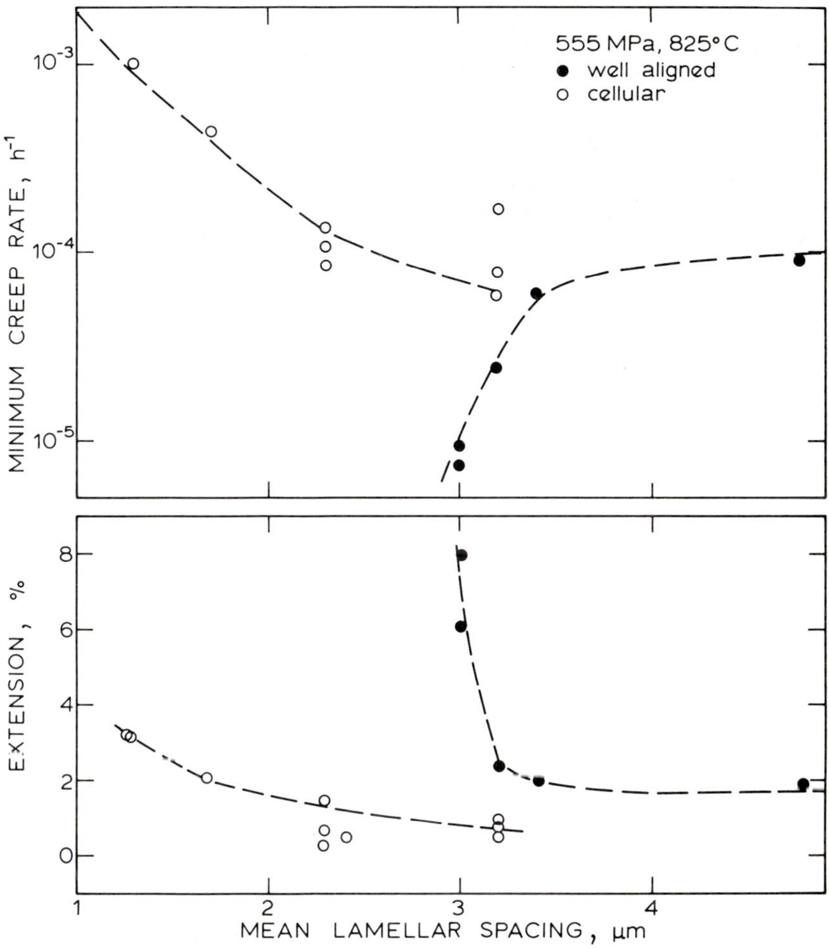


**Figure 6.19 Variation in creep behaviour of $Ni_3Al$–$Ni_3Nb$ with solidification rate for a standard creep testing condition (850°C, 555 MPa). The different microstructures obtained are indicated**

the strain rate parallel to the fibres is

$$\dot{\varepsilon}_{zz} = A_1(\sigma \cos^2 \phi)^n \tag{6.36}$$

There will be an associated strain rate normal to the fibres to satisfy volume conservation

$$\dot{\varepsilon}_{xx} = -\tfrac{1}{2}\dot{\varepsilon}_{zz} \tag{6.37}$$

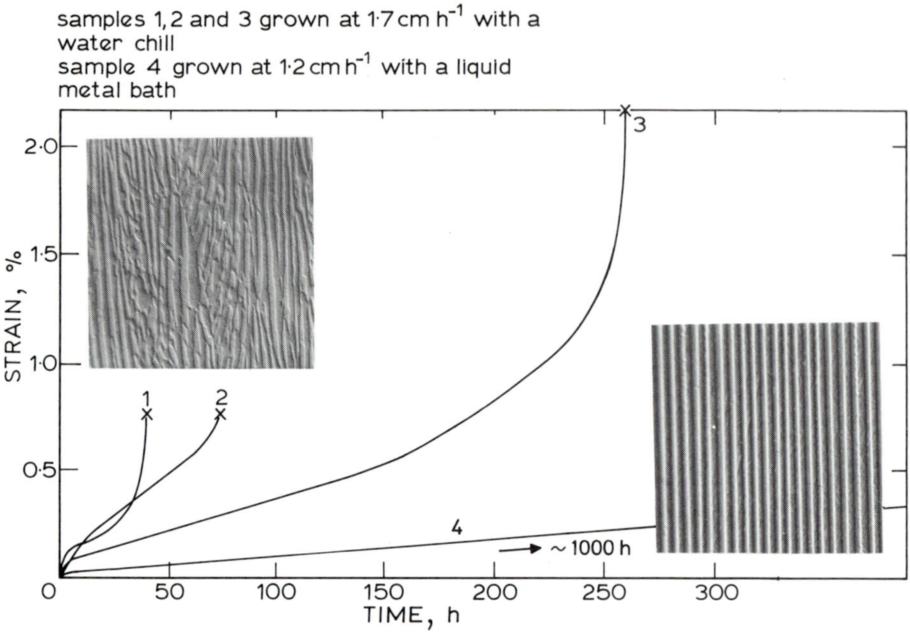


**Figure 6.20 Creep curves for $Ni_3Al$–$Ni_3Nb$ prepared in two temperature gradients and tested under nominally identical conditions. The insets show the microstructures of the relevant test specimens [555 MPa, 850°C]**

Together these lead to an effective strain rate parallel to the applied stress

$$\dot{\varepsilon}_1 = \dot{\varepsilon}_{zz} \cos^2 \phi + \dot{\varepsilon}_{xx} \sin^2 \phi$$

$$= \dot{\varepsilon}_c[\cos^{2(n+1)} \phi - \tfrac{1}{2} \cos^{2n} \phi \sin^2 \phi] \qquad (6.38)$$

where $\dot{\varepsilon}_c = A_1\sigma^n$ is the composite creep rate when the stress $\sigma$ is applied parallel to the fibres.

(2) The shear stress $\sigma_{xz} = \sigma \sin \phi \cos \phi$ leads to deformation of the matrix by shear parallel to the fibres and does not necessarily involve deformation of the fibres. If *matrix* creep is represented by $\dot{\varepsilon}_m = A_2\sigma^m$, then

$$\dot{\varepsilon}_{xz} = f_1 A_2(\sigma \sin \phi \cos \phi)^m \qquad (6.39)$$

where $f_1 = 3^{(m+1)/2}/2$ is a numerical constant relating shear to tensile deformation. The contribution of this mechanism to the creep rate parallel to the applied stress is given by

$$\dot{\varepsilon}_2 = f_1 \dot{\varepsilon}_m(\sin \phi \cos \phi)^{m+1} \qquad (6.40)$$

(3) $\sigma_{xx} = \sigma \sin^2 \phi$ also leads to deformation by matrix creep alone and does not necessarily involve fibre creep. However, transverse deformation by this mechanism will be very anisotropic since the fibres will inhibit matrix creep parallel to the fibres whereas relatively unconstrained flow can occur normal to the fibres, i.e.

$$\dot{\varepsilon}_{xx} = A_2(\sigma \sin^2 \phi)^m = \dot{\varepsilon}_{yy} \tag{6.41}$$

$$\dot{\varepsilon}_{zz} = 0$$

Thus the contribution to the strain rate parallel to the applied stress $\dot{\varepsilon}_3$ is

$$\dot{\varepsilon}_3 = \dot{\varepsilon}_m \sin^{2(m+1)} \phi \tag{6.42}$$

The three contributions are additive, so the total strain rate $\dot{\varepsilon}$ parallel to the applied stress is given by the expression

$$\begin{aligned}\dot{\varepsilon} &= \dot{\varepsilon}_1 + \dot{\varepsilon}_2 + \dot{\varepsilon}_3 \\ &= \dot{\varepsilon}_c[\cos^{2(n+1)} \phi - \tfrac{1}{2} \cos^{2n} \phi \sin^2 \phi] \\ &\quad + f_1 \dot{\varepsilon}_m(\sin \phi \cos \phi)^{m+1} + \dot{\varepsilon}_m \sin^{2(m+1)} \phi\end{aligned} \tag{6.43}$$

Miles *et al.*[18] have studied the creep behaviour of the (Co,Cr)–$Cr_7C_3$ *in situ* composite under conditions of anisotropic loading for $0 \leqslant \phi \leqslant 90°$ and have compared the results with the above theory. Their data are shown in Fig. 6.21 and the solid line is a plot of eqn (6.43) using values of the parameters $\dot{\varepsilon}_c$ and $\dot{\varepsilon}_m$ evaluated from the data points for $\phi = 0$ and $\phi = 90°$. The value of $f = 35$ is computed from eqn (6.39) using the measured value of stress exponent $m = 5{\cdot}5$. The calculated curve gives a good fit to the experimental data. The particular alloy considered has a very weak matrix and these considerations point to the need for a strong matrix in addition to fibre strengthening. The results of a more rigorous theory by Johnson[19] are also indicated in Fig. 6.21.

The discussion on off-axis loading has been restricted to fibre reinforced systems. However, similar but more severe effects are likely to occur in lamellar composites. We have shown that in the fibre-reinforced alloy transverse tensile stresses can lead to large transverse elongations by flow of the matrix between the fibres (Fig. 6.22*a*). This cannot occur with lamellar reinforced alloys since the 'soft' elements are essentially unconnected and must deform in isolation. They are constrained by the reinforcement in two dimensions (Fig. 6.22*b*). Thus lamellar eutectics such as $\gamma$–$\gamma'$–$\delta$ have much less inherent transverse ductility than the fibrous eutectics; indeed, this has been a major factor in abandoning certain alloys that initially appeared to have promising properties (e.g. $\gamma$–$\gamma'$–$\delta$). Not all fibrous systems have high transverse ductility; for example, some of the NITAC alloys displayed virtually zero transverse ductility in creep conditions. However, this has been identified as being due to decohesion at grain boundaries[20] within the matrix and not

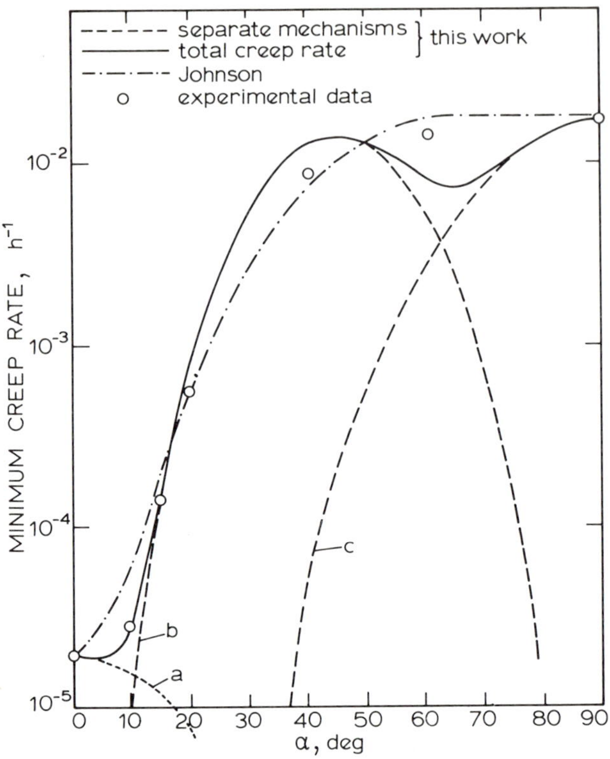


**Figure 6.21 Minimum creep rate of (Co, Cr)–$Cr_7C_3$ tested at 825°C with a stress of 200 MPa applied at various angles $\phi$ to the direction of the aligned fibres. The theoretical curves computed from eqn (6.43) and from Johnson[19] are included [Ref. 18]**

because of an intrinsic lack of ductility of the composite structure. This problem is amenable to solution by the approaches adopted for superalloys and, of course, could be completely eliminated by seeding single eutectic crystals.

In turbine blade applications the stresses at the root present a particular problem for all directional microstructures including *in situ* composites. The centrifugal force on the aerofoil creates a pure shear stress in the fibre direction at the fir-tree roots where the blade is anchored. For fibrous systems creep is fully described by eqn (6.40) and is determined by only the matrix behaviour. In lamellar-reinforced composites shear by sliding of the interphase boundary is possible. Consequently alloys with weak matrices and phase boundaries are

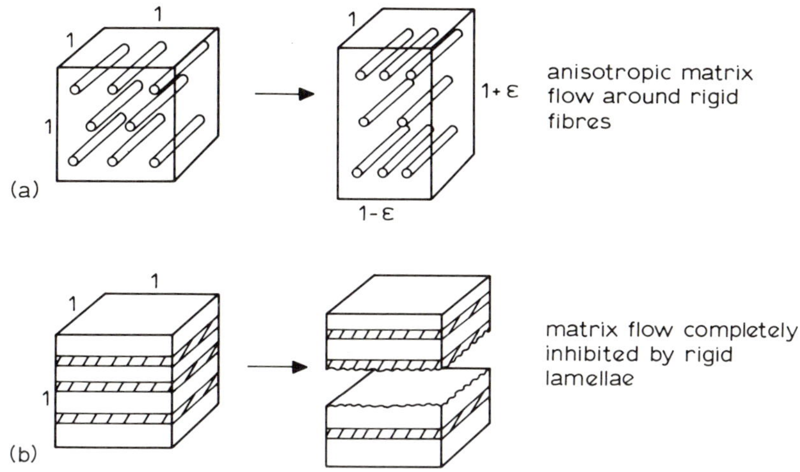


**Figure 6.22 Schematic illustration of the transverse deformation of (a) fibre- and (b) lamellar-reinforced composites**

particularly vulnerable to failure by this mechanism and are unsuitable for turbine blade applications even although they may have high strength when stressed parallel to the reinforcement.

## 6.4 FATIGUE

The high cycle fatigue behaviour of most *in situ* composites has been shown to be outstandingly good. A selection of data is shown in Table 6.2 and Fig. 6.23; results for superalloys are included for comparison. The ratio of the stress for rupture in $10^7$ cycles to the tensile strength of 0·5–0·6 for the composites is about twice that reported for superalloys, and bearing in mind that the tensile strengths of eutectics are superior to those of superalloys this amounts to a very large improvement over current materials. The mode of failure depends on the stress- (or strain-) range $\Delta\sigma$ of the test. In high-cycle/low $\Delta\sigma$ conditions cracking nucleates within the matrix and extends by the fracture of fibres when they are encountered by the crack tip. In low-cycle/high $\Delta\sigma$ conditions the initial failure is due to fibre fracture and further fibre segmentation ahead of an advancing crack weakens the composite.

Several factors appear to contribute to this high fatigue resistance. Probably the most important is due to the high moduli of the composites that lead to small strain cycles for a given fatigue loading. A plausible stress/strain cycle for a composite consisting of elastic fibres and plastic non-hardening matrix is shown in Fig. 6.24; the behaviours of the fibre and matrix are also indicated. Whereas the fibres follow the straight elastic line, the area of the

**Table 6.2 Comparison of fatigue limits of *in situ* composites**

| Alloy | Temp., °C | Condition $V_I$, cm $h^{-1}$ | Notes | $\sigma_{UTS}$, MPa | $\Delta\sigma_{10^7}$, MPa | $\Delta\sigma_{10^7}/\sigma_{UTS}$ |
|---|---|---|---|---|---|---|
| NITAC | 25 | 0·6 | DS | 1304 | 518 | 0·42 |
| NITAC | 25 | 0·6 | soln | 1249 | 483 | 0·41 |
| NITAC | 25 | 0·6 | aged | 1228 | — | — |
| COTAC | 25 | 0·6 | DS | 963 | 518 | 0·57 |
| COTAC | 25 | 2·5 | DS | 1042 | 690 | 0·70 |
| $\gamma$–$\gamma'$–$\delta$ (0% Cr) | 25 | 3 | DS | 1193 | 966 | 0·84 |
| $\gamma$–$\gamma'$–$\delta$ (0% Cr) | 25 | 5·7 | DS | — | 966 | — |
| $\gamma$–$\gamma'$–$\delta$ (6% Cr) | 25 | 3 | DS | 1187 | 656 | 0·58 |
| Ni–Cr | 25 | ? | DS | 828 | 587 | 0·71 |
| $\gamma$–$\delta$ | 25 | 4·7 | DS | 745 | 414 | 0·56 |
| $\gamma'$–$\delta$ | 25 | ? | DS | 1173 | 690 | 0·59 |
| Ni–W | 25 | 3·0 | quenched | 920 | 350 | 0·38 |
| COTAC 74 | 700 | ? | DS | 1240 | 700 | 0·56 |
| COTAC 74 | 800 | ? | DS | 960–1100 | 500 | 0·45–0·52 |
| COTAC | 825 | 0·6 | DS | ? | 330 | 0·75 |
| COTAC | 825 | 0·6 | aged | ? | 485 | 0·65 |
| NITAC | 825 | 0·6 | DS | ? | 400 | ? |

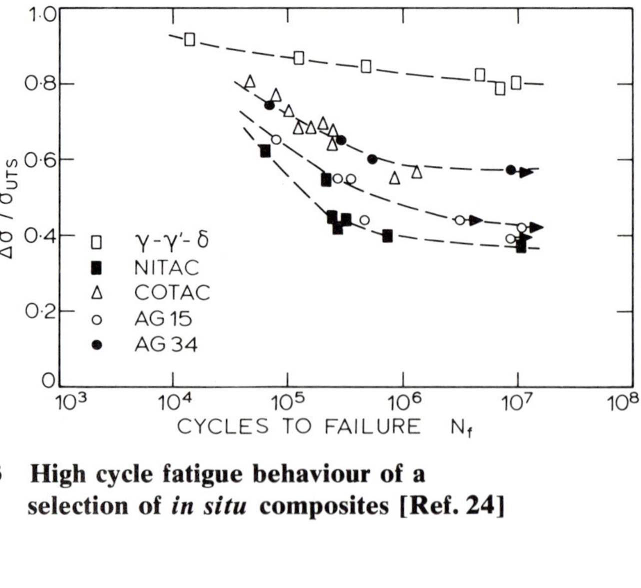


**Figure 6.23 High cycle fatigue behaviour of a selection of *in situ* composites [Ref. 24]**

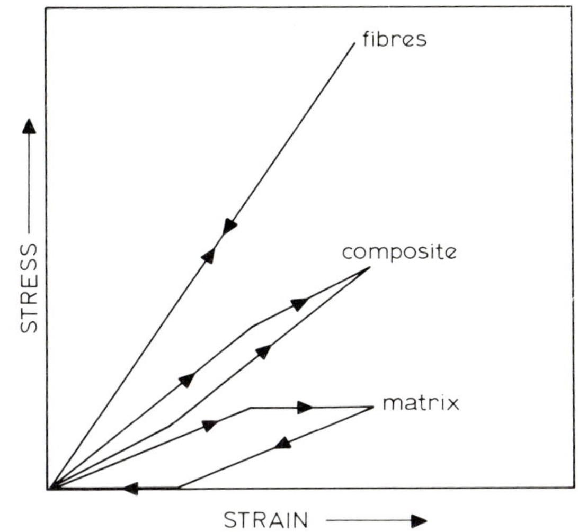


**Figure 6.24 Schematic stress/strain cycle for an imaginary composite consisting of elastic fibres and plastic matrix. The responses of fibres and matrix are indicated**

matrix loop is a measure of the plastic deformation and damage. Further improvements should result from measures such as precipitation and solid solution strengthening, designed to increase the matrix yield stress of the alloys, and this expectation has been largely confirmed by the work of Bibring *et al.*[2] in the evolution of the COTAC alloys.

In the case of the fibre reinforced alloys it is not clear whether the good fatigue properties are due to delayed crack nucleation or to slow crack growth (or indeed to a combination of both). However, for lamellar eutectics such as $\gamma'$–$\delta$ and $\gamma$–$\gamma'$–$\delta$ advancing cracks are diverted along debonded interphase boundaries, thus blunting them and reducing the crack growth rate. This phenomenon appears to be less important for the fibre reinforced materials; studies of fatigue crack propagation[21] show cracking perpendicular to the fibre direction to be qualitatively similar to that in superalloys, whereas crack growth parallel to the fibres is manifested by multiple crack branching which leads to a very unusual negative gradient in the $\Delta a/\Delta n$ versus $\Delta K$ plot shown in Fig. 6.25. (This simply means that the 'real' $\Delta K$ is reduced as the crack grows and cannot be identified with the apparent crack length.)

The fatigue behaviours of *in situ* composites are much less sensitive to the morphology and scale of the microstructure than, for example, creep. The detailed response varies for different alloys as demonstrated by the particularly full experimental studies of Stoloff[22,23] and co-workers. Thus, in $\gamma$–$\gamma'$–$\delta$[22] fibre

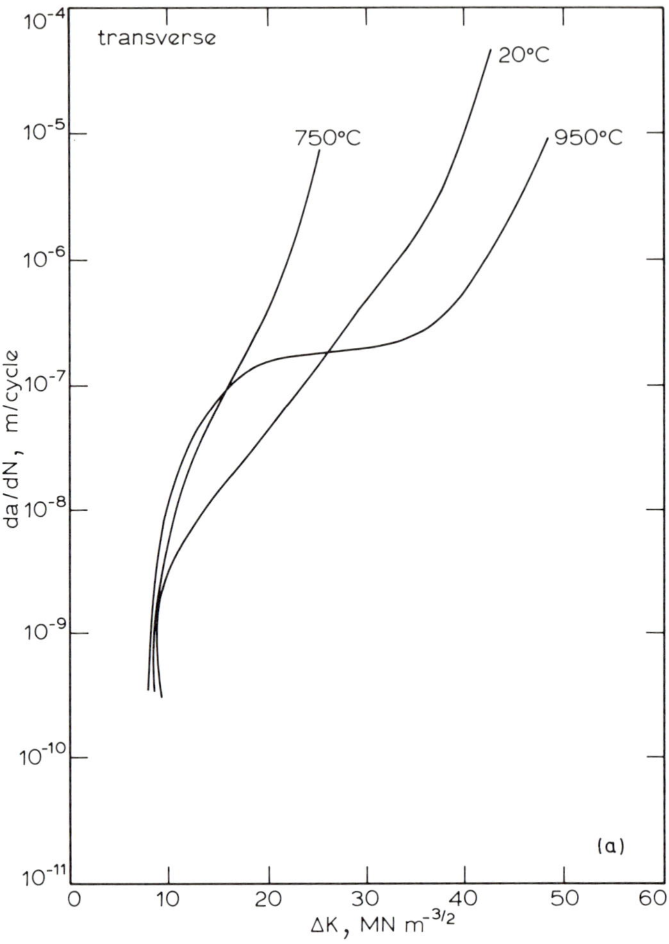


**Figure 6.25 Growth rate of fatigue cracks $\Delta a/\Delta N$ in (Co, Cr)–$Cr_7C_3$ as a function of the stress intensity factor $\Delta K$ [Ref. 21] (a) transverse to solidification direction and . . .**

refinement increases the fatigue resistance only at high stress levels whereas in a COTAC alloy[23] there are improvements at all stresses. Attempts to assess the role of fibre length through prestrain experiments to fragment the fibres have been inconclusive although they suggest a tendency to reduced fatigue resistance particularly at high temperatures.

Very little information on strain-controlled low cycle fatigue is available and consequently the interaction between creep and fatigue is not understood.

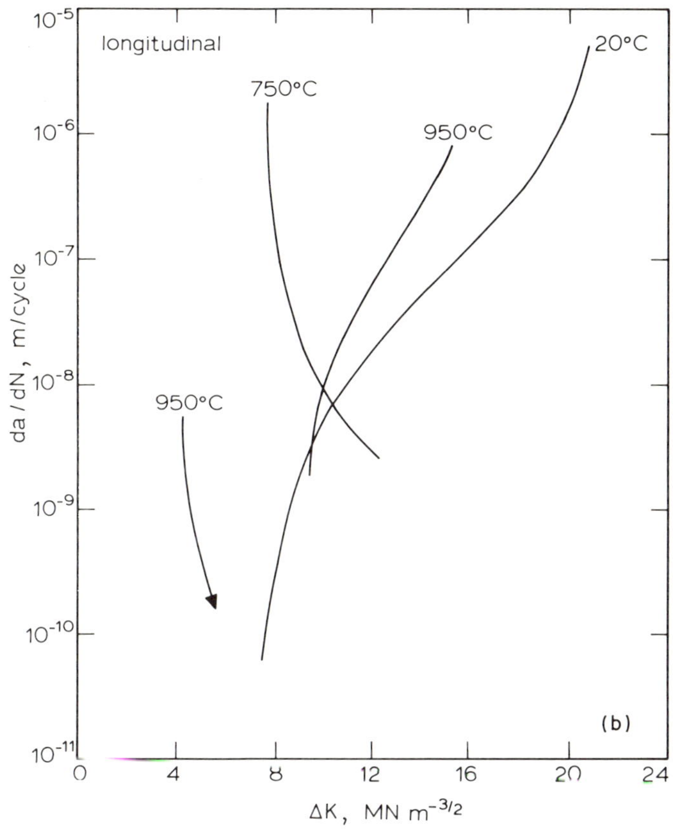


**Figure 6.25 ... (b) parallel to solidification direction**

However, the high elastic moduli must militate against good strain-controlled creep resistance. The available data on $\gamma$–$\gamma'$–$\delta$ and on a NITAC alloy are compatible with a Coffin–Manson relationship between strain range $\Delta\varepsilon_p$ and the number of cycles to failure $N_f$.[24]

$$N_f \Delta\varepsilon_p^{1/2} = \text{constant} \tag{6.44}$$

The somewhat related problem of cyclic creep has been discussed by Henderson and McLean[25] in terms of a detailed examination of creep transients during stress cycling. The principal features of the transient strain during unloading/loading sequences observed on experiments on $\gamma$–$\gamma'$–$Cr_3C_2$ are shown in Fig. 6.26 for the two extreme cases where (i) the lower stress is maintained for sufficient time to allow complete recovery and (ii) reloading to the original stress immediately follows unloading. The one stage observed that is common to both cycles is:

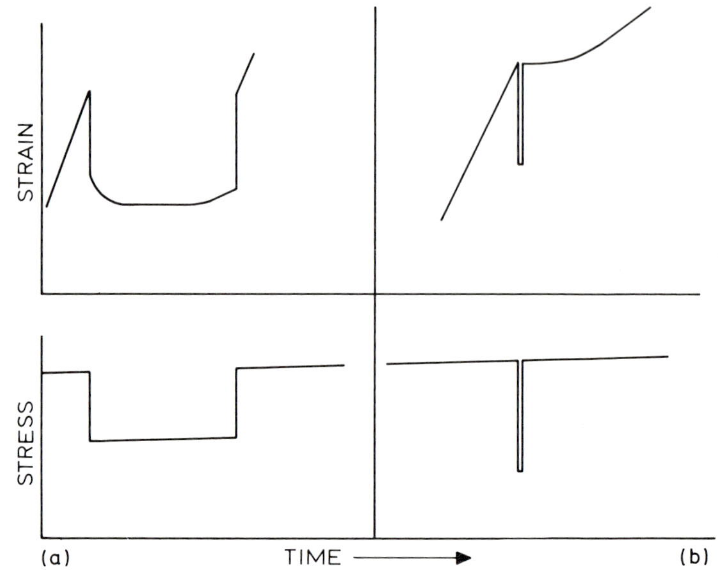


**Figure 6.26 Schematic illustration indicating the various stages in the transient creep observed following stress changes for $\gamma$–$\gamma'$–$Cr_3C_2$: (a) hold period at low stress and (b) no-hold period at low stress**

(a) the time-independent contraction $\Delta\varepsilon_i$ that exceeds the expected elastic contraction $\Delta\varepsilon_e = \Delta\sigma/E_c$.

If the lower stress is maintained for a long time (a) is followed by:

(b) a time-dependent contraction,
(c) a period $\Delta t$ during which no creep is observed (cf. Chapter 5), and
(d) the development of a new forward creep rate $\dot{\varepsilon}_2'$ that is smaller than the creep rate $\dot{\varepsilon}_2$ that would be obtained by initially loading to the reduced stress.

Stages (b), (c) and (d) are suppressed if reloading immediately follows the stress reduction. The major difference in the transients occurs on reloading to the original stress. When recovery has been allowed at the lower stress the reloading response consists of

(e) an immediate strain $\Delta\varepsilon_i'$ that exceeds $\Delta\varepsilon_e$ and also (slightly) $\Delta\varepsilon_i$,
(f) a pseudo-primary creep regime of decreasing creep rate with increasing time, and
(g) re-establishment of the original creep rate.

The equivalent reloading response when recovery has been prevented are

(e′) an immediate strain $\Delta\varepsilon''_i = \Delta\varepsilon_i > \Delta\varepsilon_e$,
(f′) a forward creep behaviour that often consists of a delay period $\Delta t'$ of no creep followed by a slowly increasing creep rate, and
(g′) re-establishment of the original creep rate.

The most important differences are in stages (f) and (f′) and their consequences to the accumulation of creep strain under conditions of stress cycling are shown in Fig. 6.27. Clearly the effects of such reloading delay periods will be to cause lower strains than would be expected by applying the simple creep equation to the total time on the high stress; moreover, the deviation will be greatest for the smallest cycle period (or rather for the smallest periods on the lowest stress). This is exactly the behaviour that has been reported for superalloys although relevant information is not available for *in situ* composites. The interpretation of Henderson and McLean[25] appeals to composite theory where the transients are due to the interaction between material elements (here fibre and matrix) with different elastic and plastic characteristics. They speculate further that the consistency of this formalism with the results for superalloys suggests that the transient effects there are due to the interaction of grains and/or dendrites rather than due to dislocation interactions.

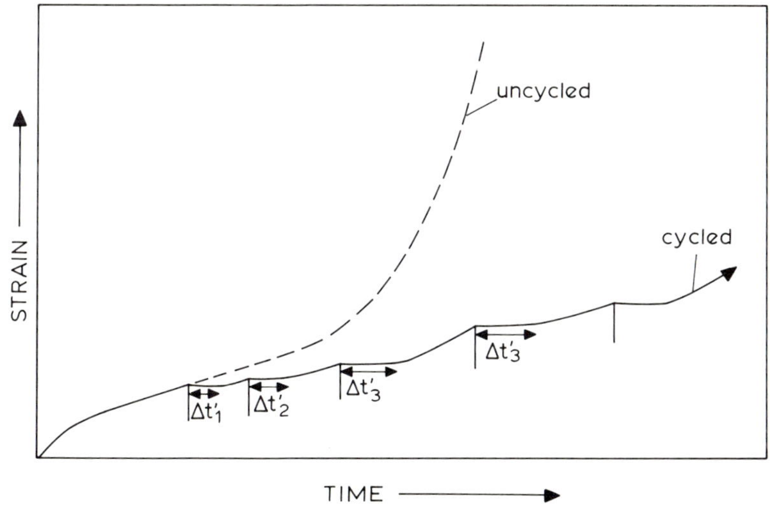


**Figure 6.27 Consequences of the incubation period on reloading to the cumulative strain during the period at high stress for different stress cycling periods**

The total creep strain $\varepsilon_t$ that occurs when a high stress $\sigma_1$ is applied for a period $t_1$ followed by $\sigma_2$ for a time $t_2$ may be expressed

$$\varepsilon_t = \dot{\varepsilon}_1 t_1 + \dot{\varepsilon}_2 t_2 + \Delta\varepsilon_{r1} + \Delta\varepsilon_{r2} + \Delta\varepsilon_h \tag{6.45}$$

where $\dot{\varepsilon}_1$, $\dot{\varepsilon}_2$ are the steady state creep rates with stresses $\sigma_1$ and $\sigma_2$, $\Delta\varepsilon_{r1}$, $\Delta\varepsilon_{r2}$ are strain perturbations attributable to the delay periods $\Delta t$, $\Delta t'$ on unloading and reloading respectively and $\Delta\varepsilon_h$ is the discontinuous strain that occurs on reloading (i.e. $\Delta\varepsilon_{r1} = -\dot{\varepsilon}_1\Delta t$ and $\Delta\varepsilon_{r2} = -\dot{\varepsilon}_2\Delta t'$). The expression for $\Delta t$ on unloading, developed in Chapter 5 (i.e. eqn (5.24)) remains valid in this context.

$$\Delta t = \frac{\mu^2\mathbf{b}^2}{K}[(\sigma_2 - \sigma_0)^{-2} - (\sigma_1 - \sigma_0)^{-2}] \tag{6.46}$$

$\Delta t$ is simply the time required for the dislocation substructure established at the higher stress to coarsen sufficiently to allow yield with the lower stress. A similar concept may be used to model $\Delta t'$ where the plastic strain during cycling may be thought of creating a dislocation substructure characteristic of a stress greater than $\sigma_1$, $\sigma_1 + \Delta\sigma'$ say. Then

$$\Delta t' = \frac{\mu^2\mathbf{b}^2}{K}[(\sigma_1 - \sigma_0)^{-2} - (\sigma_1 + \Delta\sigma^2 - \sigma_0)^{-2}] \tag{6.47}$$

where $\Delta\sigma'$ depends on the *difference* in elastic stress of fibre and matrix and on the elastic strain of the composite $(\sigma_1 - \sigma_2)/E_c$, i.e.

$$\Delta\sigma' = 2\left[\frac{E_f V_f - E_m(1 - V_f)}{E_c}(\sigma_1 - \sigma_2)\right] \tag{6.48}$$

Thus $\Delta t'$ should increase with increasing $(\sigma_1 - \sigma_2)$ and with increasing $(\mu^2\mathbf{b}^2)/K$ (i.e. decreasing temperature). The effect would be greatest for large differences in elastic strengths. Figure 6.28 compares a set of data for $\gamma$–$\gamma'$–$Cr_3C_2$ with curves computed from eqns (6.47) and (6.48) using appropriate values of $E_f$, $E_m$, $V_f$ and $(\mu^2\mathbf{b}^2)/K$.

There is reasonable agreement for low values of $\Delta\sigma$, but for $\Delta\sigma \geqslant 100$ MPa $\Delta t'$ steadily falls whereas the theory predicts a monotonic increase. This latter effect may be understood in terms of the heterogeneous nature of the deformation in superalloys and in *in situ* composites. The theory assumes that deformation and hardening are homogeneous. However, where slip occurs in bands we may think of a competition between 'hardening' of the deforming material within the bands and material 'softening' by the creation of additional deformation bands. Thus the deviation from eqns (6.47) and (6.48) for high $(\sigma_1 - \sigma_2)$ may be due to the dominance of the 'softening' phenomenon.

The influence of $t_2$ on $\Delta\varepsilon_h$ has been investigated by Gibbons *et al.*[26] who propose the empirical relation

$\Delta\varepsilon_h = bt_2^{2/3} + c$ where $b$ and $c$ are constants.

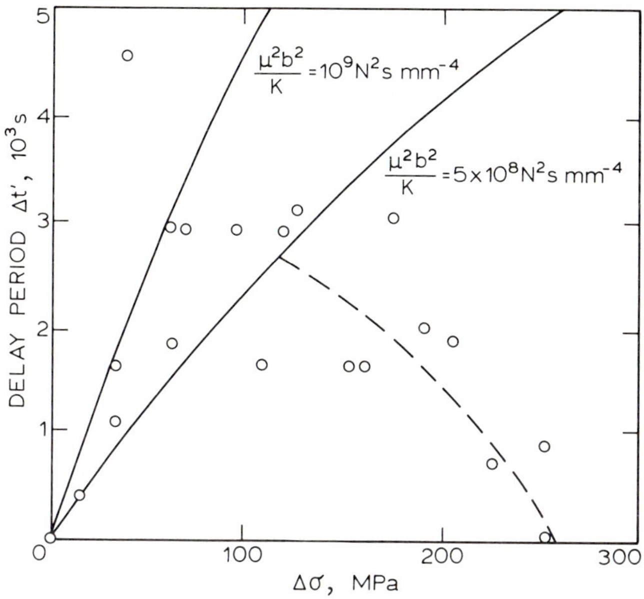


**Figure 6.28 Incubation period $\Delta t'$ on reloading after various stress changes applied to $\gamma$–$\gamma'$–$Cr_3C_2$ during creep testing at 825°C and a stress of 275 MPa. The theoretical curves computed from eqns (6.47) and (6.48) for the values of $(\mu^2 b^2)/K$ indicated are included**

## 6.5 THERMAL CYCLING

The microstructure and properties of *in situ* composites have been thought to be particularly vulnerable to degradation under temperature cycling conditions and, indeed, some of the earlier alloys such as COTAC 3 were seriously affected by changing temperatures.[27,28] However, these problems have been ameliorated by the same remedies that have improved the off-axis and fatigue behaviour, namely, alloy modification that increases the alloy yield stress and minimizes plastic flow in the matrix during temperature excursions. Composites, in general, are subject to two quite different types of damage because of their special structure. Firstly on the microscopic level, since fibres (or lamellae) and matrix have different coefficients of linear expansion then local internal stresses must be generated between the phases which can cause damage accumulation by matrix yield. Secondly, on a more macroscopic scale, temperature gradients that inevitably develop during temperature cycling may lead to anisotropic cracking of the matrix along the weak longitudinal direction.

### (a) Microscopic effects

Consider a composite loaded with a stress $\sigma$ which is assumed to be partitioned between fibre and matrix according to the Rule of Mixtures at all temperatures. If the phases were unconnected then an increase in temperature $\Delta T$ would cause the fibres and matrix to strain by different amounts as shown schematically in Fig. 6.29 ($\alpha_f \Delta T$ and $\alpha_m \Delta T$ respectively where $\alpha_f$, $\alpha_m$ are coefficients of expansion). Since the phases are coupled the composite must adopt an intermediate extension with the fibres in tension and the matrix in compression (if $\alpha_f < \alpha_m$). Figure 6.30 shows schematically the change in fibre and matrix stress for the case of a composite consisting of elastic fibres and a plastic or creeping matrix that is subject to a uniform tensile traction during a reversible temperature excursion $\Delta T$; ABC corresponds to heating and CDA to cooling. The corresponding strain versus temperature cycle is also shown; of course, for continuous composites matrix and fibres are constrained to have the same elongation. The area of the strain/temperature loop is a measure of the plastic deformation and damage that accumulates during each thermal cycle. Early

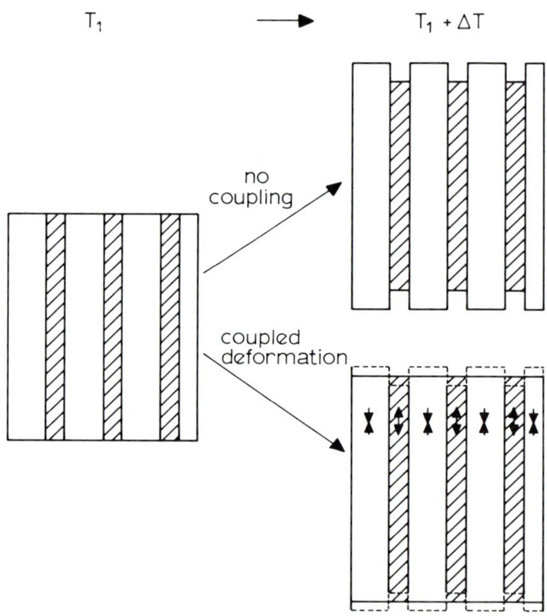


**Figure 6.29 Schematic illustrations of thermal expansion of a composite showing original structure, imaginary independent expansion of fibre and matrix, and coupled expansion of the composite**

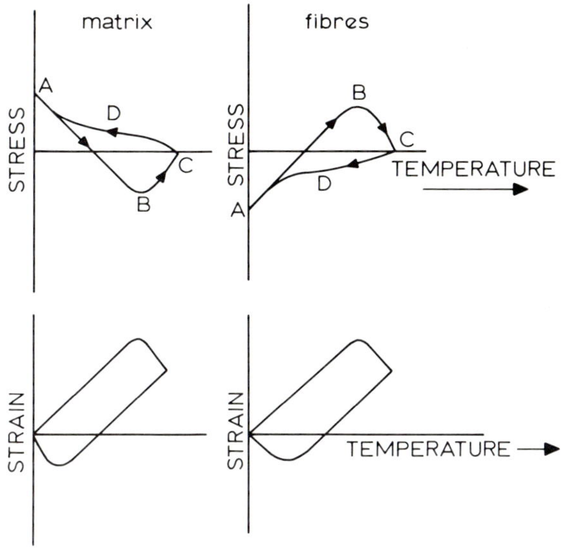


**Figure 6.30 Plots of changes in stress and strain of fibre and matrix during a reversible temperature cycle $\Delta T$ applied to a composite with a constant applied tensile stress**

yield gives a wide strain hysteresis loop and maximum damage. By increasing the yield stress the elastic zone, where no dislocations are generated, is extended.

Garmong[29,30] has developed a formal model of this mechanism that quantifies the effects described qualitatively above. He has proposed that temperature cycling under strain controlled conditions is likely to be more damaging than under stress controlled/free displacement conditions. In practical terms, stator turbine blades which are fixed at both ends to disks should be more vulnerable than rotor blades which are unconstrained at one end.

The damage accumulation in the matrix may have a variety of effects:

(i) high local dislocation densities may alter thermodynamic equilibrium and lead to changes in composite morphology,
(ii) the matrix diffusivities may be altered, or
(iii) stress concentrations may lead to premature fracture of the fibres (or lamellae).

There is now a great deal of evidence that the 'simpler' *in situ* composites that have weak matrices exhibit either a roughening of initially smooth fibres or extensive fibre fragmentation during temperature cycling where the effects are greatest with increasing maximum temperature. In both cases there is an associated deterioration in stress-rupture properties. However, the more highly alloyed composites show neither the microstructural instabilities nor the degradation in properties.[31]

### (b) Macroscopic effects

In considering the microscopic phenomenon it is assumed that the material is isothermal at any instant but that the temperature is changing. The situation in a real component, such as a turbine blade, is quite different because of non-uniform heating and cooling of different regions and these produce macroscopic stress distributions which can cause the nucleation and propagation of fatigue cracks. This is precisely the type of thermal fatigue that is studied extensively on superalloys using either sophisticated closed loop testing with control of both temperature and strain or by simulation of component shape behaviour in fluidized beds or combustion burners. The anisotropic

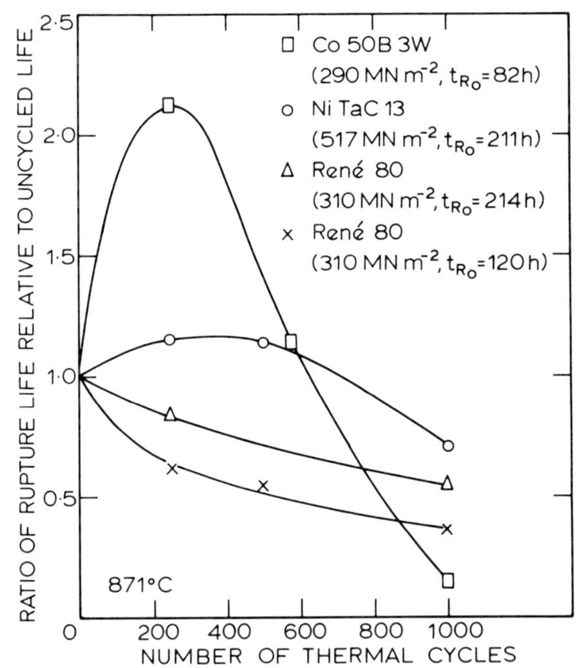


**Figure 6.31 Rupture life as a function of prior thermal cycling heat treatment for two *in situ* composites and two developmental superalloys [Ref. 32]**

nature of *in situ* composites, however, presents a special problem if the cracking characteristics in both longitudinal and transverse directions are to be determined independently. Woodford[33] has designed a suitable test specimen and testing procedure.

Woodford[32,33] has also carried out the most systematic studies of both types of thermal fatigue behaviour. The effects of microscopic damage have been determined by two approaches:

(i) prior thermal cycling followed by normal stress rupture testing, and
(ii) thermal cycling under load to rupture.

The effects are sensitive to the nature of the alloy and the details of the temperature cycle, and although conditions can certainly be found that degrade the mechanical properties of *in situ* composites, conventional superalloys also suffer from the same treatments. Figure 6.31 shows the effects of prior cyclic heat treatments on the rupture lives of two *in situ* composites and two developmental superalloys; it shows that there can be an element of cyclic strengthening before the properties begin to degrade. The composite that is most severely affected is that with the weakest matrix and it appears that the problem has been largely eliminated in the more highly alloyed systems where the matrices have been strengthened. Woodford suggests that the most damaging cycle consists of rapid changes over a wide temperature range with a long hold time at the maximum temperature.

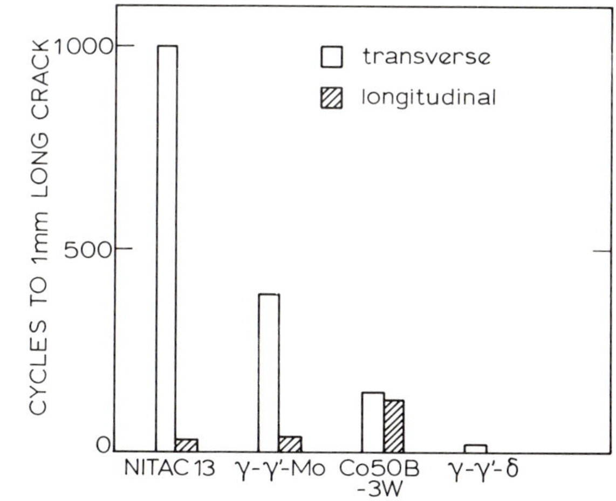


**Figure 6.32 Comparison of the number of cycles to produce a 1 mm crack length in the longitudinal and transverse directions during thermal fatigue of selected *in situ* composites [Ref. 33]**

Studies of macroscopic cracking[33] show most *in situ* composites to have outstanding resistance to the initiation and growth of cracks perpendicular to the solidification direction. However, the cracking parallel to the reinforcement varies widely between alloys as is demonstrated in Fig. 6.32. The lamellar *in situ* composites such as $\gamma$–$\gamma'$–$\delta$ are particularly vulnerable to rapid cracking along longitudinal grain boundaries. Of the various alloys considered, the $\gamma$–$\gamma'$–carbide composites appear to have the most acceptable thermal fatigue properties, performing at least as well as current superalloys.

## REFERENCES CHAPTER 6


1 A. KELLY: 'Strong solids', 1973, Oxford University Press

2 H. BIBRING: Proc. Conf. *In Situ* Composites I, National Materials Advisory Board Publication 308–II, p. 1, Washington, DC, January 1973

3 E. R. BUCHANAN AND L. A. TARSHIS: *Metall. Trans.*, 1973, **4**, 1895

4 E. R. BUCHANAN AND L. A. TARSHIS: *Metall. Trans.*, 1974, **5**, 1413

5 P. ANNARUMMA AND M. TURPIN: *Metall. Trans.*, 1972, **3**, 137

6 D. E. MILES, E. BULLOCK AND M. McLEAN: Proc. Conf. *In Situ* Composites II, eds. M. R. Jackson *et al.*, p. 539, Xerox Indvidualized Publishing, Lexington, Mass., 1976

7 E. R. THOMPSON, F. D. GEORGE AND E. M. BREINAN: Proc. Conf. *In Situ* Composites I, National Materials Advisory Board Publication 308–II, p. 71, Washington, DC, 1973

8 G. A. CHADWICK: *Acta Metall.*, 1976, **24**, 1137

9 R. W. ARMSTRONG, F. D. LEMKEY AND E. R. THOMPSON: Proc. Conf. *In Situ* Composites II, p. 476, eds. M. R. Jackson *et al.*, Xerox Individualized Publishing, Lexington, Mass, 1976

10 M. DUPEUX AND F. DURAND: *Metall. Trans.* 1975, **6A**, 2143

11 D. A. WOODFORD: *Metall. Trans.*, 1977, **8A**, 639

12 T. KHAN, J. F. STOHR AND H. BIBRING in 'Superalloys 1980', Proc. 4th Int. Symp. Superalloys, eds. J. K. Tien *et al.*, p. 531, 1980, Metals Park, Ohio, ASM

13 D. McDANELS, R. A. SIGNORELLI AND J. W. WEETON: NASA Report No. TND-4173, 1967

14 E. BULLOCK, P. N. QUESTED AND M. McLEAN: *Acta Metall.*, 1977, **5**, 333

15 K. D. SHEFFLER, R. H. BARKALOW AND A. YUEN: Proc. Conf. *In Situ* Composites II, eds. M. R. Jackson *et al.*, Xerox Individualized Publishing, Lexington, Mass., 1976

16 J. L. WALTER: Proc. Conf. *In Situ* Composites IV, 16–17 November 1981, Boston, Mass., eds. F. D. Lemkey, H. E. Cline and M. McLean, New York–Oxford, Elsevier North Holland, in press

17 P. N. QUESTED, E. BULLOCK AND M. McLEAN: 'Solidification and casting of metals', 466, 1979, London, Metals Society

18 D. E. MILES AND M. McLEAN: *Met. Sci.*, 1977, **11**, 563

19 A. F. JOHNSON: *J. Mech. Phys. Solids*, 1977, **25**, 117

20 M. F. X. GIGLIOTTI, M. R. JACKSON, S. W. YANG, M. F. HENRY AND D. A. WOODFORD: Proc. Conf. *In Situ* Composites IV, Boston, Mass., 16–17 November 1981, eds. F. D. Lemkey, H. E. Cline and M. McLean, New York–Oxford, Elsevier North Holland, in press

21 R. B. SCARLIN AND M. O. SPEIDEL: Brown Boveri et Cie, Baden, Switzerland, Final Report of COST 50 Programme CH3/4, July 1976

22 J. E. GROSSMAN AND N. S. STOLOFF: *Metall. Trans.*, 1978, **9A**, 117

23 C. M. AUSTIN, N. S. STOLOFF AND D. J. DUQUETTE: *Metall. Trans.*, 1977, **8A**, 1021

24 N. S. STOLOFF: Proc. Conf. *In Situ* Composites III, eds. J. L. Walter *et al.*, Ginn and Company, Lexington, Mass., 1979

25 P. J. HENDERSON AND M. McLEAN: *Acta Metall.*, 1982, **30**, 1121

26 T. B. GIBBONS, V. LUPINC AND D. McLEAN: *Met. Sci.*, 1975, **9**, 437

27 E. M. BREINAN, E. R. THOMPSON AND F. D. LEMKEY: Proc. Conf. *In Situ* Composites I, p. 201, National Materials Advisory Board Publication 308–II, Washington, DC, 1973

28 F. M. DUNLEVEY AND J. F. WALLACE: *Metall. Trans.*, 1974, **5**, 1351

29 G. GARMONG: *Metall. Trans.*, 1974, **5**, 2183

30 G. GARMONG: Proc. Conf. *In Situ* Composites II, eds. M. R. Jackson *et al.*, Xerox Individualized Publishing, Lexington, Mass., 1976
31 M. GELL: Proc. AGARD Specialists Meeting on Directionally Solidified *In Situ* Composites, eds. E. R. Thompson and P. R. Sahm, AGARD-CP-156, Washington, DC, 1974
32 D. A. WOODFORD: Proc. Conf. *In Situ* Composites II, eds. M. R. Jackson *et al.*, p. 211, Xerox Individualized Publishing, Lexington, Mass., 1976
33 D. A. WOODFORD: Proc. Conf. *In Situ* Composites III, eds. J. L. Walter *et al.*, p. 410, Ginn and Company, Lexington, Mass., 1979
34 J. F. STOHR AND R. VALLÉ: *Philos. Mag.*, 1975, **1**, 43

# 7
# Microstructural stability of anisotropic alloys

## 7.1 INTRODUCTION

The microstructures of the directionally solidified alloys, and indeed of other anisotropic materials, are not equilibrium forms; rather the processing conditions have been contrived to produce microstructural characteristics that impart good mechanical properties. Consequently these materials are inherently thermodynamically unstable and, particularly during high temperature operation when diffusional processes are relatively rapid, they may degrade to more stable microstructural forms with less attractive properties. It is important to be able to predict the rates of such instabilities quantitatively since this will define one limit to the operating conditions of the material. All microstructural instabilities need not be detrimental (although many undoubtedly are); for example, suitable sequences of heat treatments can dissolve precipitated phases formed during the initial processing and redeposit them in a more suitable form.

It is convenient to distinguish two broad categories of instability on the basis of the nature of the driving force for the change in material structure.

(a) *Intrinsic* instabilities arise from the inherent structure of the material having a higher internal energy than necessary so that mass transport by diffusional processes leads to a gradual approach to thermodynamic equilibrium. Thus the transverse compositional gradients produced by dendritic solidification provide a chemical driving force that can, in principle, be removed by homogenizing heat treatments. On the other hand, the large areas of phase boundaries associated with dispersed phases (e.g. fibres, lamellae, $\gamma'$ particles) cause capillary driving forces, due to the interfacial energies that can lead to changes in the shape, size or position of the phases. For example, a fine dispersion in a superalloy can coarsen to produce a much larger particle size during service and a composite material can spheroidize to produce equiaxed particles from the original rod or fibrous structure.

(b) *Extrinsic* instabilities derive their driving forces from the interactions of the material with an external potential field and in this case there is no equilibrium configuration unless there is a minimum in the field potential. This type of phenomenon has been studied less extensively than, for example, particle coarsening or spheroidization; however, it is now well established that the presence of temperature and stress gradients can lead to significant changes in the phase morphologies of duplex materials that do not occur under isothermal conditions.

The relative importance of intrinsic and extrinsic instabilities depends on the magnitudes of their respective driving forces, in particular testing or service conditions. Figure 7.1 provides a simple metaphor of the two types of phenomena. The steel ball in Fig. 7.1*a* falls vertically parallel to the walls of the container through the fluid under the influence of gravity; when placed on an incline or within the influence of a magnet (Figs. 7.1*b* and *c*) the trajectory of the falling ball, with respect to the frame of reference of the container, is altered to an extent dependent on the angle of the incline or on the strength of the magnet. Here Fig. 7.1*a* represents an intrinsic instability while Figs. 7.1*b* and *c* show the influence of extrinsic effects.

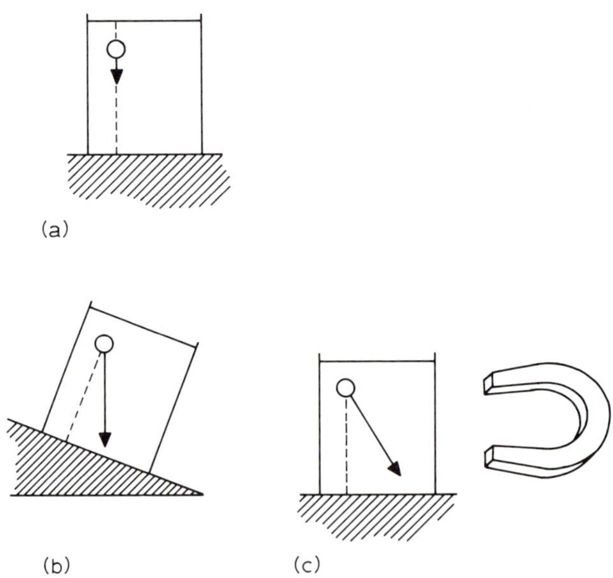


**Figure 7.1 Schematic illustration showing the change in trajectory of a steel sphere falling through a viscous fluid (a), and due to the application of an additional (b) gravitational and (c) magnetic field**

However they originate the gradients in free energy (or chemical potential) $\mu$ within the material may be regarded as giving rise to a force $\mathbf{f}$ biasing the thermal activation of each atom:

$$\mathbf{f} = -\nabla\mu \tag{7.1}$$

This leads to a drift velocity $\mathbf{v}$ of atoms in the direction of this force $\mathbf{f}$

$$\mathbf{v} = B\mathbf{f} \tag{7.2}$$

where $B$ is the mobility of the atoms which depends on the particular mass transport mechanism that is operating (e.g. volume, interface or grain boundary diffusion). If $N$ is the density of diffusing atoms, then their flux $\mathbf{j}$ is given by the expression

$$\mathbf{j} = BN\mathbf{f} = -BN\nabla\mu \tag{7.3}$$

For a single phase material this will simply lead to changes in compositional distribution. However, in two-phase alloys the requirement for conservation of matter at the interface between the matrix $\alpha$ (the diffusing medium) and a second phase $\beta$ leads to a displacement of the interface with a velocity $\mathbf{V}$ given by the expression

$$\mathbf{V} = -(\Omega\mathbf{j} \,.\, \mathbf{n})\mathbf{n} = -(B\Omega N\mathbf{f} \,.\, \mathbf{n})\mathbf{n} \tag{7.4}$$

where $\Omega$ is the atomic volume and $\mathbf{n}$ is a unit vector normal to the surface. This idealized situation is illustrated in Fig. 7.2 for diffusion of atoms of $B$ through $\alpha$ to $\beta$, which only contains $B$ atoms, by various diffusional mechanisms.

The remainder of this chapter will describe in detail various types of instability that can occur in anisotropic alloys, although many of them are also relevant to isotropic materials. Most of the phenomena described are of the intrinsic type, the evidence for and theories of extrinsic instabilities being collected in the final Section (§ 7.4).

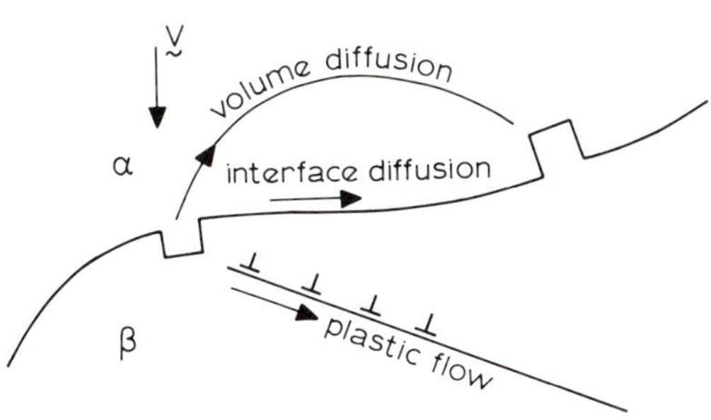


**Figure 7.2 Schematic illustration of interface velocity *V* due to various mass transport mechanisms**

## 7.2 CHEMICAL INSTABILITIES

If the processing route to produce a material leads to a partitioning of its constituent elements in a non-equilibrium form then at high temperature, when diffusion can occur, there can be significant spatial redistribution of the elements to give, for example, homogenization of dendritic segregation or the precipitation of new phases.

### 7.2.1 Homogenization of dendritic segregation

The description of dendritic segregation in Chapter 2 shows that when kinetic factors control the freezing process then a range of compositions that is determined by the nature of the relevant phase diagram is formed in a relatively regular array. In particular, directional solidification leads to aligned, columnar dendrites with compositions ranging between the equilibrium values associated with the liquidus and solidus temperatures (see Fig. 2.7); moreover, the periodicity of the segregation is sensitive to the solidification conditions, high solidification rates and temperature gradients giving finer microstructures (eqn (2.35)). The homogenization of such non-equilibrium solute distributions has been treated by Shewmon[1] and by Purdy and Kirkaldy[2] for the case of simple binary alloys.

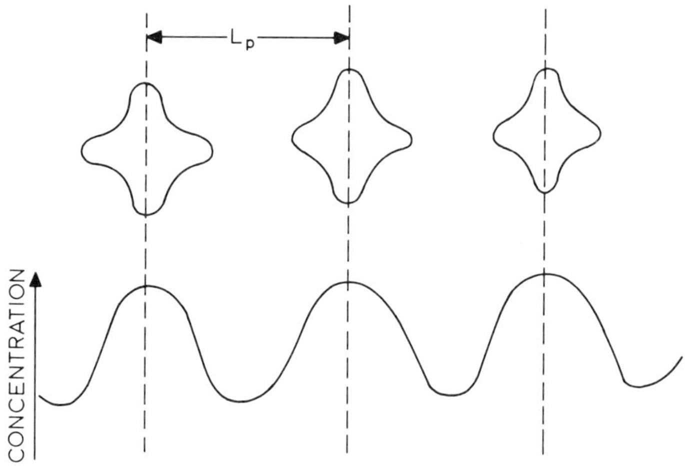


**Figure 7.3 Schematic representation of the relationship of solute concentration to dendritic microstructure; (a) shows the dendritic structure on a transverse section of directionally solidified alloys and (b) is the transverse variation in solute concentration through several dendrites**

The concentration profile across the dendrite structure may be represented by a Fourier series where the fundamental frequency is defined by the distance between the primary dendrites $L_p$. In this simple treatment we assume that the contours of constant composition are circular about the centre of each dendrite in a plane normal to the solidification axis as shown schematically in Fig. 7.3. Then, using cylindrical co-ordinates with the origin at the centre of a dendrite, the initial profile (i.e. at $t = 0$) is given by the expression:

$$C(r) = C_0 + \sum_n C_n \cos \frac{2\pi nr}{L_p} \quad \text{for } r < L_p/2 \tag{7.5}$$

where $C_0$ is the average composition of the alloy. (A more realistic profile would be obtained by taking double Fourier series in order to reproduce the non-circular concentration contours, but the simpler representation which is preferred here for clarity produces essentially the same result.) In the dilute solution approximation, the chemical potential $\mu$ is related to the concentration $C$ by the following expression:

$$\mu = \mu_0 + \mathrm{RT} \ln C(r) \tag{7.6}$$

(where $\mu_0$ is a reference potential, $R$ is the gas constant and $T$ is the absolute temperature). It follows from eqn (7.3) that the flux of solute atoms $\mathbf{j}$ has the form

$$\begin{aligned} \mathbf{j} &= -BN\frac{d\mu}{dr} \\ &= -\frac{BNRT}{C(r)}\frac{dC(r)}{dr} \\ &= -\frac{BRNT}{C(r)}\frac{2\pi}{L_p}\sum_n nC_n \sin\frac{2\pi nr}{L_p} \end{aligned} \tag{7.7}$$

This will clearly lead to a reduction in the amplitude of the various Fourier components; moreover, since the contributions to $\mathbf{j}$ from the various components of the Fourier series are proportional to $n$, the higher order components will decay rapidly leaving the fundamental component ($n = 1$). It follows that it is sufficient to consider only the fundamental component in eqn (7.5) when considering the overall extent of homogenization.

During diffusion the variation of $C(r)$ with time (i.e. $C(r, t)$) can be obtained by solution of Laplace's Equation,

$$D\frac{d^2C}{dr^2} = \frac{dC}{dt} \tag{7.8}$$

where $D$ is the appropriate diffusivity, subject to the boundary condition that eqn (7.5) is satisfied at $t = 0$. Then

$$\frac{dC}{dt}(r, t) = -\sum_n \frac{4\pi^2 n^2 D}{L_p^2} C(r, t)$$

or

$$\frac{d}{dt} \ln C(r, t) = -\sum_n \frac{4\pi^2 n^2 D}{L_p^2}$$

This has the solution

$$C(r, t) = C_0 + \sum_n C_n \cos \frac{2\pi n r}{L_p} \exp\left(-\frac{t}{\tau_n}\right) \tag{7.9}$$

where $\tau_n = L_p^2/(4\pi^2 n^2 D)$ is the time constant for decay of the $n$th harmonic. Clearly the higher harmonics decay relatively rapidly since $\tau_n$ decreases as $n$ increases. Consequently, even the most complex initial compositional profile will rapidly reduce to a relatively simple dependence on the fundamental frequency ($n = 1$).

Inspection of eqn (7.9) shows that a uniform composition $C_0$ is approached exponentially with time at a rate dependent on both the dendrite spacing $L_p$ and the diffusion coefficient. Clearly since $C_0$ can never actually be attained, there is some arbitrariness in defining when homogenization has been completed. For the alloys considered in this book, little segregation remains after $t = 3\tau$ and this time is plotted as a function of $L_p$ for various values of $D$ in Fig. 7.4. This clearly indicates the need for high heat treatment temperatures (i.e. high $D$) and fine dendrite spacings to facilitate homogenization. In many materials thermomechanical treatments can be adopted to reduce the dendrite dimensions by mechanical working prior to heat treatment. However, this is not possible for complex superalloys that are used in the cast component form.

Complete homogenization is rarely achieved in cast superalloys in either the conventionally cast or directionally solidified forms. Because of the high degree of alloying, the melting ranges of these materials are often very wide so that the highest practicable heat treatment temperatures (i.e. before incipient melting) are too low to either allow sufficient solutioning of the $\gamma'$ precipitate or give fast enough diffusion for homogenization to occur in an acceptable time scale. A major advantage of the new generation of single crystal superalloys is that the melting ranges have been reduced by removing grain boundary strengthening elements; this allows heat treatments to be carried out at considerably higher temperatures than are possible with more conventional superalloys. Figure 7.5 shows the compositional profiles obtained by Gell *et al.*[2a] using microprobe analysis across a dendritic structure of the Pratt and Whitney single crystal alloy 454 after two heat treatments of 4 h at 1204 and 1288°C respectively. At the first temperature, which is typical

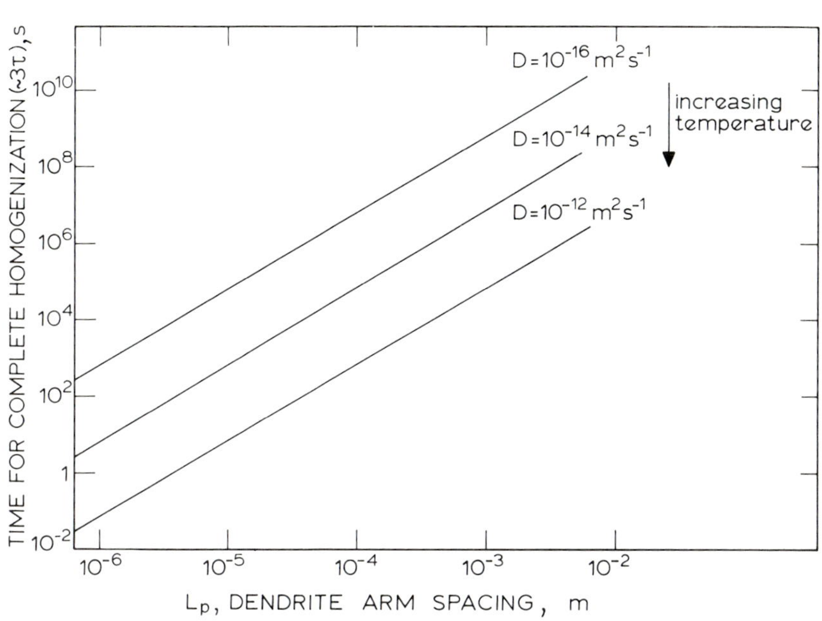


**Figure 7.4 Time for complete homogenization, taken as 3$\tau$, of dendritic segregation as a function of dendrite arm spacing $L_p$ and diffusivity $D$**

of the heat treatment conditions of conventional and directionally solidified superalloys, considerable segregation remains, whereas this has almost entirely been removed at the higher temperature. Used in combination with the finer dendrites possible through solidification in high temperature gradients, even faster heat treatment rates would be possible.

### 7.2.2 Phase changes

During service at high temperatures the nature and distribution of phases produced during processing can be changed radically either by the deposition of new phases or by changes in the distribution, chemistry or structure of the original phases. Both of these effects will occur when the equilibrium phases at the relatively low service temperatures differ from those produced at the higher processing temperatures and when the temperatures are sufficiently high to cause diffusional mass transport. Because such precipitation is specific to the detailed phase equilibria of each alloy it is not possible to generalize the behaviour of all superalloys (for example). Moreover, since diffusion coefficients are extremely sensitive to temperature the rate at which any phase transition occurs will be dramatically changed by fairly small modifications

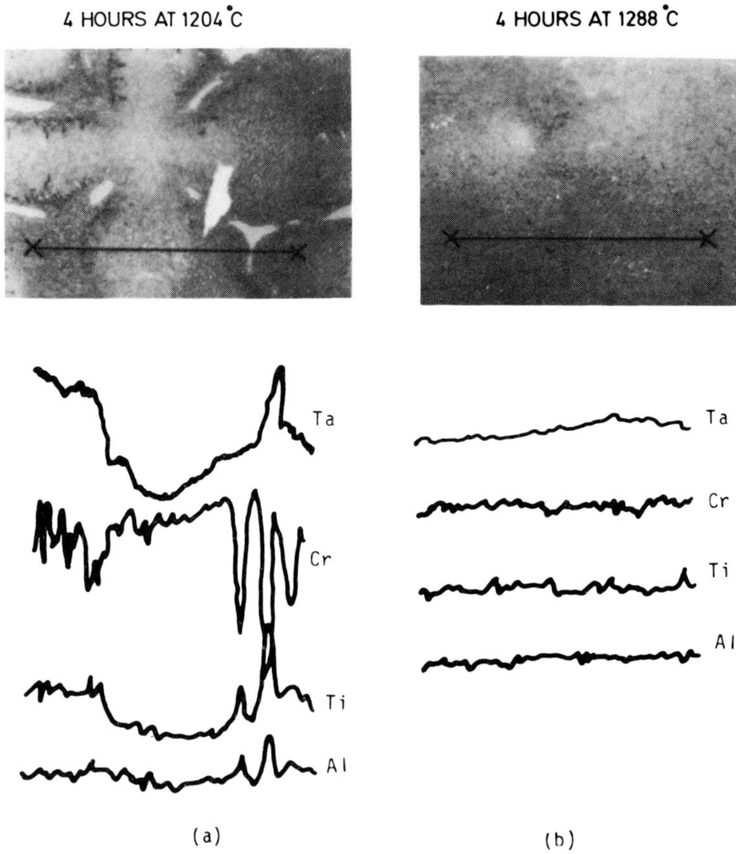


**Figure 7.5 Comparison of dendritic segregation in Pratt and Whitney single crystal alloy 454 after heat treatments of 4 h at (a) 1204°C and (b) 1288°C [Gell *et al.*[2a]]**

to the transition equilibrium temperatures. Consequently, changes in phase chemistry will only be of practical significance if they are thermodynamically possible and if they can occur at a sufficiently fast rate to influence the life of a component.

Considerable concern is expressed over the deposition of topologically-close-packed (tcp) phases which are reputed to cause severe reductions in the creep ductility and rupture lives of some superalloys.[3] The high chromium superalloys that are particularly vulnerable to this type of instability are precisely the materials that are intended for use over very long periods in industrial gas turbines. It is therefore necessary to be able to predict the vulnerability of such materials to the formation of potentially detrimental phases. In the absence of reliable equilibrium diagrams for multicomponent alloys, empirical methods have evolved for predicting the occurrence of tcp phases, of which the PHACOMP scheme has been particularly useful. This approach derives from a suggestion by Rideout *et al.*[4] that the average electron vacancy number $n_v$ of a solid solution alloy of transition metals should provide a guide to its stability, particularly with regard to $\sigma$-phase. If $n_v$ exceeds an experimentally determined critical value, the alloy would be prone to the formation of $\sigma$. These concepts have been developed over the years by various authors to deal with multiphase materials such as superalloys (see, for example, Stickler[5] or Restall and Toulson[6]). The first stage is to determine the composition of the solid solution matrix by an auditing procedure; i.e. by estimating the volume fractions and compositions of the various precipitate phases and deducting these from the total alloy composition. The mean value of $\bar{n}_v$ for the remainder is then calculated and compared with an experimentally determined critical value. The various stages in this procedure are shown as a flow diagram in Fig. 7.6. The loci of constant $n_v$ numbers in the Ni–Co–Cr ternary system are shown in Fig. 7.7, together with the compositions and results of

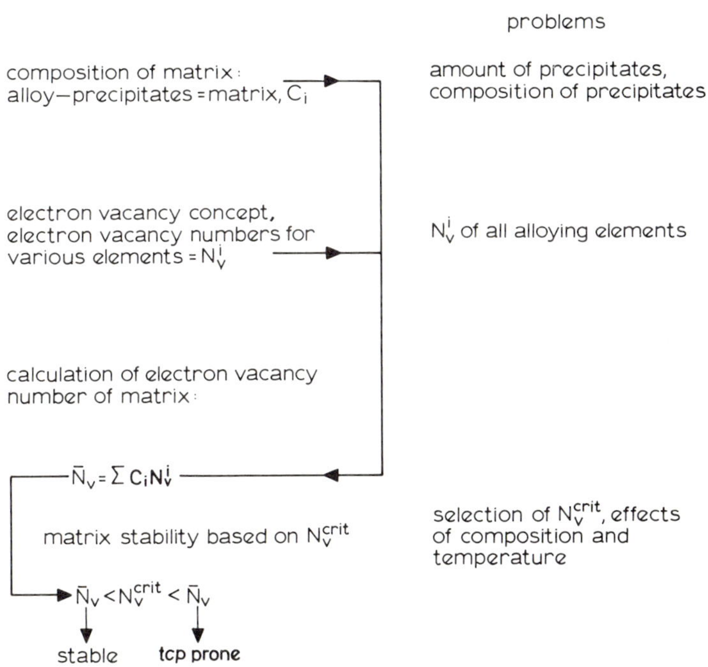


**Figure 7.6 Flow chart showing the various stages in a typical PHACOMP calculation [Stickler[5]]**

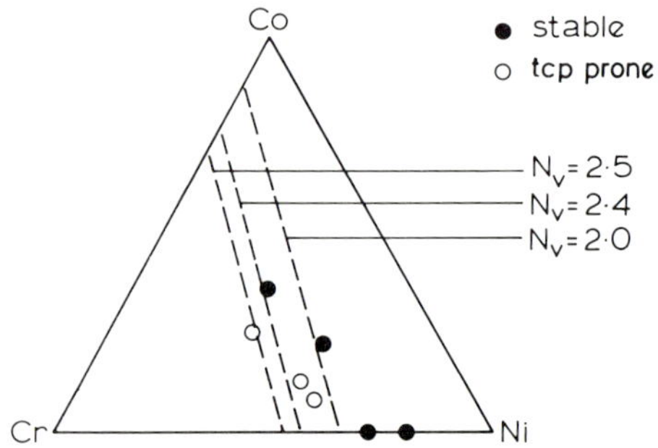


**Figure 7.7 Ternary composition diagram showing the stability of commercial alloys in relation to computed electron vacancy numbers [Stickler[5]]**

phase stability assessments on various commercial alloys. From this type of study the threshold of $\sigma$ stability is estimated to occur for $\bar{n}_v \geqslant 2{\cdot}30$ although opinion varies over the exact value. No modern commercial superalloys are tcp-unstable since such calculations figure prominently in their specification. However, the requirement for corrosion resistant alloys with high chromium contents giving high values of $\bar{n}_v$ inevitably means that alloy compositions close to the stability boundary must be used; consequently, small deviations from the specified compositions may lead to tcp instabilities.

Changes in the composition and stoichiometry of precipitated phases, particularly of carbides, are frequently observed in both superalloys and *in situ* composites, and these can be accompanied by reductions in the strength of the alloys. For example, the distribution of various carbides in the directionally solidified superalloy IN738LC depends on the processing, heat treatment and service conditions. When the alloy is directionally solidified in a high temperature gradient the only carbide apparent is of the blocky MC type (shown as the dark shade in Fig. 7.8*a*). However, $M_{23}C_6$, imaged as an intermediate shade, occurs both along the grain boundaries and at the surfaces of MC particles in material that has either been directionally solidified in a low temperature gradient (Fig. 7.8*b*) or subjected to heat treatments (Fig. 7.8*c*). This indicates a reversal from MC being stable at high temperatures (indeed it is thought to resist remelting in many alloys) to $M_{23}C_6$ being preferred at lower temperatures and its formation being associated with depletion of the $\gamma$ matrix of the carbide forming elements (e.g. W, Ti, Ta, Nb) that may contribute to solid solution strengthening, i.e.

$$6MC + 17M \rightarrow M_{23}C_6$$

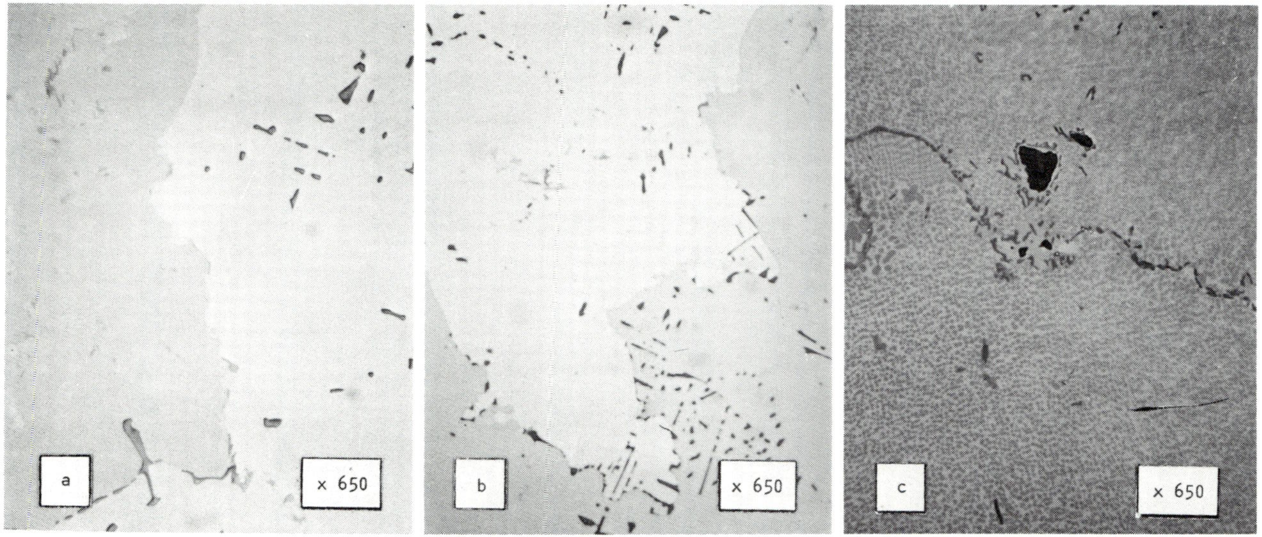


**Figure 7.8** Micrographs of directionally solidified IN738LC showing distribution of carbides to be dependent on processing conditions (colour contrast by interference film microscopy, here reproduced in monochrome; grey levels: MC—dark, $M_{23}C_6$—intermediate)
(a) $G = 13\ \text{K mm}^{-1}$, $V_I = 60\ \text{mm h}^{-1}$—no heat treatment
(b) $G = 3\ \text{K mm}^{-1}$, $V_I = 200\ \text{mm h}^{-1}$—no heat treatment
(c) $G = 13\ \text{K mm}^{-1}$, $V_I = 60\ \text{mm h}^{-1}$—heat treated

Similar effects have been observed in certain *in situ* composites that depend on carbide reinforcement. Thus (Co,Cr)–$Cr_7C_3$ and (Fe,Cr)–$Cr_7C_3$ alloys are intrinsically unstable with respect to the $M_7C_3 \rightarrow M_{23}C_6$ transition below about 900°C.[7] In contrast the MC reinforced alloys, such as NITAC and COTAC, exhibit no such changes. Figure 7.9 demonstrates the existence of two types of carbide in the (Co,Cr)–$Cr_7C_3$ *in situ* composite that has been aged at 825°C, showing both changes in the structure of the fibre and deposition of equiaxed carbides in the matrix. As a consequence of this type of instability, alloys such as (Co,Cr)–$Cr_7C_3$ often do not exhibit the intrinsic strengthening associated with the microstructural refinement of composites that was discussed in Chapter 6. Rather, the loss of strength due to degradation of the reinforcement can dominate, making coarser microstructures preferred; however, the *in situ* composites showing this behaviour are relatively rare. It is worth observing in passing that chemical and thermodynamic incompatibilities of fibres and matrices create particularly severe problems for many synthetic composites.

## 7.3 INSTABILITIES DUE TO CAPILLARITY

The free energy associated with the many types of interface that occur particularly in multiphase alloys such as superalloys and composite materials can provide a driving force for several types of morphological instability. The most important, which are a consequence of the progressive reduction in the total interfacial free energy of the system, are usually, but not always, manifested as reductions in the total interfacial area of the various interfaces

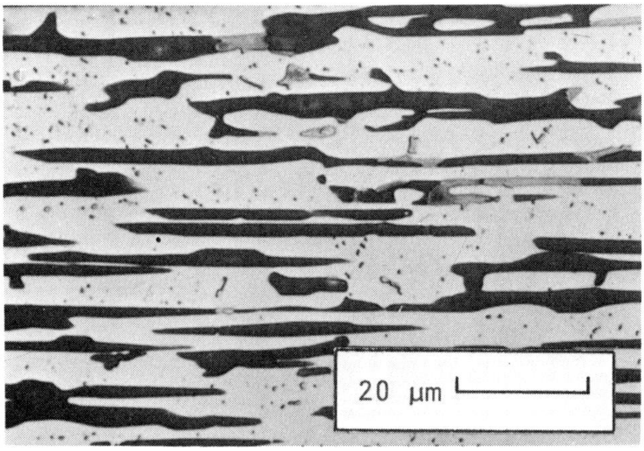


**Figure 7.9 Development of $M_{23}C_6$ on the surface and between $M_7C_3$ fibres during heat treatment of the (Co,Cr)–$Cr_7C_3$ *in situ* composite**

(e.g. grain boundaries, precipitate boundaries etc.) Exceptionally increases in boundary area may occur where there are substantial anisotropies in interfacial energy and surfaces of low energy are being produced. Such effects can lead to changes in size, shape or position of precipitated phases by diffusional processes; these possibilities will be discussed separately later, but first a general description of capillarity motivated instabilities that is sufficient for many purposes will be considered.

### 7.3.1 Herring scaling laws

The driving force **F** for capillary instabilities derives from the relevant interfacial energy; the rate of development of the instability depends on which kinetic process is operable. All shape changes of this type can be related to the transfer of material from one boundary region to another. Consider the boundary between two phases $\alpha$ and $\beta$, shown schematically in Fig. 7.10*a*,

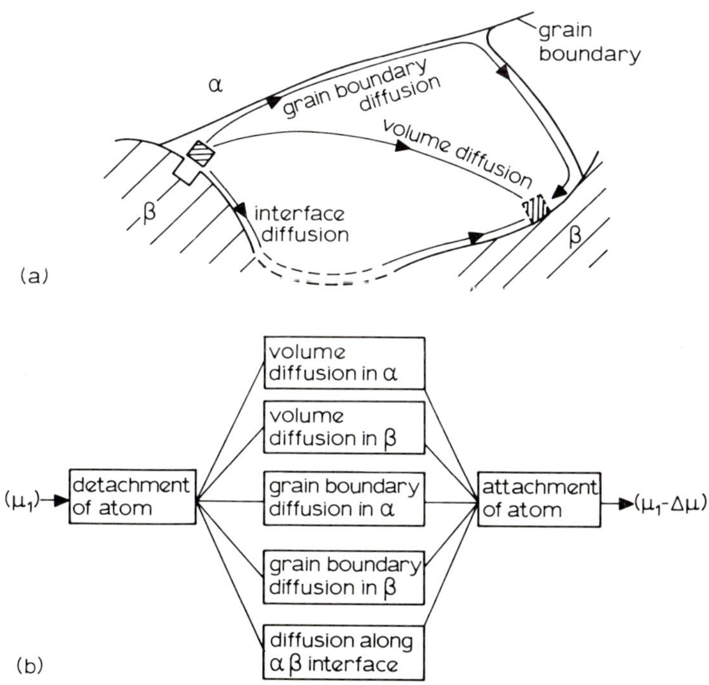


**Figure 7.10 (a) Schematic illustration of changes in phase morphology due to various mass transport mechanisms and (b) indication of various stages leading to reduced chemical potential by the mass transport process**

which may or may not be continuous. Matter can flow from one region to another first by detachment of an atom and then by transport by a variety of parallel mechanisms; the mechanism leading to the maximum amount of mass transport dominates. The possible flow mechanisms indicated schematically in Fig. 7.10*b* are:

(a) diffusion through one or both phases
(b) diffusion along the $\alpha\beta$ phase boundary (which is only significant if the phase boundary is continuous)
(c) diffusion along grain boundaries in either phase
(d) plastic flow in either phase.

The driving force for this mass transport results from the curvature of the $\alpha\beta$ boundary; the chemical potential of a solute atom on the $\alpha\beta$ interface is related to the curvature of the boundary $K$ by the Gibbs–Thompson equation:

$$\mu = \mu_0 + \gamma \Omega K \tag{7.10}$$

where $\Omega$ is the atomic volume and $\gamma$ is the interfacial energy.

Application of eqn (7.3) shows that the diffusional flux leading to mass transport depends on differences in curvature of different regions of the interface.

$$\mathbf{j} = -BN\gamma\Omega \frac{\mathrm{d}K}{\mathrm{d}s} \tag{7.11}$$

where $s$ is a distance along the boundary.

Herring,[8] by a method of dimensional analysis, has established a series of scaling laws that successfully describe the *forms* of the equations for the evolution of all capillary instabilities. He has shown that a characteristic linear dimension $d$ describing the scale of a microstructural feature is given by an expression of the form

$$d^{\mathrm{n}} = d_0^{\mathrm{n}} + \alpha A_{\mathrm{i}} t \tag{7.12}$$

where $\alpha$ is a dimensionless geometrical constant of proportionality, $A_{\mathrm{i}}$ is a parameter characterizing the kinetics of the appropriate mass transport mode, and the exponent $n$ takes the following integral values depending on the mechanism controlling the kinetics of the process:

$n = 1$ for viscous or plastic flow in either phase
$n = 2$ for interface reaction control (i.e. when the diffusional step is more rapid than the detachment/attachment of the mobile atom at the boundary)
$n = 3$ for volume diffusion through one or both phases
$n = 4$ for two-dimensional diffusion either along the $\alpha\beta$ phase boundary or on grain boundaries within either phase
$n = 5$ for single dimensional diffusion along dislocations.

Expressions for $A_i$ are tabulated in Table 7.1 for different mechanisms; more detailed analyses of different types of instability are available that derive explicit expressions for $\alpha$ for various specific geometries and which often account for the statistical evolution of precipitate distributions. Some of these are described below.

### 7.3.2 Spheroidization

The shape that minimizes the surface area of a fixed volume is a sphere and, when the interfacial energy is isotropic, it follows that there will be a driving force for the relaxation in shape of non-spherical inclusions to their equilibrium shape configuration. The rod and lamellar precipitate morphologies that are critical to the performance of *in situ* composites are particularly potentially vulnerable to microstructural degradation since they are far from the low energy ultimate configurations.

Most treatments of spheroidization derive from the work of Lord Rayleigh[9] who showed that an infinitely long cylinder of fluid of radius $a$ was unstable with respect to periodic shape perturbations, $\delta \cos(2\pi z/\lambda)$, for a certain wavelength of the perturbation, $\lambda$. The configuration is shown schematically in Fig. 7.11. The total surface area of the distorted cylinder $E$ relative to that of the perfect cylinder $E_0$ is

$$E - E_0 = \frac{\pi\delta^2}{2a}\left(\frac{4\pi^2 a^2}{\lambda^2} - 1\right) \tag{7.13}$$

**Table 7.1 Mass transport kinetic constants for different mechanisms**

| Mechanism | |
|---|---|
| Volume diffusion | $A_v = \dfrac{C_0 D_v \gamma \Omega^2}{kT}$ |
| Interface diffusion | $A_I = \dfrac{\nu \gamma D_I \Omega^2}{kT}$ |
| Interface reaction control | $A_{IC} = \dfrac{\beta C_0 \gamma \Omega^2}{kT}$ |

$C_0$ = solute volume concentration; $D_v$, $D_I$ are solute volume and interface diffusivities; $\gamma$ = interfacial energy; $\nu$ and $\Omega$ are the area and volume of diffusing atoms; $k$ = Boltzmann's constant; $T$ = temperature in K; $\beta$ = constant

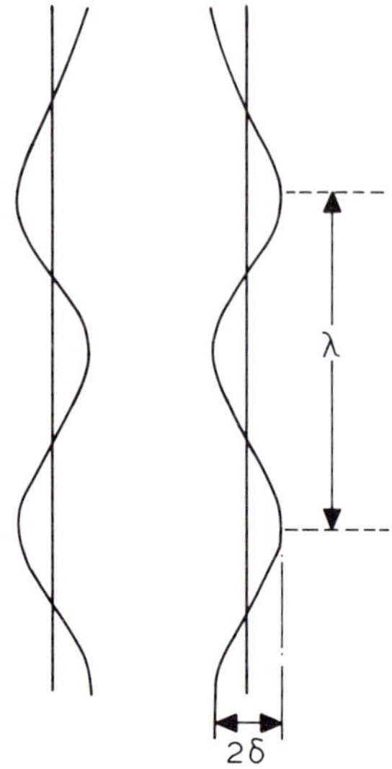


**Figure 7.11 Profile of an idealized shape perturbation on an initially cylindrical shape**

Thus for wavelengths $\lambda < 2\pi a$ the perturbation increases the energy of the configuration and is therefore unstable and decays. However, if $\lambda > 2\pi a$ then there is a decrease in surface area (and therefore of surface energy) and the perturbation is stable and will grow. This will lead eventually to the cylinder breaking up into a string of discrete spheres of the same total volume as the original cylinder. The size of spheres formed depends on the kinetics of the operative diffusional mechanism which determines the rate at which perturbations of a given wavelength grow.

The original Rayleigh analysis referred to the stability of a water jet; the theory has been extended independently by Nichols and Mullins[10] and by Cline[11] to the solid state where the different mass transport mechanisms described in § 7.3.1 can operate. Inspection of eqn (7.13) shows that the driving force for the growth of an instability increases monotonically with $\lambda$; however, increasing $\lambda$ increases the diffusion distances necessary to sustain the growth of the perturbation so that longer wavelengths tend to grow more slowly. Taken together, these two effects lead to maximum growth rates at particular wavelengths which are characteristic of the operative mass transport mechanisms. The results of Nichols and Mullins[10] calculations are shown in Table 7.2.

The complete spheroidization of infinite fibres by solid state diffusional processes to the point of rupture of the rods has been modelled and calculations made by finite difference techniques by Marinis and Sekerka[12,13] for the case

**Table 7.2 Fastest growing wavelengths $\lambda_M$ of perturbations to a cylinder of radius $a$ for different mass transport mechanisms**

| | $\lambda_M/a$ |
|---|---|
| Interface diffusion | 8·89 |
| Volume diffusion (inside cylinder) | 9·02 |
| Volume diffusion (outside cylinder) | 12·96 |

of interface diffusion. Their results show a slight difference in fastest growing wavelength ($\lambda_M/a = 9{\cdot}43$ rather than 8·89) from that derived by Nichols and Mullins. More importantly the later calculations indicate the pinching-off process accelerates faster than suggested by the earlier linear approximation. Figure 7.12 shows the calculated evolutions of such perturbations of initial amplitude $0{\cdot}1a$ and wavelengths $7{\cdot}85a$ and $2{\cdot}22a$. The first perturbation is unstable and grows while the second decays. The curves are in a dimensionless form where lengths are normalized to the fibre radius and the time parameter $\tau$ is given by the expression

$$\tau = \frac{D_s \gamma \nu \Omega^2 t}{kTa^4} \tag{7.14}$$

where $\nu$ = surface density of diffusing atoms, $\gamma$ is the interfacial energy, $D_s$ is the interfacial diffusivity, $\Omega$ is the atomic volume, $kT$ is the thermal energy and $t$ is time. This is quite compatible with the Herring Scaling Equation (eqn (7.12)); the pinch-off time for a cylinder of radius $a$ is

$$a^4 = \frac{D_s \gamma \nu \Omega^2}{5{\cdot}607 kT} t \tag{7.15}$$

There are mechanisms other than the Rayleigh instability by which cylindrical inclusions can degrade and these are shown schematically in Fig. 7.13. Apart from the Rayleigh-type instability described above there are two more rapid types of spheroidization.

(a) Rapid spheroidization of rods of finite length can occur by the detachment of droplets from the fibre ends. Similar calculations have been carried out as described above for infinite fibres. When the aspect ratio of the fibres (length/diameter) is less than 7·2 shape relaxation to a single sphere is predicted; otherwise spheroidization to a string of spheres occurs, although with a smaller $\lambda_M/a$ ratio (8·5 rather than 8·89 for interface diffusion[14]). Of more importance is the fact that the rate of pinch-off is considerably

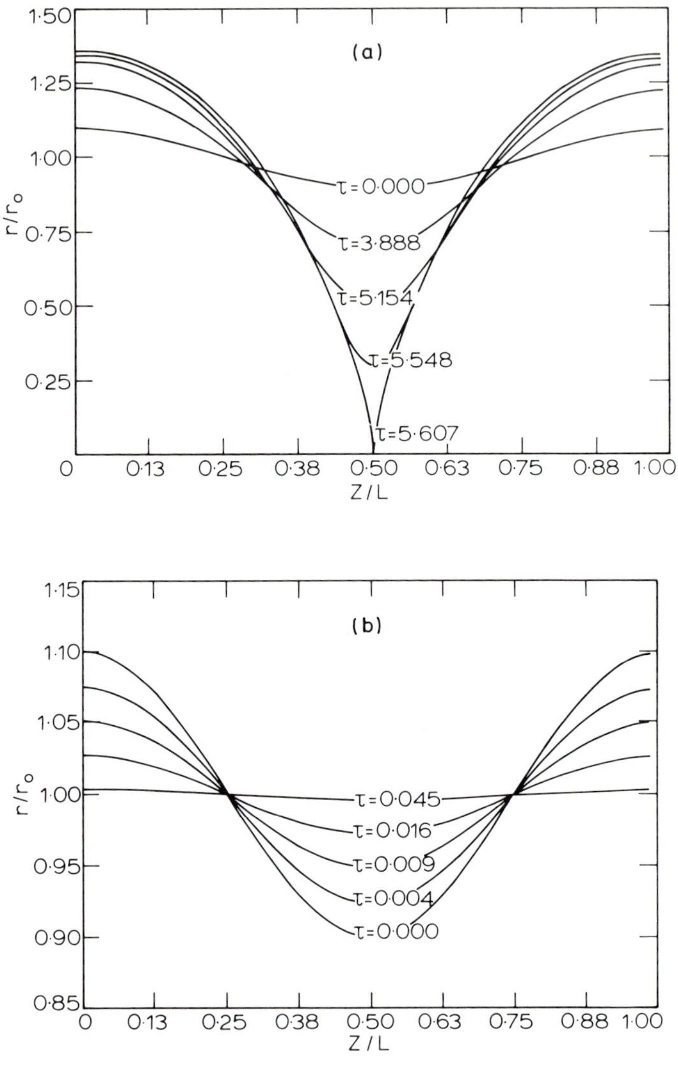


**Figure 7.12** **Surface profiles of cylinder segments plotted as the instantaneous local radius relative to the initial cylinder radius for different values of a dimensionless time parameter $\tau$ after introduction of shape perturbations of different wavelengths $\lambda$ and amplitude $0{\cdot}1a$: (a) $\lambda = 7{\cdot}85a$, (b) $\lambda = 2{\cdot}22a$ [Marinis and Sekerka[12]]**

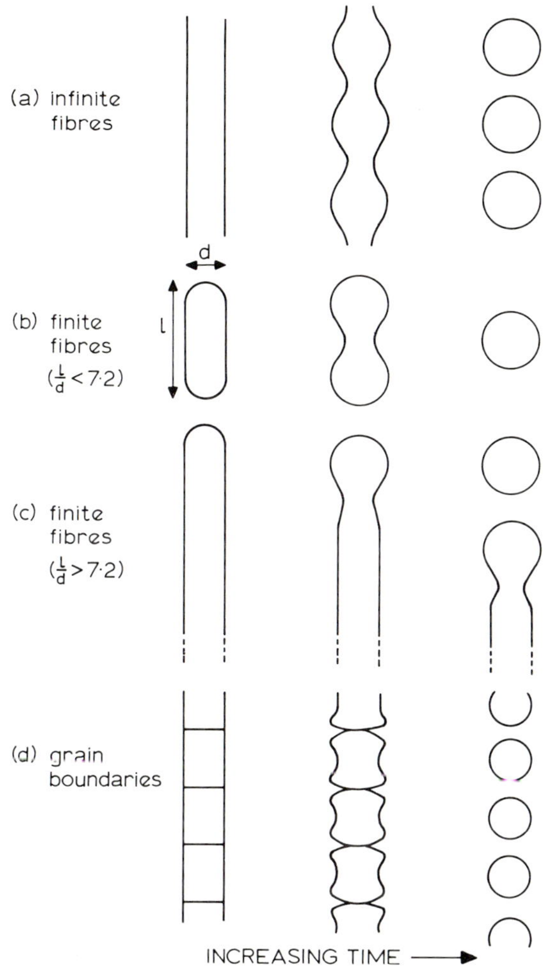


**Figure 7.13 Schematic diagram showing different modes of spheroidization of cylindrical inclusions**

more rapid from such free fibre ends, being described by eqn (7.15) with a different numerical constant.

(b) If the cylinder contains a grain boundary or is intersected by a matrix grain boundary then a groove can develop at the junction of the phase and grain boundaries that will grow and eventually penetrate the cylinder. Mullins[15] has shown that the depth $d$ of such a grain boundary groove is described by Herring-type equations which are listed in Table 7.3. The eventual spheroidized microstructure in this case is determined by the density of the grain boundaries rather than by the geometry of the cylinder.

**Table 7.3 Growth of groove dimensions by different mass transport mechanisms**

| | Groove depth $d$ | Groove width $w$ |
|---|---|---|
| Interface control | $1{\cdot}13m(At)^{1/2}$ | — |
| Interface diffusion | $0{\cdot}974m(Bt)^{1/4}$ | $4{\cdot}6(Bt)^{1/4}$ |
| Volume diffusion | $1{\cdot}01m(Ct)^{1/3}$ | $5.0(Ct)^{1/3}$ |

By contrast with the behaviour of rods, infinite plates and lamellae are inherently stable with respect to shape perturbations. This means that perturbations of *all* wavelengths, when applied to plates, decay and no spheroidization should occur. The break-up of planar phases must therefore be associated with plate edges and defects. Firstly, the grain boundary grooving mechanism is generally valid to all phase shapes. Nichols and Mullins[10] observed that the retraction of plate edges would lead to a thickened, toroidal rim which would be susceptible to the types of spheroidization described above for rods.

A central assumption of all of the theories of spheroidization outlined above is that the interfacial energies associated with the phase boundaries are independent of crystallographic orientation. Although there are few reliable measurements of these interfacial energies there is now substantial evidence that there can be large anisotropies. This will tend to stabilize fibres or lamellae that have low energy facets; any perturbation would require to create surfaces with higher interfacial energies. Nevertheless, the mechanisms we have described have all been observed to varying degrees.

Most *in situ* composites are remarkably resistant to spheroidization at realistic service temperatures although they tend to coarsen (this will be dealt with below). At very high homologous temperatures where entropy effects tend to reduce the anisotropy of interfacial energy and the diffusivity is much higher, there is ample evidence of the occurrence of Rayleigh instabilities and of the pinching-off of globules from free ends. Thus Marinis and Sekerka[13] have complemented their theoretical work by observations on an antimony–manganese model *in-situ* composite (Sb–MnSb). Figure 7.14 shows sequences from hot stage electron microscope experiments on two alloys. In Fig. 7.14*a* the liquid lead fibres entrained in solid aluminium spheroidize in the expected manner and detailed analysis shows that mass transport is by volume diffusion through the liquid. Similar microstructures are apparent for the solid $Al_3Ni$ fibre in Al shown in Figure 7.14*b*; however, in this case mass transport is by interface diffusion. It is noteworthy that the highly alloyed composites proposed for high temperature application do not appear to exhibit these effects even at the highest temperatures.

There are also many indirect observations that indicate that spheroidization can occur immediately after solidification as the material cools. Clearly the extent of mass transport during this intrinsic cooling stage depends on the cooling rate, which is governed by the temperature gradient $G_S$ and the solidification rate $V_I$; this can account for different solidification morphologies (but, of course, not all the differences). Thus Marich[16] and Marich and Jaffrey[17] have used quenching experiments to show that the transition from globular to rod morphologies as the growth rate of Fe–FeS and Cu–CuS eutectics are increased are due to spheroidization. It is likely that similar transitions in monotectic transformations are due to the same phenomenon.[18,19]

Perhaps of more worry in relation to the performance of high temperature composites intended to be used in gas turbines are the potential dangers of recrystallization of the matrix or reinforcement and the subsequent damage caused by the grain boundary grooving mechanism. Weatherly and co-workers[20,21] have shown that this can indeed be a potent spheroidization mechanism on a series of low temperature model alloys; Fig. 7.15 shows the development of such grain boundary grooving in the Al–$Al_2Cu$ lamellar composite. Careful analysis has shown the kinetics of groove growth to be well described by the Mullins theory and to indicate that mass transport occurs by an interface diffusion mechanism; indeed, this provides a powerful method of measuring interfacial diffusivities which is difficult by other methods.

### 7.3.3 Phase coarsening

Whatever the shapes of the particles, the total surface area associated with a finely divided precipitate is greater than the same volume distributed as larger particles. Thus there is a tendency for the microstructure to coarsen by diffusional processes in order to reduce the total surface area (i.e. surface energy). We have seen in Chapters 5 and 6 that the dimensions both of $\gamma'$ in superalloys and of the reinforcing phases in *in situ* composites have an important effect on the mechanical behaviours of these materials so that such phase coarsening may lead to substantial weakening. The spheroidization and phase coarsening do not operate independently; rather, a given system will degrade by a combination of all possible mechanisms in order to achieve the maximum reduction in free energy. The balance between the two will depend on the relative kinetics of the two processes, which may occur by quite different diffusional mechanisms. Thus spheroidization normally involves interface diffusion whereas phase coarsening almost invariably requires transport of matter between unconnected particles by volume diffusion. Although the phenomena compete, it is convenient to consider them separately.

The isotropic growth of a dispersed phase of constant volume entrained in a matrix which has long been identified in fluid systems is known as Ostwald ripening. The process has been analysed independently by both Wagner[22]

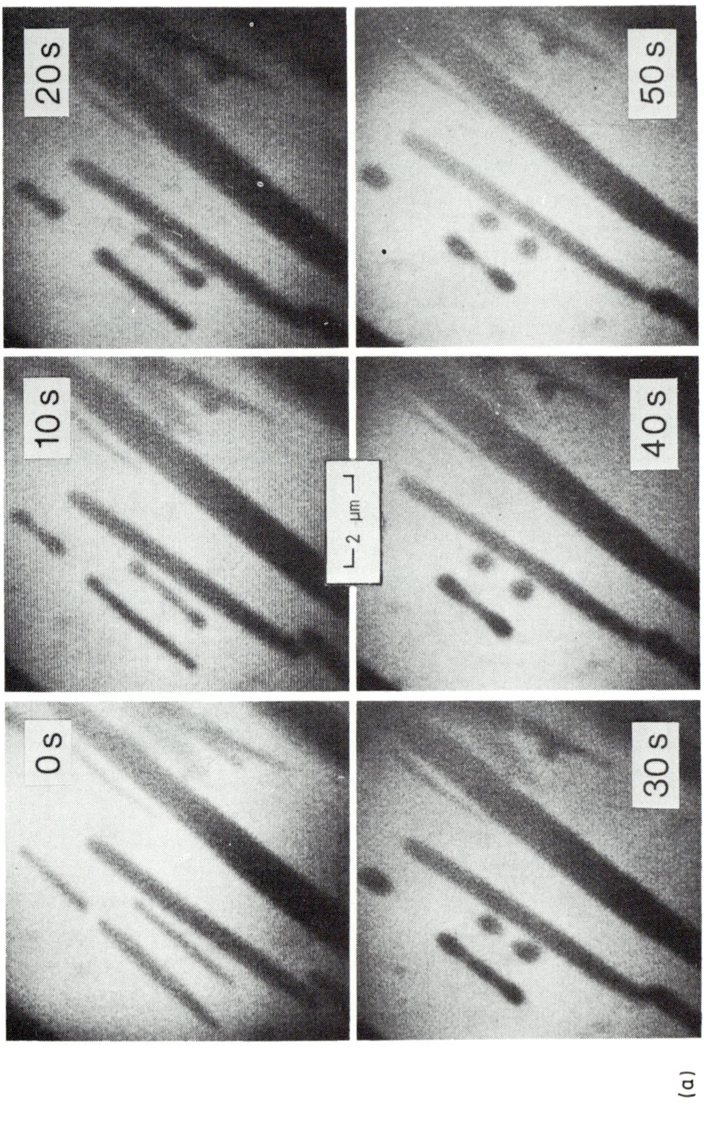


(a)

**Figure 7.14** Sequences of electron micrographs for *in situ* hot stage experiments showing progressive spheroidization in (a) Al–Pb alloys at 600°C—the Pb fibres are liquid . . .

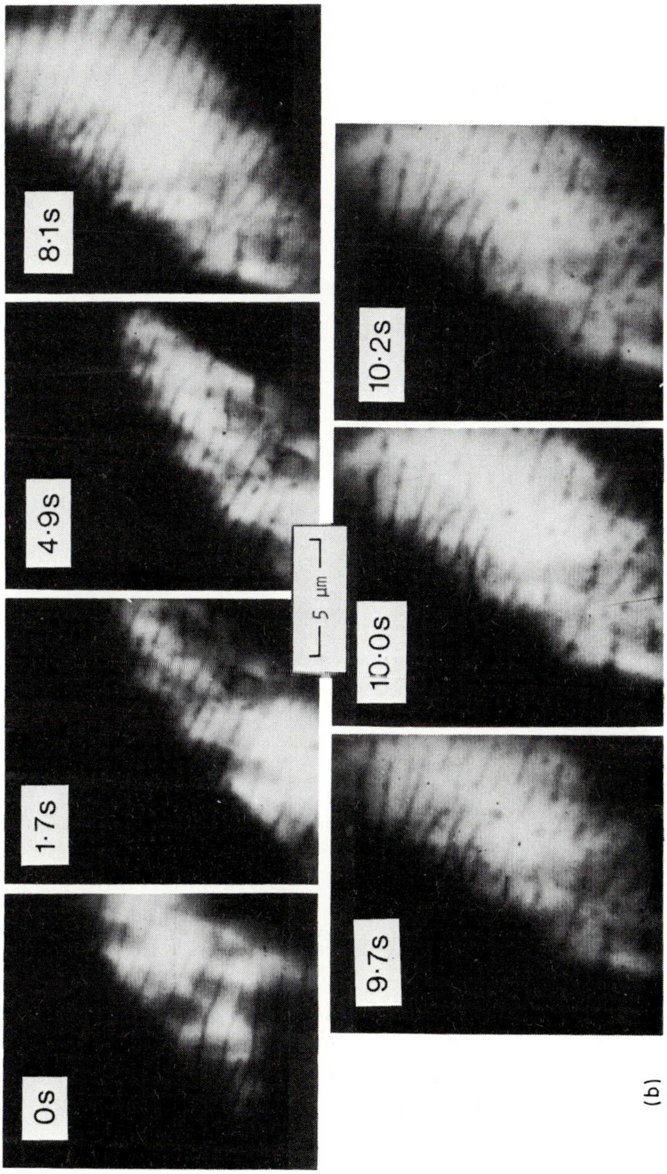


(b)

**Figure 7.14 . . . (b) Al–$Al_3Ni$ eutectic at 600°C [McLean and Loveday,[38] McLean[32]]**

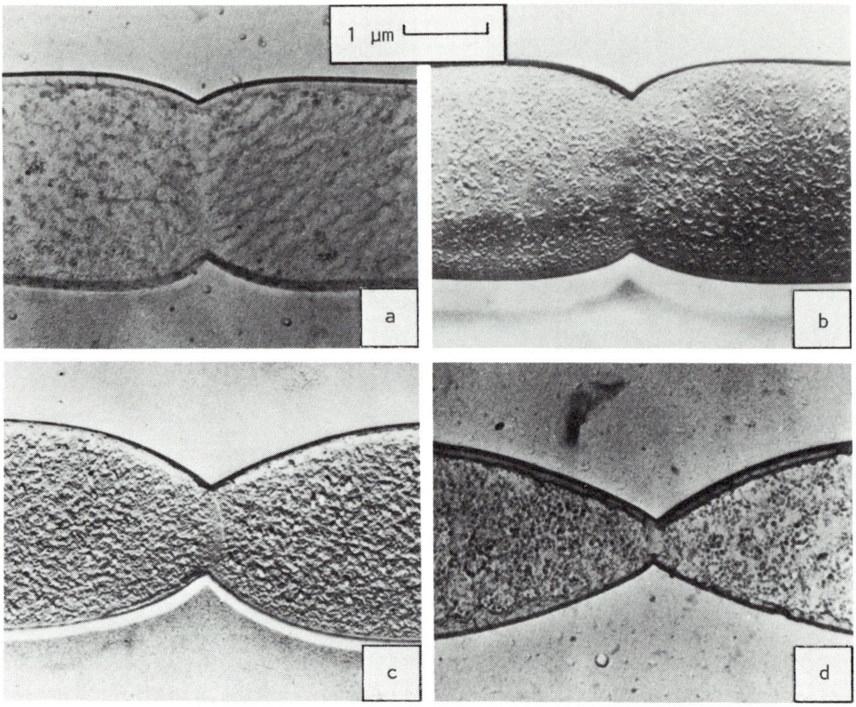


**Figure 7.15 Grain boundary grooves caused by heat treatment of deformed Al–$Al_2Cu$ [Ho and Weatherly[20]]**

and Lifshitz and Slyozov[23] for the solid state with virtually the same result. Both treatments assume spherical particles but Greenwood[24] has shown that the basic assumptions remain valid for faceted particles (assumed to retain a constant shape) where the radius of the spheres is replaced by a characteristic dimension of the particles. The models assume that the solute concentration $C_r$ adjacent to a particle of radius $r$ is given by the Gibbs–Thompson equation:

$$\frac{C_r}{C} = \exp\left(\frac{2\gamma\Omega}{kTr}\right) \simeq \left(1 + \frac{2\gamma\Omega}{kTr}\right) \tag{7.16}$$

where $C$ is the solute concentration at a particle of infinite radius. By calculating the solute flux through the matrix and requiring the particle to shrink or grow at the rate needed to supply the solute (assuming constant volume fraction) the following differential equation is obtained:

$$\frac{\mathrm{d}r}{\mathrm{d}t} = -\frac{2C\gamma\Omega^2}{kTr}\left(\frac{1}{\bar{r}} - \frac{1}{r}\right) \tag{7.17}$$

where $\bar{r}$ is the average particle radius. A schematic plot of $dr/dt$ versus $r$ (Fig. 7.16*a*) shows that particles with $r = \bar{r}$ are neither shrinking nor growing but that the equilibrium is unstable. Particles with $r < \bar{r}$ shrink at a rate that increases with decreasing size, while there is a maximum growth rate for particles with $r = 2\bar{r}$. This gives a particle size distribution that has a sharp cut-off at $r \sim 2\bar{r}$.

The solution to eqn (7.17) with appropriate boundary conditions gives the expression

$$\bar{r}^3 = \bar{r}_0^3 + \alpha \frac{C\Omega^2 D_v \gamma}{kT} t \tag{7.18}$$

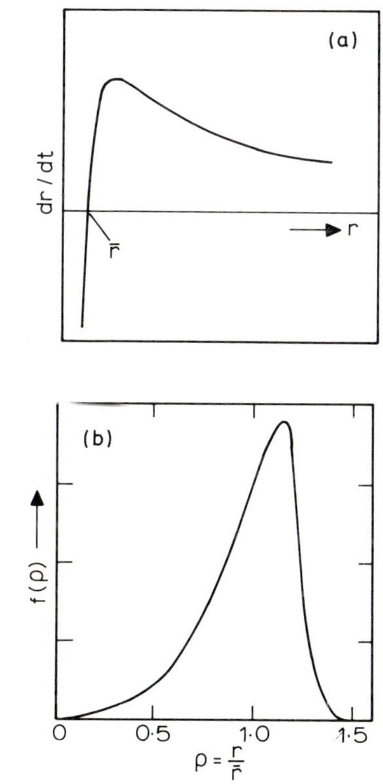


**Figure 7.16** **(a) Variation in growth rate $dr/dt$ with particle radius $r$ during the isotropic coarsening of an array of particles of mean radius $\bar{r}$**
**(b) Steady state particle size distribution $f(\rho)$ predicted by the theory of isotropic coarsening**

for the case of volume diffusion where $\bar{r}_0$ is the average particle size at $t=0$. The dimensionless constant $\alpha$ depends on the shape of the inclusions ($\alpha = 8/9$ for spheres, $=0{\cdot}47$ for rods). The theories also predict that a steady state distribution of sizes $f(r/\bar{r})$ is approached irrespective of the initial size distribution. The distribution associated with volume diffusion is shown in Fig. 7.16*b*.

Expressions analogous to eqn (7.18) have been developed for the cases of interface reaction control kinetics[22] and for coarsening by grain boundary diffusion.[25,26]

*interface control* $$\bar{r}^2 = \bar{r}_0^2 + \frac{64}{81}\frac{AC\gamma\Omega^2}{kT}t \tag{7.19}$$

*grain boundary diffusion* $$\bar{r}^4 = \bar{r}_0^4 + \frac{4}{3}\frac{CD_{gb}\gamma\Omega^2\delta}{kT}t \tag{7.20}$$

where $\delta$ is the interface thickness. We note that eqns (7.18–20) are quite compatible with the Herring scaling laws.

One critical assumption of the Lifshitz–Slyozov–Wagner theory that the diffusion fields around each particle are spherically symmetrical is only strictly valid in the limit of zero precipitate volume fraction $V_f$. As $V_f$ increases the diffusion fields will interfere and introduce errors into the simple theory. Ardell[27] has introduced a modification that accounts for finite volume fractions and retains the basic form of eqn (7.18):

$$\bar{r}^3 = \bar{r}_0^3 + f(V_f)Bt \tag{7.21}$$

where $B = (C\Omega^2 D_v\gamma)/kT$. The correction factor $f(V_f)$ and the changing particle size distributions predicted are shown in Fig. 7.17*a* and *b*. The effects of volume fraction are of particular importance for both superalloys, where $\gamma'$ volume fractions up to ~60% may be present, and *in situ* composites where up to 30% by volume of fibres occurs in some alloys.

There have been many studies of phase coarsening in a variety of alloys that are consistent with the basic form of eqn (7.18). Two typical sets of data relevant to the materials discussed in this book are shown in Fig. 7.18. Thus Fig. 7.18*a* shows the growth in mean size of $\gamma'$ particles in the cast superalloy IN738LC while the coarsening of fibres in the COTAC 74 *in situ* composite is shown in Fig. 7.18*b*.[28] In both cases analyses of the temperature dependences of such plots indicate activation energies for coarsening that are compatible with those associated with diffusion. However, there is less confirmation of the predicted steady state distribution of particle sizes. Data for IN738LC directionally solidified in a high temperature gradient at a low rate in order to produce a dendrite-free homogeneous microstructure are shown in Fig. 7.19; the single population of cuboidal $\gamma'$ particles has a much more symmetrical distribution than the theoretical one and exhibits no sharp cut-off. This is thought to be due to the coalescence of large particles.

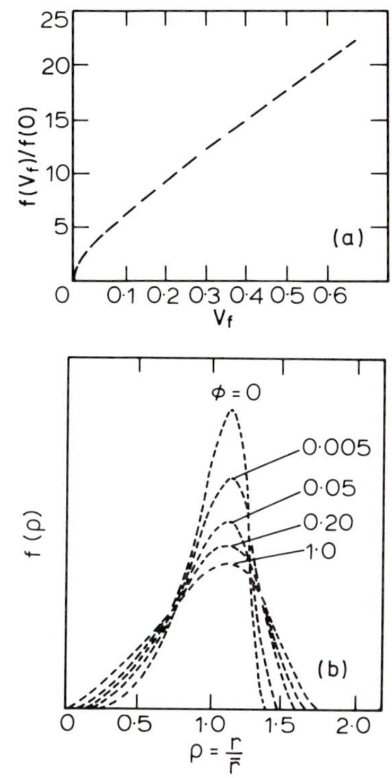


**Figure 7.17 (a) Correction factor to theory of isotropic coarsening and (b) particle size distributions predicted by the theory of isotropic coarsening for systems containing various volume fractions of dispersed phase [Ardell[27]]**

In practice the initial distribution of $\gamma'$ in cast nickel-base superalloys is very complex. In addition to the compositional variations across the dendrites that give modulations in $\gamma'$ characteristics (i.e. volume fraction, morphology, solution temperature, composition) with periods of 100 μm or so, locally there are often two (or more) distinct populations of particles, on quite different scales, which are produced by complex heat treatments in order to optimize certain properties. For example, the standard commercial heat treatment for IN738LC (i.e. 2 h at 1393 K + 16 h at 1118 K) gives two distinct shapes of $\gamma'$ particles, namely 20% by volume of cuboids of average diameter 880 nm and 25% by volume of spheroids with mean diameters of 89 nm. Both populations coarsen independently in accord with eqn (7.18) but at the same time there is a change in the relative amounts of each while maintaining a constant

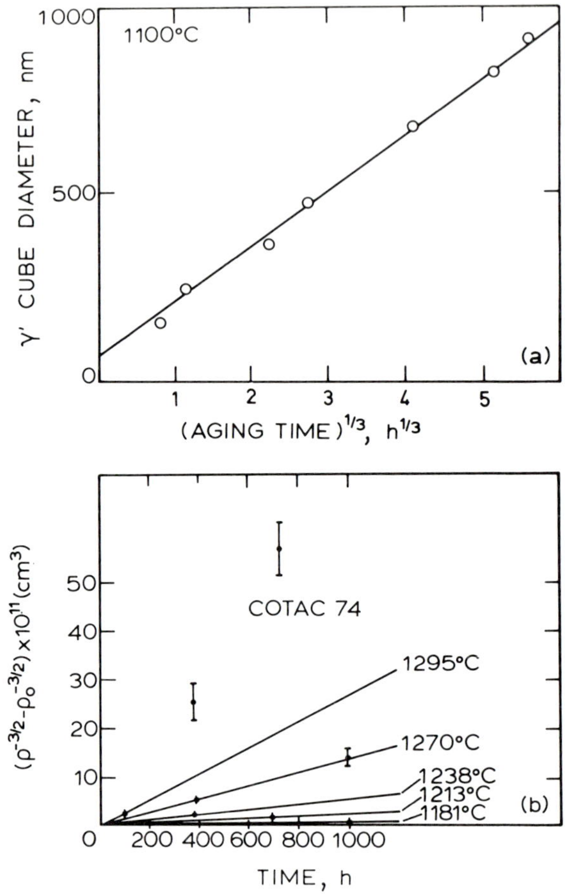


**Figure 7.18 Growth of particle and fibre sizes during coarsening heat treatments: (a) $\gamma'$ cuboids in IN738LC at 1100°C; (b) NbC fibres in COTAC 74 [Stohr[28]]**

volume fraction; the smaller spheroids gradually dissolve while the cuboids grow. It is noteworthy that no significant variations in the kinetic constants occur as the relative volume fractions vary although eqn (7.21) would predict quite large effects. Indeed, experimental data supporting a variation in coarsening rate with precipitate volume fraction have been reported on only one alloy.[29] Further work on this topic is required.

The aspects of isotropic coarsening described above largely involve volume diffusion through a matrix. However, more heterogeneous particle growth can occur due to a local rapid diffusion along a two dimensional

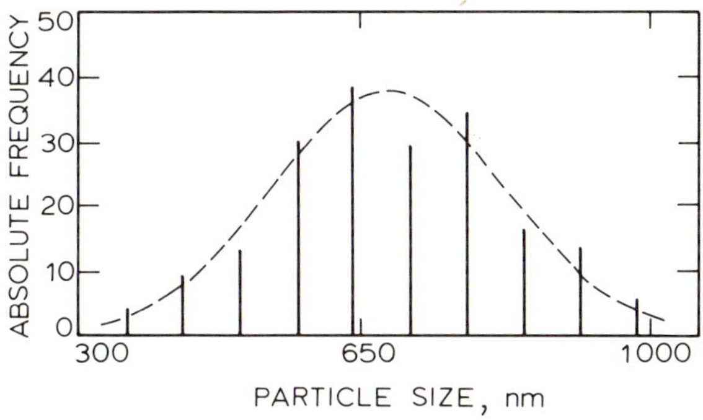


**Figure 7.19 Distribution of $\gamma'$ cuboids after heat treatment of IN738LC at 1100°C for 68 h**

interface such as the particle/matrix boundary or a matrix grain boundary. At its simplest, this may only be the accelerated growth of particles or fibres situated at static grain boundaries, as shown in Fig. 7.20, without radically affecting their shapes or distributions. In other circumstances quite different microstructural features can develop because of interfacial diffusion and two of the most important phenomena are described in some detail below.

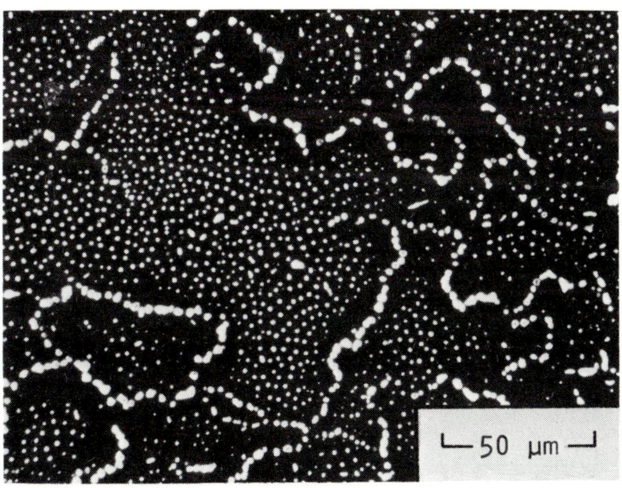


**Figure 7.20 Micrograph of *in situ* composite COTAC 74 after heat treatment at 1200°C showing preferential growth of fibres situated at grain boundaries [Stohr[28]]**

(*a*) *Fault retraction in composites*

Most *in situ* composites do not have a perfectly aligned microstructure of infinitely long fibres or lamellae but rather contain varying densities of faults and fibre (or lamellar) terminations as shown schematically in Fig. 7.21. As we have already discussed in the context of fibre spheroidization, the high chemical potentials at the ends of the fibre (or lamellar) segments and at the saddle points connecting the branches relative to the chemical potential along the length of the fibre provide a driving force for interfacial diffusion that is manifested as a retraction of the fibre end or of an advance of the saddle point. The velocity $dz/dt$ of the fibre termination is given by the expression

$$\frac{dz}{dt} = -\frac{\alpha \nu D_i \gamma \Omega^2}{kTa^3} \tag{7.22}$$

where $a$ is the fibre radius, $\alpha$ is a constant and the other parameters are described above.

When adjacent faults moving in opposite directions meet, as illustrated in Fig. 7.21, they annihilate and thus reduce the mean fibre (or lamellar)

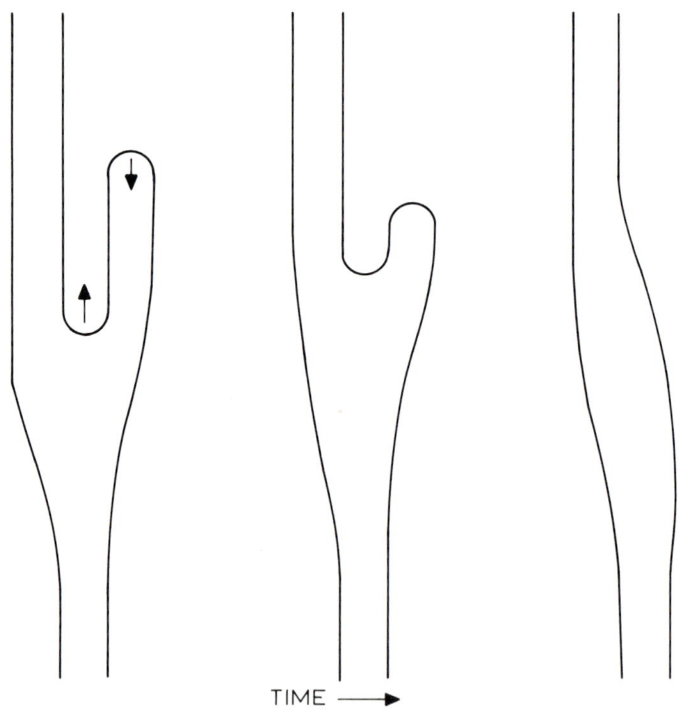


**Figure 7.21 Schematic illustration of the mechanism of fibre coarsening by fault retraction**

density and increase the fibre diameter or lamellar thickness. Nakagawa and Weatherly[30] have considered the contribution that termination migration can make to fibre coarsening in some detail. They assume that the density of fibres $N$ per unit area measured on a transverse section of a composite consists of $N_0$ continuous fibres and $n$ of mean length $\bar{z}$ per unit volume:

$$N = N_0 + n\bar{z} \tag{7.23}$$

Since $N_0$ is unaffected by fault migration it is assumed to be invariant, although it will change by isotropic coarsening. Then

$$\frac{dN}{dt} = n\frac{d\bar{z}}{dt} + \bar{z}\frac{dn}{dt} = n\frac{d\bar{z}}{dt} + \frac{N}{n}\frac{dn}{dt} \tag{7.24}$$

if $N_0 \ll N$ which is observed experimentally for many systems. By assuming random distributions of both rod lengths and terminations, Nakagawa and Weatherly[30] show that

$$\frac{N_0}{N} = (1 + n_0 K t)^2 \tag{7.25}$$

where $K$ is a mass transport rate constant and $n_0$, $N_0$ refer to $t = 0$.

Nakagawa and Weatherly[30] showed that the anomalously high coarsening rates observed by Bayles *et al.*[31] in Al–$Al_3Ni$, relative to the values associated with Ostwald ripening, were consistent with fault migration by interface diffusion. Indeed, McLean[32] has recently directly observed fault migration in this system by hot-stage electron microscopy.

(*b*) *Discontinuous coarsening*

There have been many observations of localized coarsening in biphase materials associated with the migration of grain boundaries. These effects appear to be particularly important in materials containing high volume fractions of a fine scale feature (e.g. eutectoids or $\gamma$–$\gamma'$ microstructures in superalloys). A migrating matrix grain boundary can interact with the dispersed phases in a variety of ways which are shown schematically in Fig. 7.22:

(i) the particles may be immobile and anchor the boundary until it bows sufficiently to break away
(ii) the particles may be dragged by the boundary
(iii) the grain boundary passes through both phases
(iv) the precipitate may be dissolved at the advancing grain boundary and reprecipitated in the new grain as it passes
(v) the precipitate particles in the advancing grain may grow in a discontinuous fashion at the expense of the particles in the shrinking grain.

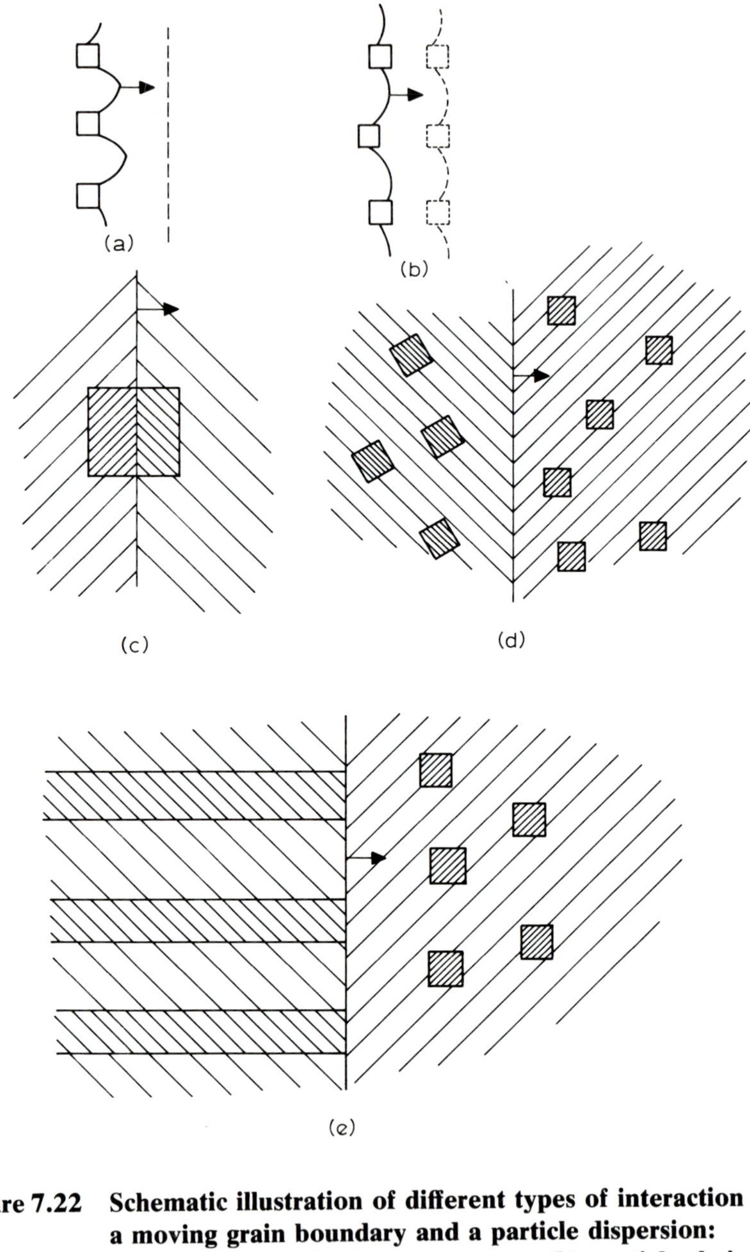

**Figure 7.22** **Schematic illustration of different types of interaction between a moving grain boundary and a particle dispersion: (a) boundary anchoring and escape; (b) particles being dragged; (c) grain boundary passing through particle; (d) particles dissolved at boundary and reprecipitated in second grain; (e) discontinuous coarsening at grain boundary**

Mechanisms (iv) and (v) appear commonly in many superalloys although (iv) is restricted to those alloys having low $\gamma'$ volume fractions.[33] Discontinuous precipitation dominates in the high performance superalloys relevant to directional solidification. Figure 7.23*a* shows a typical example in directionally solidified IN738LC. Examination along the lengths of ingots that were suddenly quenched during the course of directional solidification confirms that the feature develops by solid state diffusion at grain boundaries in interdendritic regions and shows that they often nucleate on small volumes of solid $\gamma$–$\gamma'$ eutectic. The $\gamma$–$\gamma'$ coherency is retained during such coarsening. A similar effect in the Co–Si eutectoid is shown in Fig. 7.23*b*.

Livingston and Cahn[34] have modelled the phenomenon of discontinuous coarsening for the idealized case of lamellae of thickness $\lambda_1$, increasing to $\lambda_2$ by grain boundary diffusion (Fig. 7.24). The driving force for the process derives entirely from the reduction in the total energy of interphase interfaces. The theory has much in common with that of eutectic solidification in that the simple requirements of flux and matter conservation are insufficient to solve the appropriate diffusion equations. The authors consider the consequences of two alternative postulates: (1) an extremum condition that requires that the velocity $V$ of the grain boundary is maximized and (2) a requirement that the rate of entropy production is a maximum. They find that, for growth by grain boundary diffusion,

$$\lambda_2 = 1{\cdot}5\lambda_1 \qquad \text{for maximum } V$$

$$= 2\lambda_1 \qquad \text{for maximum entropy production}$$

This result is in broad agreement with their experimental observations on the Co–Si eutectoid. However, there is no equivalent theory available to describe the behaviour of $\gamma'$ in superalloys.

## 7.4 INSTABILITIES DUE TO EXTERNAL POTENTIAL GRADIENTS

Materials in service rarely experience the uniform isothermal conditions normally used in the laboratory to characterize the microstructural stabilities of alloys. Variations in some environmental parameters (such as temperature, stress or electric fields) can cause potential energy gradients within the material, and these provide a driving force for diffusion that can lead to microstructural instabilities in polyphase alloys. In practice the chemical and capillary instabilities described above will occur at the same time as the extrinsic effects that are the subject of this Section. Here we will attempt to establish a formalism that will allow the delineation of the conditions where these latter effects can either be dominant or significantly modify the aging characteristics normally associated with capillary driving forces.

The most detailed work available concerns the effects of temperature gradients. The following treatment will be largely restricted to the theoretical

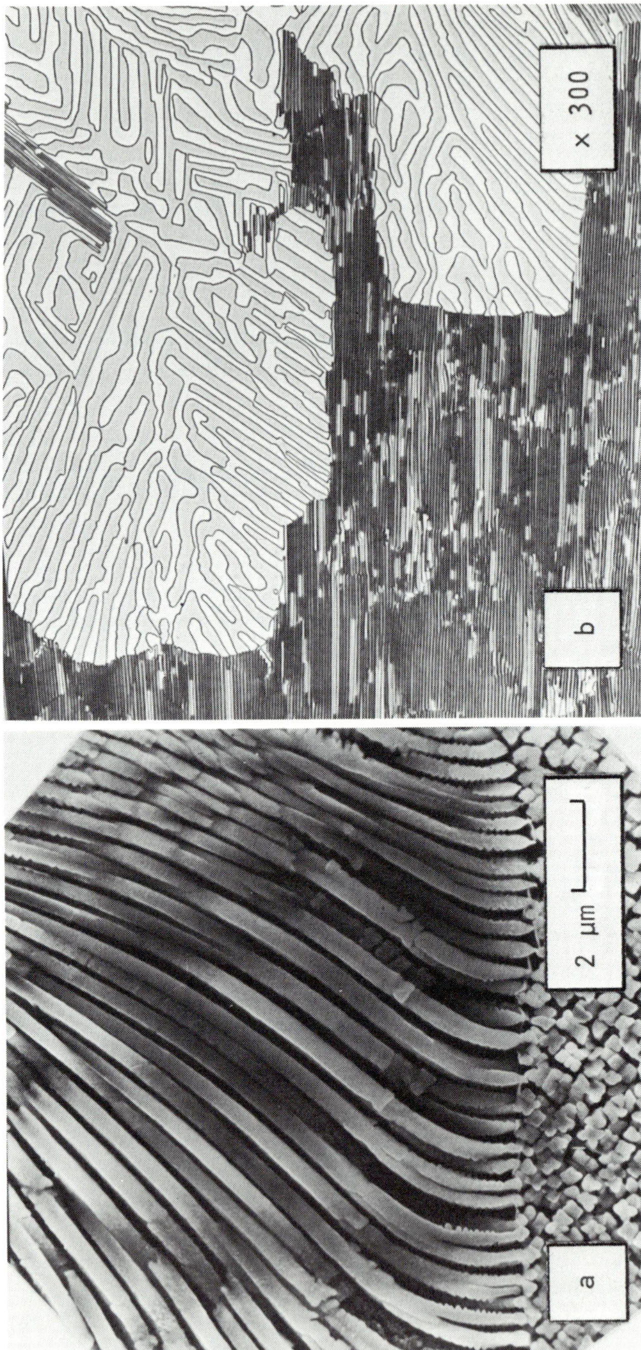


**Figure 7.23 Discontinuous coarsening (a) of $\gamma'$ in IN738LC [Rosenthal and West[57]) and (b) of the Co–Si eutectoid [Livingston and Cahn[(34)]]**

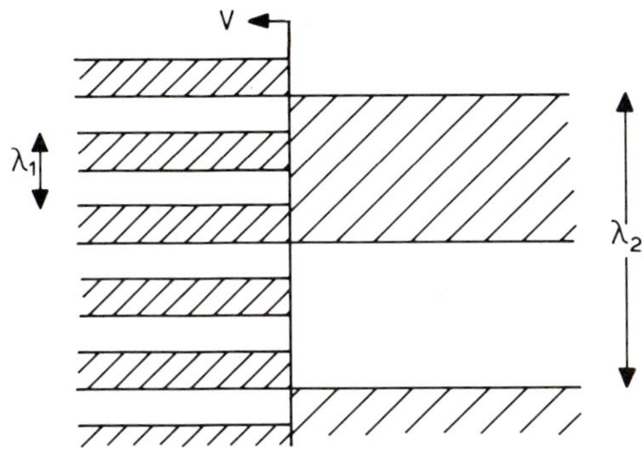


**Figure 7.24 Schematic diagram defining the parameters in the theory of discontinuous coarsening**

developments and experimental testing of temperature gradient phenomena. However, much of the formalism and many of the conclusions have quite direct analogues with respect to other potential gradients (e.g. electric or magnetic fields.)

### 7.4.1 Temperature gradients

Here we consider a solid precipitated phase $\beta$ entrained in a solid matrix $\alpha$; $\beta$ consists entirely of $B$ atoms and $\alpha$ is a dilute solid solution of $B$ in a matrix of $A$ atoms. In the presence of a temperature gradient $G$ each solute atom in the matrix experiences a force $\mathbf{f}$ which leads to a diffusive drift of $B$ through the matrix. There are two components of $\mathbf{f}$ caused by quite different effects; the first tends to dominate in the liquid state (e.g. in zone refining) while the second is most important when both phases are solid.

(a) The solubility of $B$ in $\alpha$ is generally temperature-dependent. If the inclusions are finely dispersed and equilibrium is attained at the particle/matrix interface then the local equilibrium solute concentrations will be maintained throughout the matrix. The concentration gradients of $B$ in $\alpha$ can then be simply related to the slope of the $\alpha$ solvus of the $\alpha\beta$ phase diagram ($m$) and to $G$ as shown schematically in Fig. 7.25. For a linear temperature gradient

$$\frac{\mathrm{d}C}{\mathrm{d}x}=\frac{G}{m} \tag{7.26}$$

where $C$ is the solute concentration. Taking $\mu=\mu_0+kT\ln C$ we may write eqn (7.1) in the form

$$\mathbf{f}_1=\frac{kT}{C}\frac{G}{m} \tag{7.27}$$

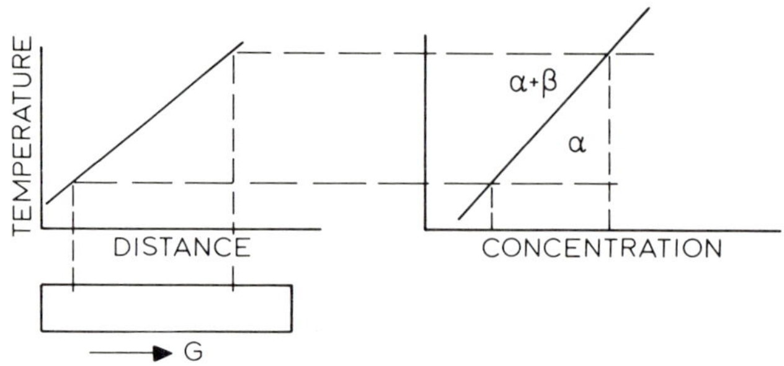


**Figure 7.25 Relationship of solute concentration to temperature gradient due to variation in equilibrium solute concentration**

(b) Heat and matter flow are not independent variables although in most circumstances their interaction is not detected. However, in certain circumstances a heat flow caused by a temperature gradient can generate a flux of matter. The phenomenon, which is known as the Soret effect, is characterized by an interaction term $Q^*$ which is known as the heat of transport.

$$\mathbf{f}_2 = -\frac{Q^*}{T}G \tag{7.28}$$

There remains some debate over the full physical significance of $Q^*$ and there are few reliable measurements on systems of practical interest. Indeed, it is difficult to anticipate either the sign or magnitude of $Q^*$.

The diffusive flux $\mathbf{j}_B$ of solute atoms through $\alpha$ resulting from this driving force interacts with phase boundaries to give changes in position, size or shape of the inclusions. For a linear temperature gradient eqn (7.3) may be expressed as a simple analogue to Ohm's law:

$$\mathbf{j}_B = -BN\frac{\Delta\mu}{\Delta x} \tag{7.29}$$

where $\Delta\mu = -\mathbf{f}\Delta x$ is a potential difference and $R = \Delta x/BN$ is a resistance to mass transport. Although quite rigorous solutions to the diffusion equations are available, the main features of temperature gradient instabilities can be illustrated in this approximation.

Figure 7.26 shows schematically an equivalent resistance circuit associated with the transfer of an atom from position 1 to position 2 on an interface (that need not be connected) as illustrated in Fig. 7.10. This involves (a) the detachment of the atom from its original site, (b) diffusion of the atom by several competitive mechanisms and (c) incorporation of the atom in its final

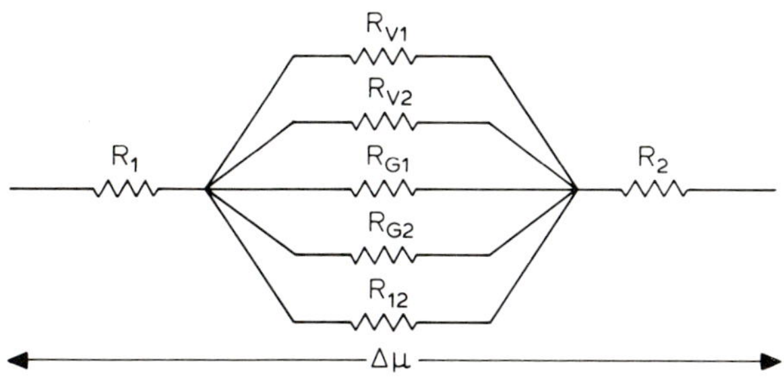


**Figure 7.26 Resistance network analogue to mass transport illustrated in Fig. 7.10**

site as illustrated in Fig. 7.10*b*. Since the chemical potential drop $\Delta\mu = -\mathbf{f}\Delta x$ is defined by the initial and final configurations, the diffusion flux and the associated rate of change of the interface position are determined by the effective mass transport resistance for all the possible mechanisms that are indicated schematically in Fig. 7.26.

$$R_{\text{eff}} = R_1 + R_2 + \left(\frac{1}{R_{v1}} + \frac{1}{R_{v2}} + \frac{1}{R_{G1}} + \frac{1}{R_{G2}} + \frac{1}{R_{12}}\right)^{-1} \tag{7.30}$$

The various resistance terms of importance to high temperature alloys are listed in Table 7.4.

A spherical inclusion at radius $a$ in such a flux field will move in the opposite sense due to dissolution of atoms at the high potential side and redeposition of them at the low potential side as illustrated in Fig. 7.27. The

**Table 7.4 Expressions for the resistance to mass transport due to various mass transport modes**

| Mass transport mode | Concentration of diffusing atoms | Resistance to mass transport |
|---|---|---|
| Interface reaction | — | $A$ (constant) |
| Volume diffusion | $c$ | $\phi \dfrac{kT}{cD_v} a$ |
| Interface diffusion | $\dfrac{\delta c}{2a}$ | $\phi \dfrac{kT}{cD_i} \cdot \dfrac{2a^2}{\delta}$ |

$\delta$ = interface thickness

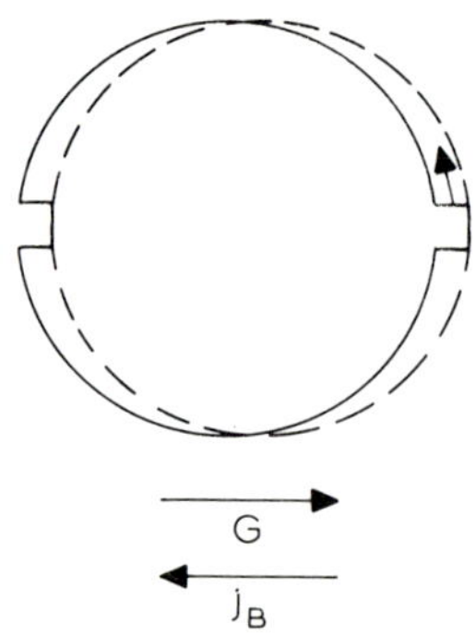


**Figure 7.27 Particle migration due to atomic diffusion in a temperature gradient**

inclusion velocity is given by

$$\frac{dx}{dt} = -\Omega \mathbf{j}_B = \frac{1}{R_{eff}} \cdot \phi \mathbf{f} a \tag{7.31}$$

where $\Omega$ is the atomic volume and $\phi$ is a geometrical constant. Substitution of the mass transport resistance terms from Table 7.4 shows that the variation of particle migration velocity with particle size depends on the dominant mode of mass transport

*interface reaction kinetics*

$$\frac{dx}{dt} = \frac{\phi \mathbf{f} \Omega}{A} a \quad \text{increases with increasing } a \tag{7.32a}$$

*volume diffusion*

$$\frac{dx}{dt} = \frac{\phi C D_v \Omega \mathbf{f}}{kT} \quad \text{independent of } a \tag{7.32b}$$

*interface diffusion*

$$\frac{dx}{dt} = \frac{\phi C D_i \Omega \mathbf{f}}{2kTa} \quad \text{decreases with increasing } a \tag{7.32c}$$

This result was originally derived by Shewmon.[35]

When one phase is liquid the term due to the temperature dependence of solubility dominates. There have been many observations of the movement of liquid inclusions or zones through a solid phase;[36] indeed, this forms the basis of temperature gradient zone refining.[37] The study of the migration of liquid lead inclusions through solid aluminium by McLean and Loveday[38] is fairly typical of this type of behaviour. The plot of particle velocity versus inclusion size shown in Fig. 7.28 suggests a transition between interface

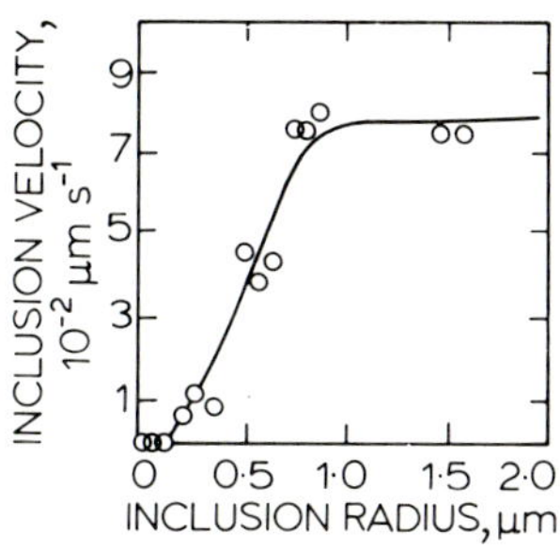


**Figure 7.28 Migration velocity of liquid lead inclusions through solid Al at 600°C in a temperature gradient of 185 K mm$^{-1}$ as a function of particle size [McLean and Loveday[38]]**

reaction kinetics and volume diffusion when compared with eqn (7.32). However, Doherty and Strutt[39] observed that this interpretation requires much larger undercoolings of the solid/liquid interface than are thought to be associated with atomically rough metal interfaces; they have proposed that dislocations in the solid matrix can provide an anchoring to the motion of the inclusions and this does in fact rationalize the lack of motion of the smallest particles. There is much less evidence of thermal migration in the solid state where the Soret contribution appears to dominate. Probably the most important observations are associated with the migration of nuclear fuels in the relatively high temperature gradients associated with nuclear reactors. Thus, Lindemer and Pearson[40] studied the migration of various HTGR reactor fuels through protective coatings of pyrolytic carbon and silicon carbide; similarly, Gulden[41] has examined the thermal migration of $U_2C_2$ through pyrolytic carbon.

The recognition that very high temperature gradients ($\sim$100 K mm$^{-1}$) can be generated in certain high temperature components such as air-cooled turbine blades has led to some concern that temperature gradient induced instabilities could be of some importance in the operation of these components—in particular, that *in situ* composites might be vulnerable to loss of strength through migration of the strengthening phases.

Perhaps of more practical importance is whether temperature gradient diffusion can lead to an acceleration of the aging mechanisms normally associated with isothermal heat treatment (e.g. particle coarsening). The available evidence is confusing indicating, often on the same system, that the temperature gradient can

(a) increase
(b) not affect } the rate of particle coarsening.
(c) decrease

The variability of reported results is indicated in Fig. 7.29 which shows results of independent comparisons of isothermal and temperature gradient coarsening in the Pb–Ag and Al–$Al_3Ni$ model *in situ* composites. Thus Jones[42] and Jones and May[43] show coarsening in both alloys to be accelerated by the presence of a temperature gradient while Collins *et al.*[44] find no difference for Pb–Ag and Davies *et al.*[45] show coarsening of Al–$Al_3Ni$ to be smaller when a temperature gradient is applied.

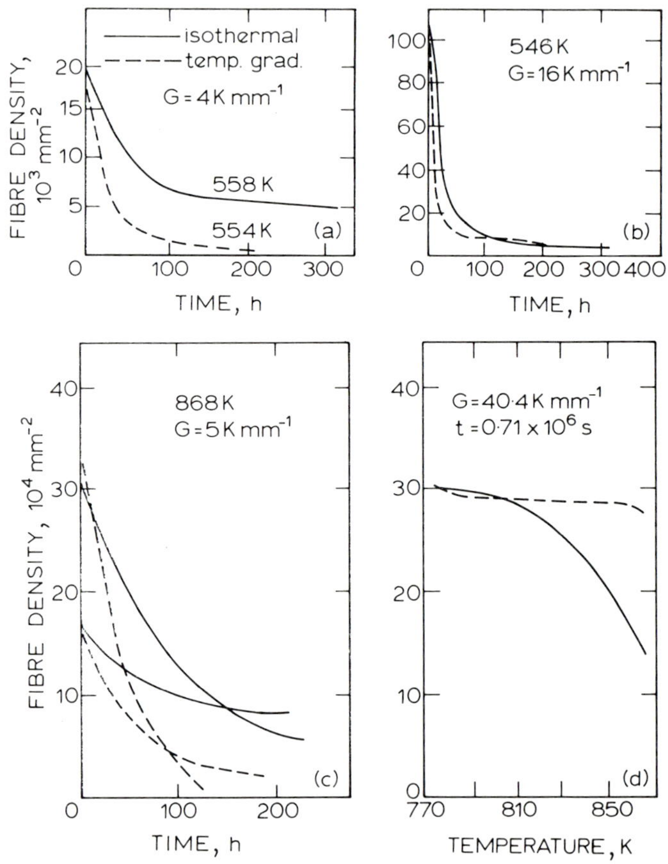


**Figure 7.29** **Changes in fibre density with time during both isothermal and temperature gradient annealing for two *in situ* composites reported by different authors:**
**Ag–Pb (a) Jones,[42] (b) Collins *et al.*[44]**
**Al–$Al_3Ni$ (c) Jones and May,[43] (d) Davies *et al.*[45]**

Two rival models have been proposed to account for the possible influence of thermal migration on particle coarsening.

(a) Jones[46] proposes that during phase migration, collision and sintering of inclusions can increase the size
(b) McLean[47] proposes that there will be a coarsening due to the front and back interfaces of an inclusion (at different temperatures) moving with slightly different velocities.

The two models are not entirely incompatible—although they might seem so at first sight. If we require that a small volume element (Fig. 7.30) must be maintained in thermodynamic equilibrium, then there must be a balance between (a) the difference in atomic flux down the temperature gradient across the hot and cold surfaces of the element and (b) the matter lost by different numbers of inclusions migrating up the temperature gradient across the hot and cold surfaces. This leads to a reduction in the density of inclusions and thermodynamic equilibrium is achieved by 'plating out' the excess solute atoms on the reduced number of inclusions to maintain a constant volume

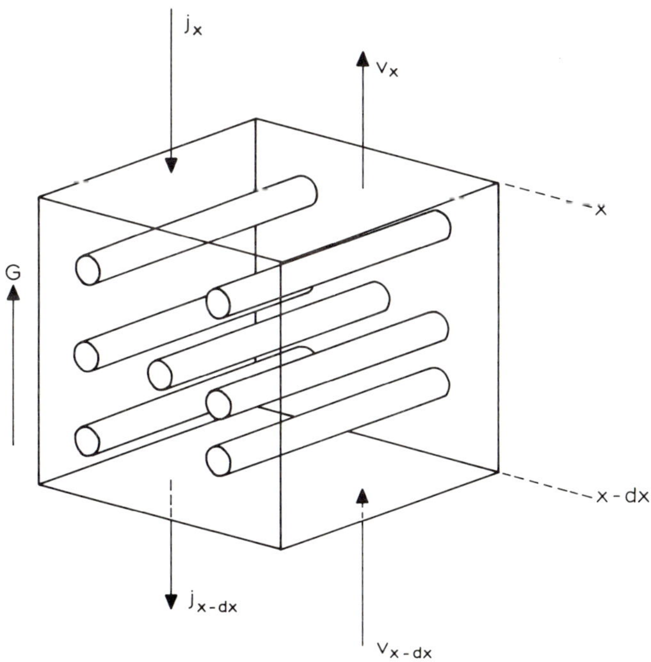


**Figure 7.30 Schematic diagram indicating the solute flux and migration of fibres into and out of a volume element of fibre reinforced alloy**

fraction—i.e. coarsening occurs.

$$(v_x - v_{x-\mathrm{d}x})\rho \, \mathrm{d}t = \mathrm{d}\rho \, \mathrm{d}x$$

i.e.

$$\frac{1}{\rho}\frac{\mathrm{d}\rho}{\mathrm{d}t} = \frac{\mathrm{d}v}{\mathrm{d}x} = \frac{\mathrm{d}v}{\mathrm{d}T}\frac{\mathrm{d}T}{\mathrm{d}x} = G\frac{\mathrm{d}v}{\mathrm{d}T}$$

where $v$ is the velocity and $\rho$ the density of inclusions. Expressing eqn (7.32$b$) for volume diffusion as

$$v = \frac{\alpha G}{T} \exp\left(-\frac{Q}{kT}\right)$$

where $\alpha$ is a constant and $Q$ is the activation energy for diffusion, then

$$\frac{\rho}{\rho_0} = \exp\left[-\frac{QGvt}{kT^2}\right] \sim \left[1 - \frac{QGvt}{kT^2}\right]$$

or

$$\frac{r}{r_0} = 1 + \frac{1}{2}\frac{QGvt}{kT^2}$$

where $r$, $r_0$ are the current and initial fibre radii. This model therefore predicts that $\rho$ should decrease and $r$ increase linearly with $t$; the results do not seem to indicate this although they are generally not sufficiently sensitive to discriminate between the various models. Material heat treated in a temperature gradient certainly coarsens more rapidly at the hot end than at the cold end; but is that due to the temperature gradient or simply because the kinetics are faster there?

The two models give spatial distributions

$$\frac{\mathrm{d}}{\mathrm{d}z} \ln\left[r(z)^3 - r_0^3\right] = -\frac{QG}{kT^2} + \frac{G}{T} \quad \text{for Ostwald ripening}$$

$$\frac{\mathrm{d}}{\mathrm{d}z} \ln\left[r(z) - r_0\right] = -\frac{QG}{kT^2} + \frac{G}{T} \quad \text{for temperature gradient coarsening}$$

Three such analyses show Ostwald ripening dominant for Al–$Al_3Ni$[32] and Pb/Ag[44] and temperature gradient effects dominant for (Co,Cr)–$Cr_7C_3$.[48] Although, in principle, solid phases should migrate in a temperature gradient albeit at a lower rate than for liquids, attempts to observe temperature gradient migration directly by hot stage microscopy[32] by the method successfully applied to Pb/Al alloys[38] have not been successful. Instead, they have confirmed the dominance of capillarity rather than temperature gradient effects. This may indicate the operation of a dislocation anchoring mechanism of the type proposed by Doherty and Strutt[39] to explain the inhibition of motion of liquid lead droplets in solid aluminium.

The few observations of fibre migration, by McLean[48] and by Jones and co-workers,[49] have been on alloys that have been heated to within a few degrees of their melting points. For example, Fig. 7.31 shows that an initially

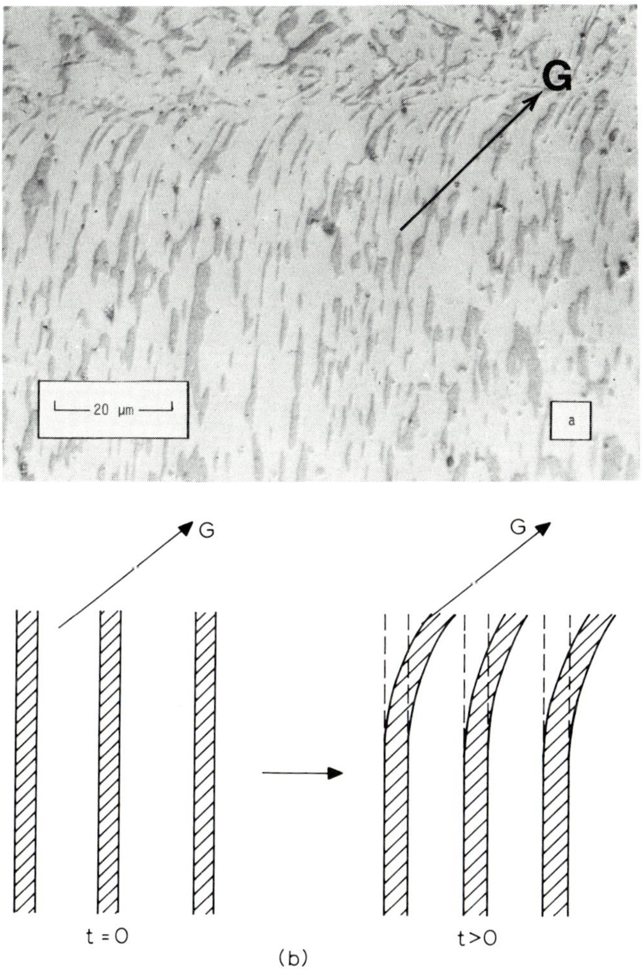


**Figure 7.31 (a) Micrograph of bent $Cr_7C_3$ fibre in the (Co,Cr)–$Cr_7C_3$ *in situ* composite due to differential migration of different parts of the fibre in an oblique temperature gradient as illustrated in (b) [McLean[48]]**

straight carbide fibre in the (Co,Cr)–$Cr_7C_3$ *in situ* composite has bent in a temperature gradient inclined to the fibres due to the hotter parts of the fibres migrating at a faster rate than the cooler parts.

Houghton and Jones[50] have also shown that at sufficiently high temperatures the thermal diffusion flux can lead to dramatic changes in the shapes of interfaces. Essentially the directed flux due to the temperature gradient can overcome the stabilizing effects of capillarity to produce a roughening of an initially flat planar interface with a characteristic wavelength. This phenomenon is closely related to the morphological instabilities, proposed by Mullins and Sekerka,[51] that can occur during crystal growth.

The weight of evidence now available suggests that phase migration and roughening in *in situ* composites, specifically due to temperature gradient effects, only occur above about $0{\cdot}9\ T_m$ ($T_m$ is the absolute melting point). Under projected service conditions of high temperature materials such effects should not be prevalent. Moreover, many data for phase coarsening are consistent with the forms of the capillary coarsening equations rather than with the theories of temperature gradient coarsening although it is not clear if the kinetic constants are modified by the temperature gradient. Thus the concerns over the vulnerability of *in situ* composites to temperature gradient instabilities have been proved to be unfounded. Nevertheless, conditions can be found where these instabilities are important, and they may well prove to be of significance in other applications.

### 7.4.2 Stress gradients

There is some evidence that variation in temperature with time can lead to some very dramatic effects. In particular, some *in-situ* composites degrade very rapidly in cycling temperatures whereas they are remarkably stable in isothermal conditions.[52,53] Figure 7.32 shows the type of effect that is observed where initially smooth fibres progressively roughen, clearly indicating that capillarity is not dominant.

What effect does changing the temperature have on such alloys? Assuming that the matrix and fibre are initially in equilibrium and stress free, a discontinuous change in temperature $\Delta T$ will lead to

(a) differential expansion (or contraction) of the fibre and matrix leading to a local gradient in stress
(b) a change of solubility of the fibre in the matrix.

Each will give rise to a gradient in chemical potential which will lead to an effective force on solute atoms. In addition there will be the usual capillarity term (c) arising from the Gibbs–Thompson effect.

*(a) Thermal expansion effect*

The volume mismatch $\delta V$ between a spherical inclusion of radius $a$ and the matrix (with coefficients of expansion $\alpha_i$, $\alpha_m$ respectively) due to a temperature change $\Delta T$ is given by the expression

$$\delta V = 4\pi a^3 \Delta T(\alpha_i - \alpha_m)$$

This will lead to stress distributions around the inclusions

$$\sigma_{rr} = -\frac{G\delta V}{\pi r^3}$$

where $G$ is the shear modulus. Following Eshelby[54] this gives an interaction energy with solute atoms

$$E_{int} = -\frac{1}{2}\frac{\Omega}{G}\sigma_{rr}^2$$

(where $\Omega$ is the atomic volume of the solute) and a force on the solute atoms

$$\mathbf{f}_1 = -\frac{\partial E_{int}}{\partial r} = 48 G\Omega\Delta T^2(\alpha_i - \sigma_m)^2 \cdot \frac{a^6}{r^7}$$

For the case of cylinders or fibres

$$\mathbf{f}_1 = 8G\Omega\Delta T^2(\alpha_i - \alpha_m)^2 \frac{a^4}{r^5}$$

*(b) Temperature dependence of solubility*

Under isothermal conditions the chemical potential of solute atoms is related to the equilibrium solute concentration

$$\mu = \mu_0 + kT \ln C(T)$$

When the temperature is suddenly changed to $T + \Delta T$, the solute concentration appropriate to $T$ will be retained far from the inclusion while the new equilibrium value $C(T+T)$ will rapidly be established at the inclusion/matrix interface. This gives rise to a difference in potential across the matrix.

$$\Delta\mu = \underset{\text{in the matrix}}{k(T+\Delta T)\ln C(T)} - \underset{\text{at the interface}}{k(T+\Delta T)\ln C(T+\Delta T)}$$

Writing $C(T) = C_0 \exp(-Q_s/kT)$, where $Q_s$ is the heat of solution, it follows that

$$\Delta\mu = Q_s\frac{\Delta T}{T}$$

The force on atoms due to this effect is given by an expression of the form

$$\mathbf{f}_2 = \frac{Q_s \Delta T}{\phi_1 T a}$$

where $\phi_1$ is a dimensionless constant.

(*c*) *Capillarity effect*
The gradient in chemical potential arising from the Gibbs–Thompson effect can be expressed

$$\mathbf{f}_3 = \frac{2\gamma\Omega}{\phi_2 a}\left[\frac{1}{\bar{a}} - \frac{1}{a}\right]$$

where $\bar{a}$ is the mean radius of the inclusion, $\gamma$ is the matrix/inclusion interfacial energy and $\phi_2$ is a constant.

Taking the value of **f** at $r = 2a$ which is the furthest point from fibres in most eutectics, and inserting the typical values of the materials parameters listed in Table 7.5

$$\mathbf{f}_1 \sim \mathbf{f}_3 \sim 10^{-18}\,\text{N}$$

$$\mathbf{f}_2 \sim 10^{-14}\,\text{N}$$

The cumulative effect of all three contributions will be a preferential dissolution of small fibres on heating cycles and deposition on large fibres during cooling. This may explain why some smaller fibres remain smooth while neighbouring larger fibres roughen (Fig. 7.32).

Another source of stress-induced diffusion that leads to a dramatic change in precipitate morphology can be attributed to local stress gradients due to

**Table 7.5 Values of the materials parameters used to evaluate the relative magnitudes of the various driving forces for diffusion during thermal cycling**

| Symbol | Value | Units |
|---|---|---|
| $\Omega$ | $10^{-29}$ | $\text{m}^3$ |
| $G$ | $10^{11}$ | $\text{N m}^{-2}$ |
| $\Delta T$ | 200 | K |
| $\alpha_i$ | $2\times10^{-6}$ | $\text{K}^{-1}$ |
| $\alpha_m$ | $14\times10^{-6}$ | $\text{K}^{-1}$ |
| $a$ | $10^{-6}$ | m |
| $r$ | $2\times10^{-6}$ | m |
| $\phi$ | $\sim1$ | |
| $\gamma$ | 0·1 | $\text{m J}^{-2}$ |
| $\bar{a}$ | $\gg a$ | |
| $Q_s$ | $4\times10^4$ | $\text{m mol}^{-1}$ |
| $T$ | 1000 | K |

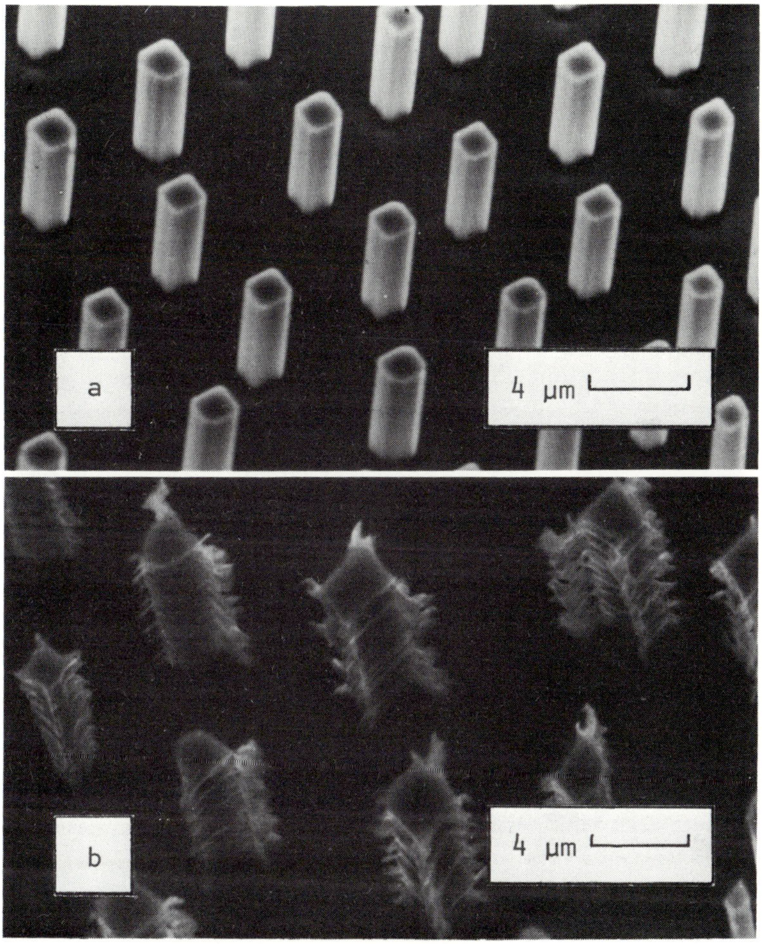


**Figure 7.32 Scanning electron micrographs of deeply etched Ni–NbC *in situ* composite after (a) isothermal and (b) temperature cycling heat treatment [Cooper and Billingham[53]]**

differences in elastic and lattice constants between matrix and precipitate in certain alloys. Tien and Copley[55] first examined this effect. Some nickel-base alloys containing cubic $\gamma'$ precipitates when aged under stress give:

(a) plates with a tensile stress
(b) fibres with a compressive stress
(c) cubes with no stress.

The effect only appears to occur in alloys having large lattice mismatch between the ordered $\gamma'$ particles and matrix. Figure 7.33 indicates how the internal stresses due to the application of both tensile and compressive loads, assuming $E_{\text{inclusion}} < E_{\text{matrix}}$ as it is for $\gamma'$ reinforced alloys at high temperature, can be relieved depending on the relative magnitudes of the particle and matrix lattice constants. In alloys where $a_{\text{inclusion}} > a_{\text{matrix}}$ then a tensile stress should lead to the development of plates with faces normal to the stress axis and a compressive stress should give rod shaped particles oriented parallel to the stress. This behaviour has indeed been observed, initially by Tien and Copely.[55] More recently Pearson *et al.*[56] have proposed that this effect might be applied deliberately to certain superalloys to inhibit the creep mechanism resulting from dislocation climb around particles.

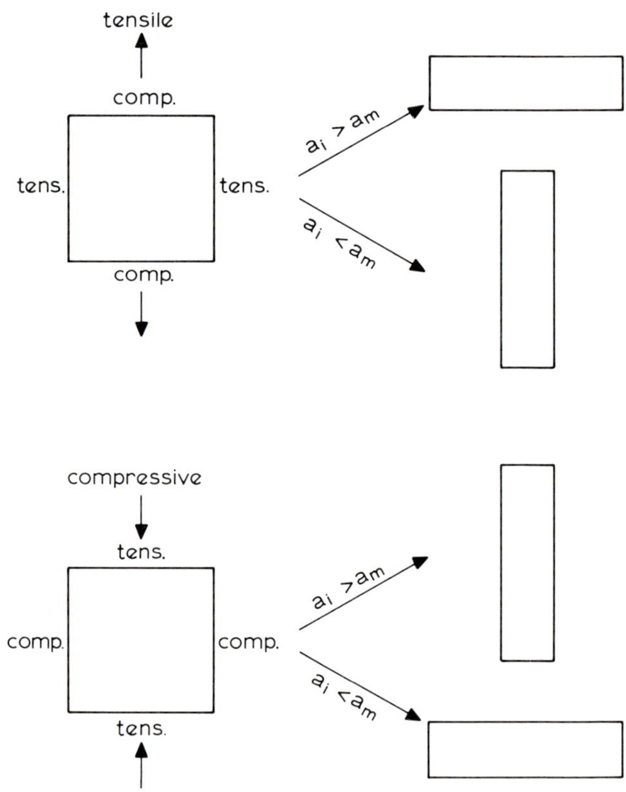


**Figure 7.33 Schematic diagram indicating the internal stress distribution around an inclusion ($E_{\text{inc}} < E_{\text{matrix}}$) in tensile and compressive loading and how it may be relieved by change in shape**

## REFERENCES CHAPTER 7


1 P. G. SHEWMON in 'Physical metallurgy', ed. R. W. Cahn, 1965, Amsterdam, North Holland

2 G. R. PURDY AND J. S. KIRKCALDY: *Metall. Trans.*, 1971, **2**, 371

2a M. GELL, D. DUHL AND M. GIAMEI: Proc. 4th Int. Symp. on Superalloys, Seven Springs, Pa, September 1980

3 C. T. SIMS in 'The superalloys', eds. C. T. Sims and W. C. Hagel, 1972, New York, Wiley Interscience

4 S. RIDEOUT, W. D. MANLEY, E. L. KAMEN, B. S. LEMENT AND P. A. BECK: *Trans. Metall. Soc. AIME*, 1951, **191**, 872

5 R. STICKLER in 'High temperature materials for gas turbines', eds. P. R. Sahm and M. O. Speidel, 1974, Amsterdam, Elsevier

6 J. E. RESTALL AND E. C. TOULSON: *Met. Mater.*, 1973, **7**, 134 and 187

7 P. N. QUESTED, E. BULLOCK AND M. McLEAN: 'Solidification and casting of metals', 466, 1979, London, Metals Society

8 C. HERRING: *J. Appl. Phys.*, 1950, **21**, 301

9 LORD RAYLEIGH: *Proc. London Math. Soc.*, 1878, **10**, 4

10 F. A. NICHOLS AND W. W. MULLINS: *Trans. Metall. Soc. AIME*, 1965, **233**, 1840

11 H. E. CLINE: *Acta Metall.*, 1971, **19**, 481

12 T. F. MARINIS AND R. F. SEKERKA: Proc. Conf. *In Situ* Composites III, eds. J. L. Walter *et al.*, Ginn and Co., Lexington, Mass., 1979

13 T. F. MARINIS AND R. F. SEKERKA: Proc. Conf. *In situ* Composites IV, Boston, Mass., 1981, eds. F. D. Lemkey, H. E. Cline and M. McLean, New York–Oxford, Elsevier North-Holland, in press

14 F. A. NICHOLS AND W. W. MULLINS: *J. Appl. Phys.* 1965, **36**, 1826

15 W. W. MULLINS in 'Metal surfaces', eds. W. D. Robertson and N. A. Gjostein, 1964, Metals Park, Ohio, ASM

16 S. MARICH: *Metall. Trans.*, 1970, **1**, 2953

17 S. MARICH AND D. JAFFREY: *Metall. Trans.*, 1971, **2**, 2681

18 A. HELLAWELL AND R. GRUGEL: Proc. Conf. *In Situ* Composites III, eds. J. L. Walter *et al.*, Ginn and Co., Lexington, Mass., 1979
19 M. H. JOHNSON AND R. A. PARR: Ref. 18
20 E. HO AND G. C. WEATHERLY: Proc. Conf. *In Situ* Composites II, eds. M. R. Jackson *et al.*, Xerox Individualized Publishing, Lexington, Mass., 1976
21 E. HO AND G. C. WEATHERLY: *Acta Metall.*, 1975, **23**, 1451
22 C. WAGNER: *Z. Elektrochem.*, 1961, **65**, 581
23 I. M. LIFSHITZ AND V. V. SLYOZOV: *J. Phys. Chem. Solids*, 1961, **19**, 35
24 G. W. GREENWOOD: Proc. Conf. 'Mechanism of phase transformation in crystalline solids,' Inst. of Metals Monograph 33, 1969
25 A. J. ARDELL: *Acta Metall.*, 1972, **20**, 61
26 H. O. K. KIRCHNER: *Metall. Trans.*, 1971, **2**, 2861
27 A. J. ARDELL: *Acta Metall.*, 1972, **20**, 601
28 J. F. STORR: Proc. Conf. *In Situ* Composites III, eds. J. L. Walter *et al.*, Ginn and Co., Lexington, Mass., 1979
29 J. D. LIVINGSTON: *Trans. Metall. Soc. AIME*, 1959, **215**, 566
30 Y. G. NAKAGAWA AND G. C. WEATHERLY: *Acta Metall.*, 1972, **20**, 345
31 J. J. BAYLES, J. A. FORD AND M. J. SALKIND: *Trans. Metall. Soc. AIME*, 1967, **239**, 844
32 M. McLEAN: *Acta Metall.*, 1977, **25**, 1209
33 A. PORTER AND B. RALPH: *J. Mater. Sci.*, 1981, **16**, 707
34 J. D. LIVINGSTON AND J. W. CAHN: *Acta Metall.*, 1974, **22**, 495
35 P. G. SHEWMON: *Trans. Metall. Soc. AIME*, 1964, **230**, 1134
36 M. McLEAN: *Met. Sci.*, 1978, **12**, 113
37 W. PFANN: 'Zone melting', 1966, New York, Wiley
38 M. McLEAN AND M. S. LOVEDAY: *J. Mater. Sci.*, 1974, **9**, 1104
39 R. DOHERTY AND T. R. STRUTT: *J. Mater. Sci.*, 1976, **11**, 2169
40 T. B. LINDERMEER AND R. L. PEARSON: *J. Amer. Ceramic Soc.*, 1977, **60**, 5
41 T. D. GULDEN: *J. Amer. Ceramic Soc.*, 1972, **55**, 14

42 D. R. H. JONES: *Mater. Sci. Eng.*, 1974, **15**, 203
43 D. R. H. JONES AND G. J. MAY: *Acta Metall.*, 1975, **23**, 29
44 G. C. COLLINS, I. A. SMITH AND G. A. CHADWICK: *Mater. Sci. Eng.*, 1978, **36**, 1
45 J. R. DAVIES, T. H. COURTNEY AND M. A. PRZYSTUPA: *Metall. Trans.*, 1980, **11A**, 323
46 D. R. H. JONES: *Met. Sci.*, 1974, **8**, 37
47 M. McLEAN: *Scr. Metall.*, 1975, **9**, 439
48 M. McLEAN: Proc. Conf. *In Situ* Composites II, eds. M. R. Jackson *et al.*, Xerox Individualized Publishing, Lexington, Mass., 1976
49 D. C. HOUGHTON AND D. R. H. JONES: *Acta Metall.*, 1978, **26**, 695
50 D. C. HOUGHTON AND D. R. H. JONES: *Acta Metall.*, 1979, **27**, 1031
51 W. W. MULLINS AND R. F. SEKERKA: *J. Appl. Phys.*, 1963, **34**, 323; 1964, **35**, 444
52 F. M. DUNLEVEY AND J. F. WALLACE: *Metall. Trans.*, 1974, **5**, 1351
53 S. P. COOPER AND J. BILLINGHAM: *Met. Sci.*, 1980, **14**, 225
54 J. ESHELBY: *Ann. Phys.* (*Leipzig*), 1958, **1**, 116
55 I K. TIEN AND S. M. COPELY: *Metall. Trans.*, 1971, **2**, 543
56 D. D. PEARSON, F. D. LEMKEY AND B. H. KEAR: Proc. 4th Int. Symp. on Superalloys, Seven Springs, Pa, 1980, Metals Park, Ohio, ASM
57 R. ROSENTHAL AND D. WEST: Imperial College, London, private communication, 1982

# 8
# Other anisotropic materials: uses and prospects

The principal attraction of the directionally solidified alloys that have been the subject of this book has been the potential of achieving various types of anisotropic microstructure in a relatively simple single stage process. However, this ease of processability is achieved at the expense of developing heterogeneous dendritic microstructures in the case of superalloys and of working with alloys optimized to have a narrow melting range for *in situ* composites. The desired end-points of having either an elongated grain structure or a composite material can be obtained by alternative multistage processes which, while probably being more expensive than the solidification route, often are not subject to the equilibrium and kinetic restrictions related to the phase diagrams of the alloys. Here we shall outline these alternative approaches to producing anisotropic materials and attempt to compare the properties and potential applications of the different materials.

## 8.1 ELONGATED GRAIN STRUCTURES BY POWDER METALLURGY

The elongated grain structures, characteristic of directionally solidified superalloys, can be obtained by solid state processing of powders. There have been several variants of this approach which follow the general sequence indicated in Fig. 8.1. The alloy is produced in powder form, by one of a variety of techniques, and is then consolidated into as dense a solid as possible by combined sintering and thermomechanical heat treatments. For low temperature applications where fatigue resistance is required the material may be produced in component form with a small grain size at this stage. However, in order to optimize the high temperature creep and stress rupture behaviour, heat treatment in a temperature gradient can often induce directional recrystallization, giving an elongated grain structure. The primary advantage of this approach is that it avoids gross dendritic segregation; a major disadvantage is that incomplete consolidation of the powder and phase depositions at the particle surfaces can lead to embrittlement of the alloys. Moreover, directional

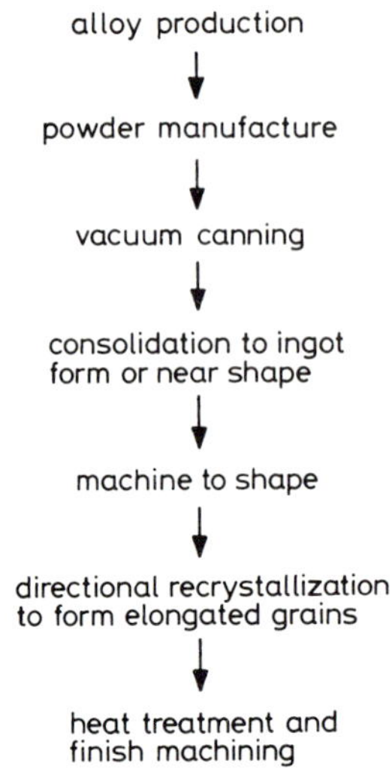


**Figure 8.1 Flow diagram of the processing sequence to produce elongated grain structures from powder**

recrystallization usually produces a preferred ⟨110⟩ texture which has neither optimum elastic nor creep properties.

Much of the process development in this field has been concerned with the production of clean powders with a fine uniform particle size distribution. Relatively simple alloys, including superalloys, have been produced from the melt by argon atomization,[1] indicated schematically in Fig. 8.2, which maintains low oxygen contents ($\leqslant$100 ppm.). More recently rapidly solidified powders of superalloys have been obtained by Pratt and Whitney by melt spinning into a helium gas coolant[2] (Fig. 8.3). The latter technique gives finer powders with nonequilibrium structures and further reduces the scale of segregation. These methods, however, are not appropriate to the production of dispersion strengthened materials where the $\gamma'$ strengthening of superalloys can be supplemented by a stable dispersion of a fine oxide such as $Y_2O_3$ or $ThO_2$. The earliest commercial dispersion strengthened alloys, such as TD nickel, were produced from solution precipitated powders;[3] in current commercial practice, however, the alloy powders are prepared in a high energy stirred ball mill where the atmosphere can be controlled[1] (Fig. 8.4).

It is essential to maintain good environmental control during the consolidation stage in order to inhibit oxide formation at the powder surface which would affect both the consolidation and subsequent recrystallization characteristics. The powders are generally enclosed in a can, under vacuum or controlled atmosphere, before hot extrusion and/or hot isostatic pressing to

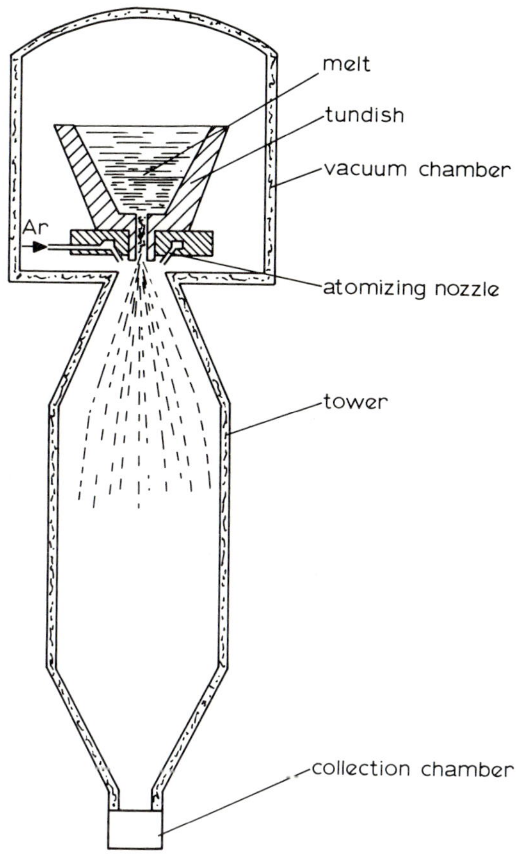


**Figure 8.2 Schematic diagram of an argon atomization rig [Ref. 1]**

give a dense microstructure. Gessinger[4] has recently reviewed the processing of oxide dispersion strengthened alloys (ODS).

Coarse elongated grain structures, which are thought to be required for optimum high temperature mechanical properties, can be obtained by suitable isothermal heat treatment of the extruded compacts. Figure 8.5[5] shows that grain growth in $Y_2O_3$ strengthened IN738LC occurs preferentially parallel to the extrusion direction for annealing temperatures greater than 1250°C. However, the effect which is probably caused by strain induced primary recrystallization is very variable. Allen[6] showed that more reliable structures could be produced by translating the compact through a high temperature gradient and that the grain aspect ratio could be related to the rate of withdrawal of the powder compact. Figure 8.6 shows that the highest grain

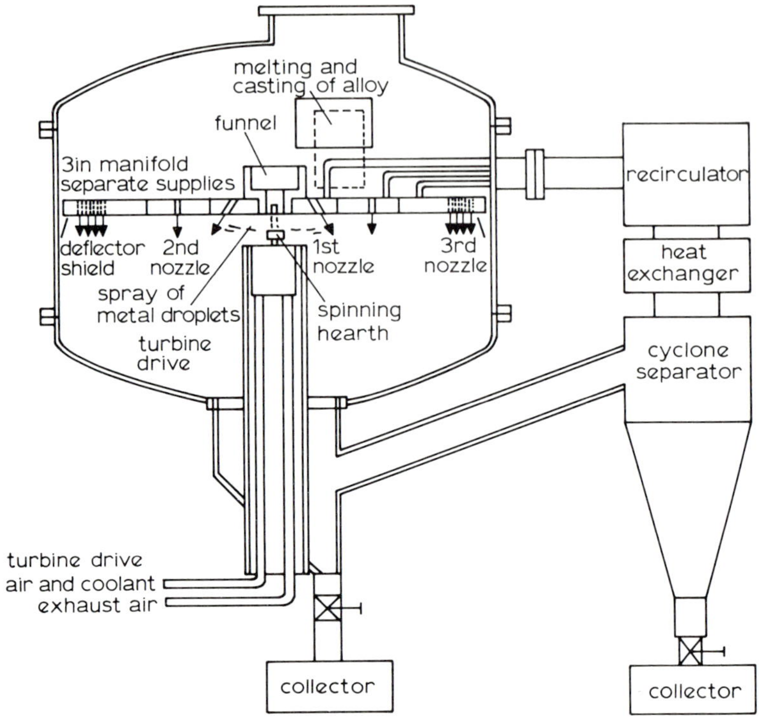


**Figure 8.3 Schematic diagram of the Pratt and Whitney apparatus for producing rapidly solidified powders by helium gas cooling [Ref. 2]**

aspect ratios are obtained at the slowest withdrawal rates.[4,7,8] Indeed, this stage of the process is identical to the directional solidification procedures but requires rather lower processing rates.

Impressive properties have been reported for both superalloys and ODS materials with elongated grains. For example, the rapidly solidified powders can be prepared from nonequilibrium compositions that permit higher proportions of solid solution and $\gamma'$ strengthening components to be retained than are possible by equilibrium solidification considerations. Thus Cox *et al.*[2] have surveyed the response of several alloy compositions to this process and typical creep curves are shown in Fig. 8.7. Data for directionally solidified MarM200 + Hf are included for comparison. From this it is clear that although rapid solidification processing offers marginal improvements to current alloy compositions at 980°C, the major advantage is in allowing the new high $\gamma'$ alloy designated RSR104 (i.e. Ni–8Al–18Mo) to be processed and heat treated.

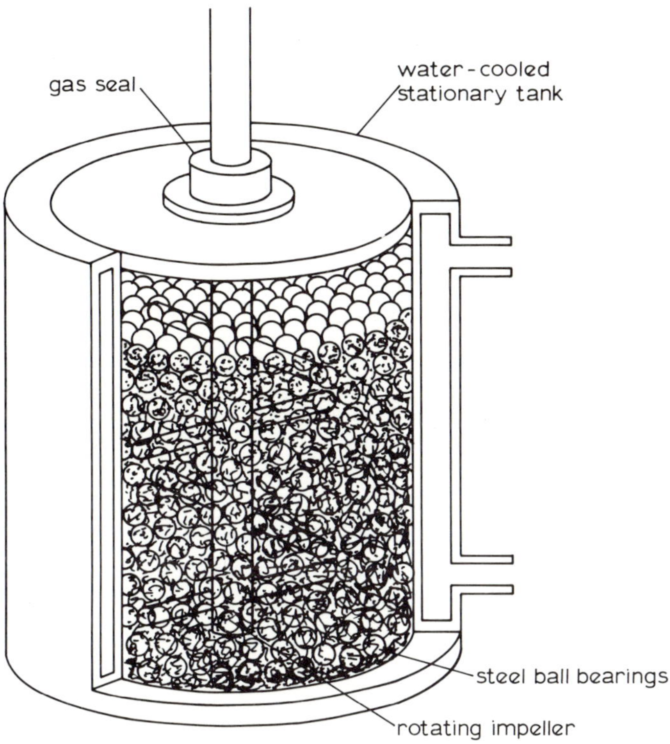


**Figure 8.4 Schematic diagram of the controlled atmosphere attritor used in the manufacture of oxide dispersion strengthened powders [Ref. 1]**

Here a life improvement of up to four times that of the best currently available directionally solidified alloy is obtained. However, when alloys like RSR104 are prepared in single crystal form they have similar stress rupture and creep properties as in the rapidly solidified form.

ODS alloys are at a more advanced stage, being now commercially exploited. They are characterized by very high stress sensitivities of both stress rupture and minimum creep rate data. For example, Wilcox and Clauer[9] have shown that when the creep data of TD nickel are represented by the Bailey–Orowan power law expression, a stress exponent of 40 (rather than the theoretical value of 4) is obtained; similarly the data yield unrealistically high values of the activation energy for creep. Typical data are shown in Fig. 8.8. The importance of elongated grains, also initially identified by Wilcox and Clauer,[10] is demonstrated by the stress rupture data shown in Fig. 8.9 where the 100 h rupture stress progressively increases as the grain aspect ratio

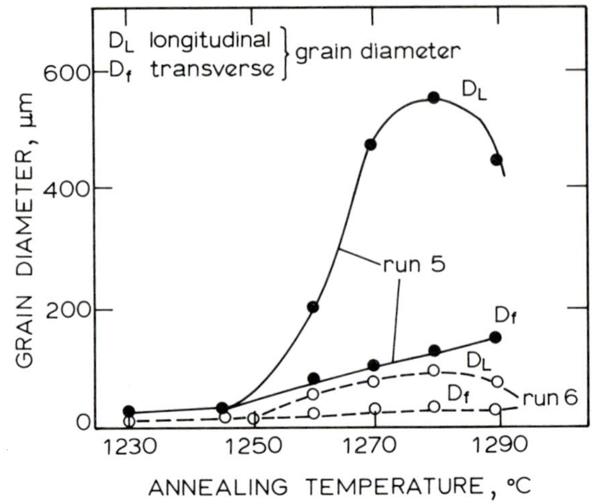


**Figure 8.5 Grain size in longitudinal and transverse directions as a function of temperature for extruded IN738 + $Y_2O_3$ [Ref. 5]**

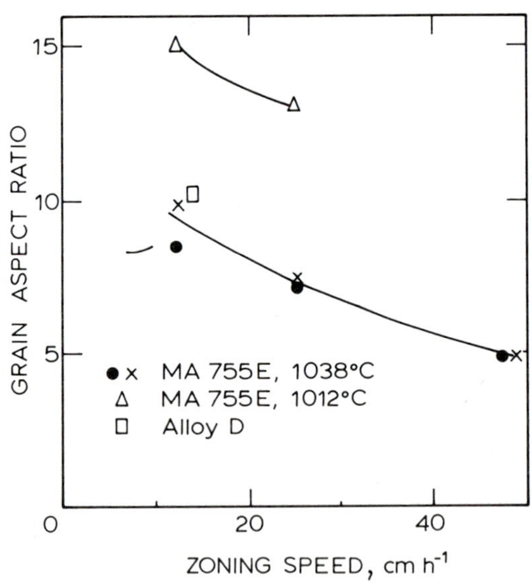


**Figure 8.6 Effect of zoning speed and extrusion temperature on grain aspect ratio [Refs. 4, 7, 8]**

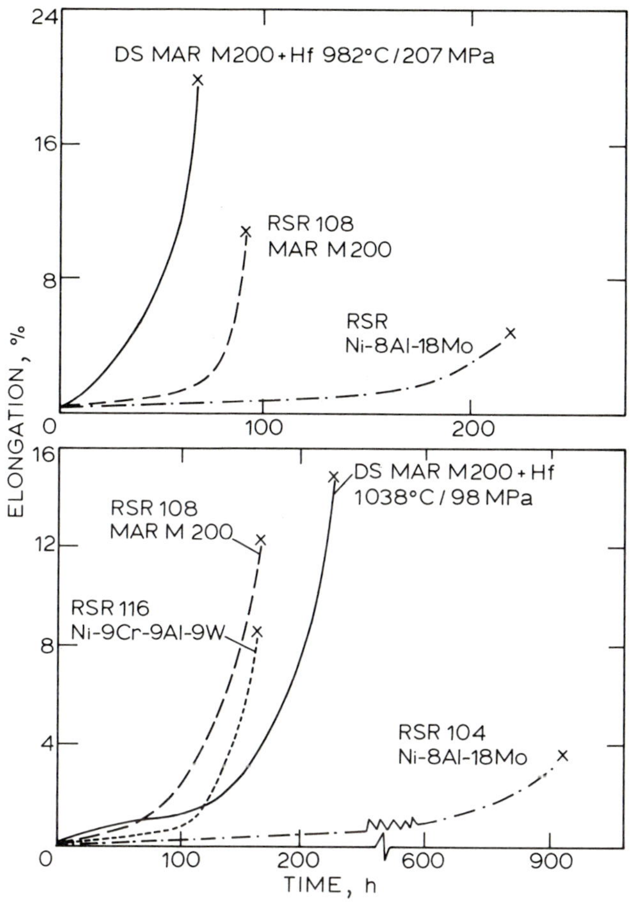


**Figure 8.7 Creep curves for various rapidly solidified alloys compared with those of directionally solidified MarM200 + Hf at 982°C/207 MPa and 1038°C/98 MPa [Ref. 2]**

approaches ~10.[11] The major recent advance has been to combine oxide dispersion strengthening which is effective at high temperatures with $\gamma'$ strengthening which introduces intermediate temperature strength. Thus the early alloys, such as TD nickel, which had no $\gamma'$ strengthening were much weaker than superalloys at temperatures below ~900°C although they retained their strength to very high temperature. The new generation of mechanically alloyed ODS materials, such as MA-6000, have much improved intermediate temperature properties, as indicated in Fig. 8.10.

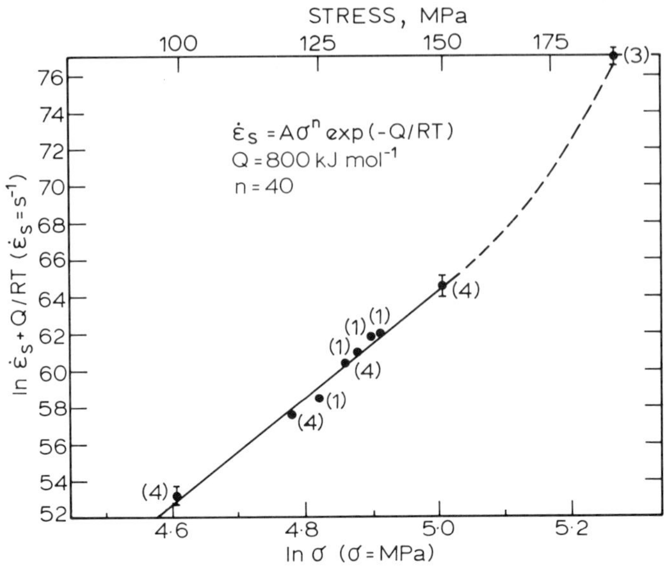


**Figure 8.8 Temperature compensated creep rate as a function of the logarithm of applied stress for TD nickel [Ref. 10]**

A great deal of work is required to characterize the behaviour of these materials in likely service conditions. In particular, the relatively low ductilities of the alloys may lead to problems in cyclic stressing (high cycle, low cycle and thermal fatigue). More immediate, however, is whether the material can be fabricated in suitable component configurations. Thus it would be necessary to retain cooling of turbine blades in order to take full advantage of the increased operating temperatures of these materials relative to superalloys.

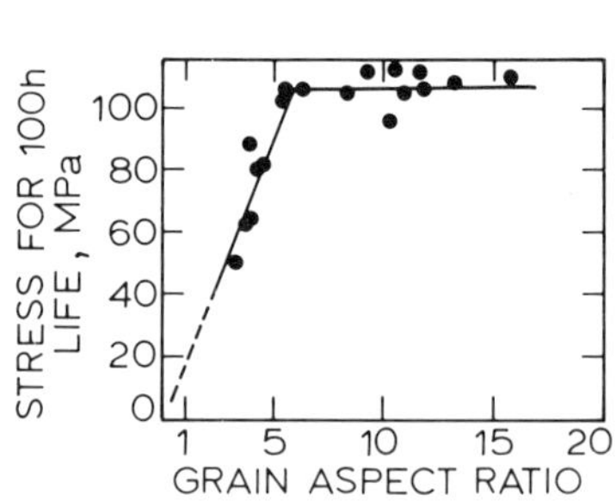


**Figure 8.9 Relationship between stress for 100 h life at 1038°C and grain aspect ratio [Ref. 11]**

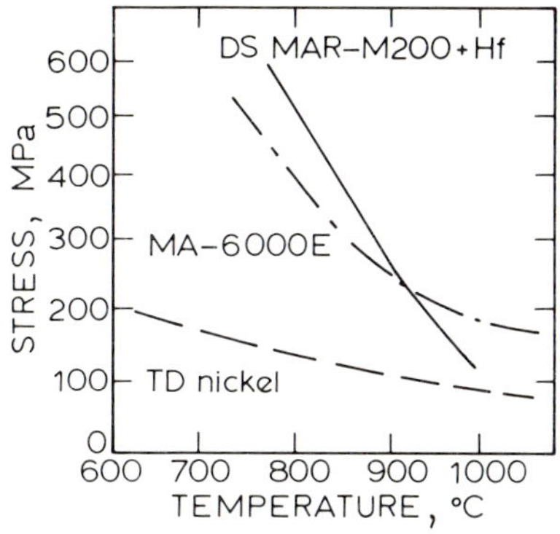


**Figure 8.10 1000 h rupture strength of MA6000E compared with directionally solidified MarM200 and TD nickel [Ref. 1]**

## 8.2 SYNTHETIC COMPOSITES

The combinations of fibre and matrix phases and the possible volume fractions of reinforcement in *in situ* composites are determined by the availability of suitable phase equilibria. In optimizing, for example, the matrix oxidation resistance there will usually be secondary effects that alter the freezing range, which affects the processability and the eutectic composition, which modifies the volume fraction of reinforcing phase of the *in situ* composite. Since there is little flexibility in altering fibre (or lamellar) volume fractions by producing composite microstructures from off-eutectic melts, the alloy designer is denied control of the major variable that controls composite strength. This limitation does not occur in synthetic composites where arbitrary volume fractions can, in principle, be incorporated in suitable matrices.

A wide range of metal–matrix composites intended for high temperature operation have been described,[12] but few continue to show promise. Attempts have been made to reinforce oxidation resistant iron-base materials and superalloys by a variety of fibres including ceramics (as both continuous fibres and submicron whiskers), boron filaments, carbon filaments and refractory metal alloy wires (such as W and Mo). Most of these systems have failed because of chemical incompatibilities between fibre and matrix. Thus boron and carbon have high solubility in suitable matrices; minor ceramic/matrix interactions lead to stress concentrations that induce premature fibre fracture, particularly during temperature cycling because of the high thermal expansion mismatches. Attempts to introduce barrier layers between fibres and matrix have not yet been successful in these systems.

The greatest degree of success appears to have been obtained by the use of refractory alloy fibres. Tungsten fibre reinforced composites have been studied extensively, particularly by Signorelli and co-workers at NASA Lewis

Research Centre.[12] Use has been made of oxidation resistant matrices such as FeCrAlY alloys and nickel-base superalloys in order to protect the high strength refractory fibres from environmental attack. However, chemical incompatibilities between W fibres and matrices remain a major problem area. Three types of fibre/matrix reactions have been identified:

(a) diffusional penetration reaction and recrystallization of the tungsten fibre
(b) precipitation of brittle intermetallic phases at the fibre/matrix boundary, and
(c) dissolution of the fibre.

The relative importance of these effects has been correlated with matrix composition and some guidelines have been evolved to minimize the effects. In general terms, superalloy compositions having high refractory element compositions are preferred on two counts: firstly they reduce the potential for further refractory element dissolution and secondly they reduce the diffusivity, and therefore the kinetics of interaction. Guidelines are now available to allow selection of chemically compatible fibres and matrices; however, this is only one criterion, other important requirements being the attainment of adequate fibre/matrix bonding and resistance to cracking in temperature cycling conditions. Iron-base matrix alloys that minimize fibre degradation are now available and it is anticipated that suitable nickel-base matrices will soon be available.

Several approaches to fabrication of synthetic composites have been proposed but they fall into two general categories:

(1) Infiltration of the reinforcing fibres by the liquid matrix. Here the melt must wet the fibres, form a good bond and yet not cause fibre/matrix degradation. Moreover, segregation of the fibres must be prevented. This approach is attractive in that current investment casting techniques can, in principle, be modified. However, lack of control of the positions of the fibres is likely to militate against this technique.
(2) Solid state processing appears to have rather more potential although it is rather slower than the casting approach. Here the fibres are sintered to the matrix which is in either powder or thin sheet form. A possible procedure to produce cored turbine blades is indicated in Fig. 8.11; composite monotapes are produced which are then assembled in component form, the cooling channels being prepared by dissolution of a suitably shaped metal core.

The fabrication procedures remain at the laboratory scale of development; major advances in processing technology would be required before these materials could be introduced into service. However, they provide the potential advantage of allowing selective reinforcement of the most highly stressed regions of components.

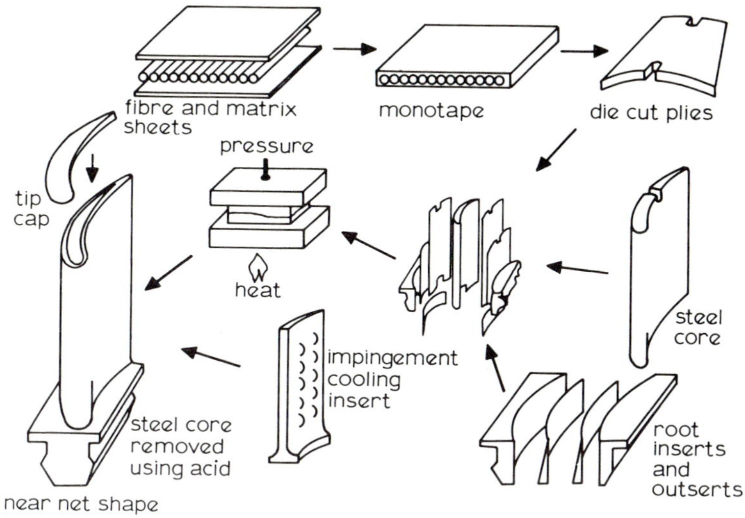


**Figure 8.11 Schematic sequence in the manufacture of cored turbine blades from synthetic tungsten fibre reinforced composites [Ref. 12]**

Significantly better stress rupture properties can be achieved with synthetic than with *in situ* composites by increasing the volume fraction of reinforcement. A typical set of data[13] is shown in Fig. 8.12; the results appear to be relatively insensitive to the nature of the matrix. The best absolute stress rupture properties have been achieved using tungsten wire reinforcement where the extent of wire/matrix interactions is minimized. The high density of tungsten, however, causes a substantial increase in density of the composite with increasing volume fraction of reinforcement as is indicated in Fig. 8.13 for both iron- and nickel-base alloy matrices. Since the likely application is as a rotating turbine blade, where the material density determines the centrifugal stress, it is necessary to introduce some density compensation into the comparison. Figure 8.14 shows the temperature at which a stress of 200 MPa gives a rupture life of $10^3$ h for tungsten wire composites as a function of volume fraction of reinforcement; data for directionally solidified MarM200 and for the $\gamma$–$\gamma'$–$\delta$ *in situ* composite are included for comparison. Although absolute improvements in stress rupture properties occur for $V_f \geqslant 0{\cdot}2$, the improvement in specific stress rupture properties of synthetic composites over either current superalloys or *in situ* composites only occurs at very high volume fractions ($\geqslant 0{\cdot}5$) where there is likely to be considerable oxidation and/or corrosion.

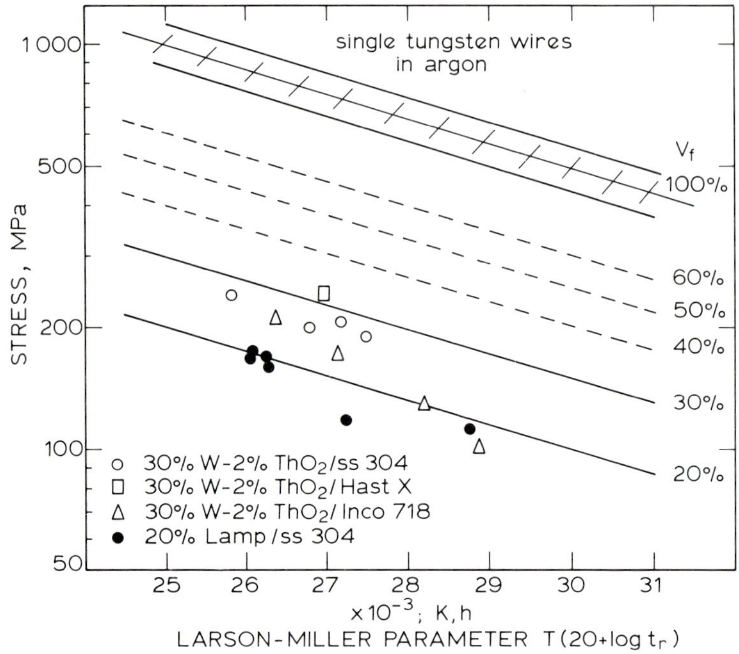


**Figure 8.12 Stress rupture data, displayed as a Larson–Miller plot, for various synthetic composites consisting of tungsten wires in various matrices [Ref. 13]**

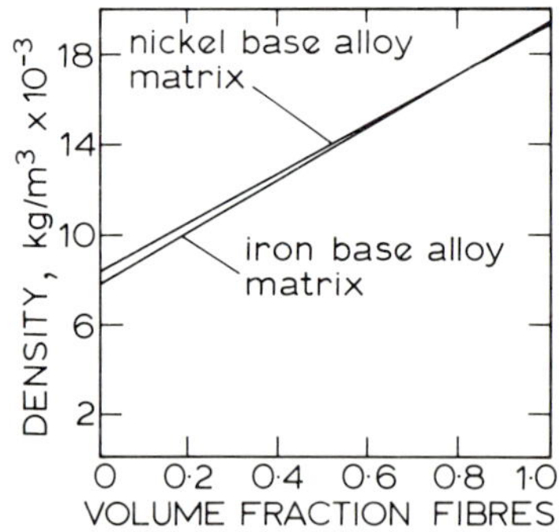


**Figure 8.13 Variations in density of nickel- and iron-base composites as a function of the volume fraction of tungsten wire reinforcement**

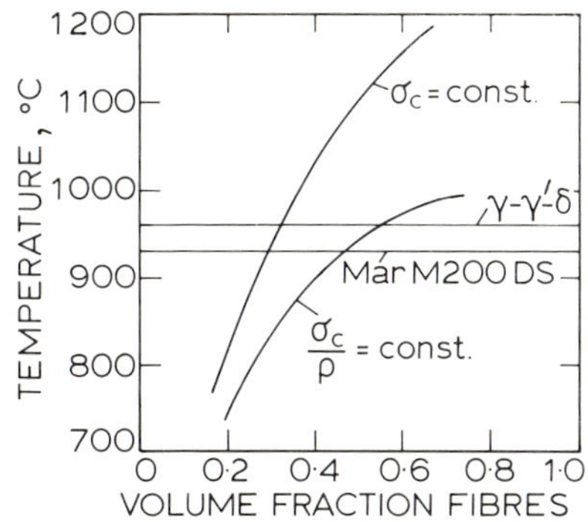


**Figure 8.14 Temperature to give $10^3$ h rupture life with a stress of 200 MPa for tungsten-wire composites as a function of the volume fraction of reinforcement. Data for directionally solidified MarM200 and the $\gamma$-$\gamma'$-$\delta$ *in situ* composite are included for comparison [Ref. 13]**

Much of the formalism developed in Chapter 6 to describe the mechanical behaviour of *in situ* composites remains valid for synthetic composites. The main difference is that the dimensions and separation of the reinforcing elements in synthetic composites are relatively large ($\gtrsim$100 μm diam.) so that the strengthening can be described adequately by 'Rule of Mixtures' concepts without recourse to additional 'dimensional strengthening' associated with the sub-micron reinforcement of *in situ* composites. However, many common problems of the two types of material may be related to their anisotropy and heterogeneity. For example, the problems of off-axis loading, particularly in pure shear, and of material degradation in temperature cycling conditions are likely to be ameliorated in synthetic composites by following the guidelines established for *in situ* composites.

Mention should also be made of the nonmetallic artificial composites such as carbon in carbon (i.e. carbon fibres in a carbon matrix) and related material such as SiC in C or SiC. These are already being used as wing leading edges in the Space Shuttle and they are likely to be further applied to nonmoving components.

## 8.3 CURRENT STATUS AND POTENTIAL

Directionally solidified and single crystal superalloys are now being commercially exploited as first stage turbine blades in advanced aero-engines for both military and commercial applications. Production and service experience appears to have been restricted to components whose lives are limited by mechanical properties rather than oxidation/corrosion resistance. Consequently, most work so far has been carried out on the highest strength superalloys such as MarM200 and Mar002 in the directionally solidified form and advanced single crystal compositions such as the Pratt and Whitney alloy 454 which offers equivalent stress rupture properties at a temperature some 30°C higher. However, in industrial gas turbines for both power generation and oil/gas pumping where long uninterrupted lives are of paramount importance, corrosion resistance is of primary concern and advantage can only be taken of improved mechanical behaviour provided the component surface stability can be maintained. This means that extension of directional solidification and single crystal technology from aeronautical to industrial applications will require developments both on different alloys and on components of much larger dimensions. There is now considerable laboratory based evidence showing that high-chromium corrosion resistant superalloys can show equivalent, or even greater, benefits from directional solidification to those achieved by aero-materials. It is inevitable that these benefits will eventually be introduced to improve the efficiency of land-based gas turbines. However, this is most likely to occur in the first instance in small ($\leqslant$500 kW) gas-pumping units or in marine propulsion gas turbines where corrosion resistance is also important. In both cases existing investment casting procedures should be adequate to supply suitable blades; extension to large power generating units will require considerable modification of casting procedures. There should be no serious technical barrier to producing such directionally solidified blades either by scaling up the withdrawal apparatus or by producing hybrid blades with directionally solidified aerofoils and conventionally cast roots.

The most advanced *in situ* and synthetic composites offer significant improvements in stress rupture properties, when stressed parallel to the reinforcement, over both directionally solidified superalloys and the most advanced experimental single-crystal material. This is indicated by the 1000 h stress rupture data shown in Fig. 8.15. Thus *in situ* composites such as the NITAC and COTAC family offer 50–100°C temperature advantage over superalloys in the low stress/high temperature regime relevant to aerofoils. Even the (Co,Cr)–$Cr_7C_3$ and $\gamma$–$\gamma'$–$Cr_3C_2$ alloys show substantial improvements over superalloys with equivalent oxidation/corrosion resistance. The potential benefits of high volume fraction synthetic composites are even greater. In spite of these apparent advantages neither type of composite has found service application, although several *in situ* composites have been assessed in engine trials.[14] Most of the early technical objections to *in situ* composites, such as poor thermal fatigue resistance and low transverse ductility, have now been overcome and this author believes that the objectives to be satisfied by *in situ* can be achieved. Technically, however, the processing

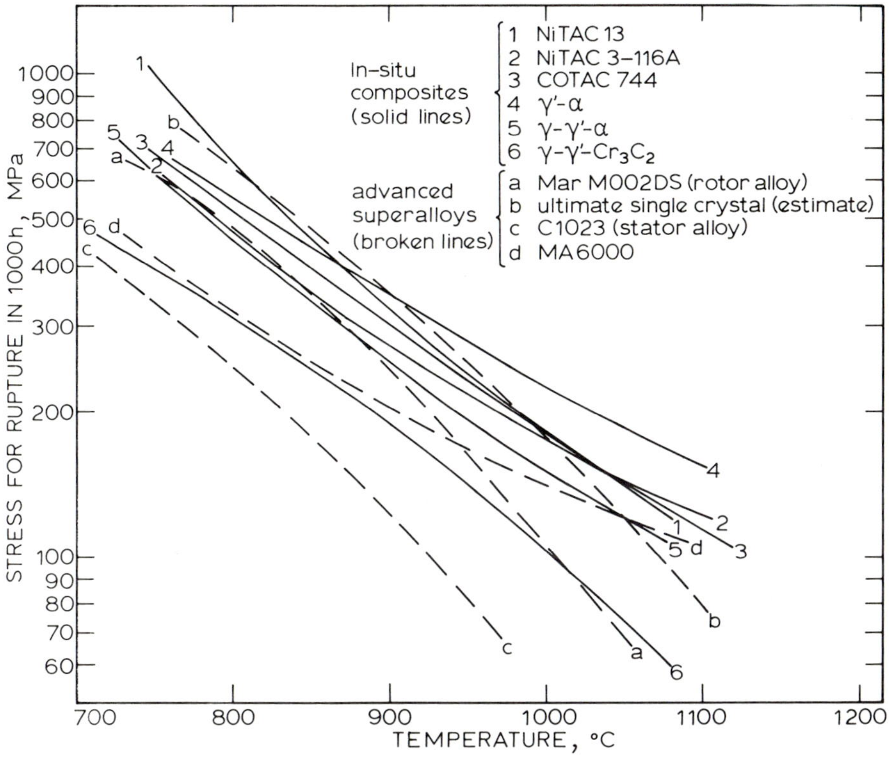


**Figure 8.15 Stress to cause rupture in 1000 h as a function of temperature for the most advanced *in situ* composites. Data are included for current rotor and stator turbine blade alloys and for an advanced mechanically alloyed material (MA6000) for comparison**

difficulties encountered in producing *in situ* components necessitate a drastic reduction in production rates relative to directionally solidified and single crystal superalloys and this presents an economic barrier to their implementation. Moreover, advances in single crystal superalloys have reduced the relative advantage of composites. There may be military applications where current *in situ* composites will be used, but commercial exploitation is likely to depend on either the development of cheaper processing routes or the development of materials more amenable to existing directional solidification technology. Processing costs are also likely to provide a barrier to the acceptance of synthetic composites.

Powder products such as ODS superalloys appear to offer a versatility in the form of the material (e.g. sheet, wire, rod) that is unlikely to be matched at equivalent cost by directionally solidified materials. Thus ODS superalloys are direct competitors with some *in situ* composites for components such as nozzle guide vanes. Moreover, ODS materials may be more easily (and cheaply) fabricated for other applications such as combuster cans in gas turbines or reaction tubes in petrochemical applications.

It seems that the task of combining and reconciling continuing improvements in performance and economy of production will require much further work on related topics. Taking up a theme introduced in Chapter 1, further impetus may arise from other applications, such as in chemical reaction plant, which may prompt innovations in the same way as heat engines have done over the past fifty years.

## REFERENCES CHAPTER 8


1 V. A. TRACEY AND C. P. CUTLER: *Powder Metall.*, 1981, **24**, 32

2 A. R. COX AND E. C. VAN REUTH: *Met. Technol.*, 1980, **7**, 238

3 F. J. ANDERS, G. B. ALEXANDER AND W. S. WARTEL: *Met. Progr.*, 1962, **88** (Dec.)

4 G. H. GESSINGER in 'High temperature alloys for gas turbines', eds. D. Coutsouradis *et al.*, p. 817, 1978, London, Applied Science Publishers

5 G. H. GESSINGER: *Metall. Trans.*, 1976, **7A**, 1203

6 R. E. ALLEN: Proc. 2nd Int. Symposium on Superalloys, Seven Springs, Pa, 1972, Metals and Ceramics Information Center, Columbus, Ohio, Battelle

7 R. L. CAIRNS, L. R. CURWICK AND J. S. BENJAMIN: *Metall. Trans.*, 1975, **6A**, 179

8 H. F. MERRICK, L. R. CURWICK AND Y. G. KIM: NASA CR-13510, Lewis Research Center, Cleveland, Jan. 1977

9 B. A. WILCOX AND A. H. CLAUER in 'The superalloys', p. 197, eds. C. T. Sims and W. C. Hagel, 1972, New York, Wiley–Interscience

10 B. A. WILCOX AND A. H. CLAUER: *Trans. Metall. Soc. AIME*, 1966, **236**, 570

11 J. S. BENJAMIN AND J. M. LARSON: 'Superalloys by powder metallurgy techniques', AIAA Aircraft Systems and Technology Meeting, Dallas, Texas, 27–29 Sept. 1976

12 D. W. PETRASEK AND R. A. SIGNORELLI: Proc. 5th Annual Conf. Composites and Advanced Ceramic Materials, Merritt Island, Florida, January 1981, NASA Technical Memorandum 82590

13 L. O. K. LARSON: Proc. 5th Int. Conf. Air Breathing Engines, 16–21 February 1981, Bangalore, India

14 Proceedings Conf. *In Situ* Composites III, eds. J. L. Walter *et al.*, Lexington, Mass., Ginn and Co., 1979

# Appendix

**Compositions of high temperature alloys that have been directionally solidified successfully (wt%)**

## (a) SUPERALLOYS

| Alloy | Ni | Co | Cr | Al | Ti | C | Ta | Mo | W | Nb | Zr | B | Others |
|---|---|---|---|---|---|---|---|---|---|---|---|---|---|
| IN713LC | bal. | — | 12·0 | 5·9 | 0·6 | 0·05 | — | 4·5 | — | 2·0 | 0·1 | 0·01 | — |
| IN713C | bal. | — | 12·5 | 6·1 | 0·8 | 0·12 | — | 4·2 | — | 2·0 | 0·1 | 0·012 | — |
| IN100 | bal. | 15·0 | 10·0 | 5·5 | 4·7 | 0·18 | — | 3·0 | — | — | 0·06 | 0·014 | 1·0 V |
| MarM200 | bal. | 10·0 | 9·0 | 5·0 | 2·0 | 0·15 | — | — | 12·5 | 1·0 | 0·05 | 0·015 | — |
| MarM002 | bal. | 10·0 | 9·0 | 5·5 | 1·5 | 0·15 | 2·5 | — | 10·0 | — | 0·05 | 0·015 | 1·5 Hf |
| MarM246 | bal. | 10·0 | 9·0 | 5·5 | 1·5 | 0·15 | 1·5 | 2·5 | 10·0 | — | 0·05 | 0·015 | — |
| B1900 | bal. | 10·0 | 8·0 | 6·0 | 1·0 | 0·10 | 4·0 | 6·0 | — | — | 0·10 | 0·015 | — |
| IN738LC | bal. | 8·5 | 16·0 | 3·5 | 3·5 | 0·11 | 1·6 | 1·75 | 2·5 | 0·7 | 0·08 | 0·008 | — |
| IN939 | bal. | 19·0 | 22·5 | 1·9 | 3·7 | 0·15 | 1·4 | — | 2·0 | 1·0 | 0·09 | 0·009 | — |
| MarM247 | bal. | 10 | 8·4 | 5·5 | 1·05 | 0·15 | 3·3 | 0·65 | 10·0 | — | 0·05 | 0·015 | 1·4 Hf |
| UTRCMMT 143 | 73·9 | — | — | 5·8 | — | — | 6·0 | 14·3 | — | — | — | — | — |
| X 40 | 10 | bal. | 25·0 | — | — | 0·5 | — | — | 7·5 | — | — | — | — |
| MarM509 | 10 | bal. | 23·5 | — | 0·2 | 0·6 | — | — | 7·0 | — | 0·5 | — | — |
| Single crystal alloys: P & W 444 | bal. | — | 8·6 | 5·1 | 1·98 | — | — | — | 11·1 | — | — | — | — |
| Single crystal alloys: P & W 454 | bal. | 5·0 | 10·0 | 5·0 | 1·5 | — | 12·0 | — | 4·0 | — | — | — | — |
| Single crystal alloys: NASAIR 100 | bal. | — | 9·0 | 5·75 | 1·2 | <0·01 | 3·3 | 1·5 | 10·5 | — | — | — | — |
| Single crystal alloys: RR SR99 | bal. | 5·0 | 8·5 | 5·5 | 2·2 | <0·015 | 2·8 | — | 9·5 | — | — | — | — |

## (b) *IN SITU* COMPOSITES

| | Structure | | Composition (wt%) | | | | | | | | | | | | |
|---|---|---|---|---|---|---|---|---|---|---|---|---|---|---|---|
| Alloy | matrix | reinforcement | Ni | Co | Cr | Al | Nb | Ta | C | W | Re | V | Mo | Y | Fe |
| NITAC3116A | $\gamma$Ni + $\gamma'Ni_3Al$ | TaC | bal. | 3·7 | 1·9 | 6·5 | — | 8·2 | 0·24 | — | 6·3 | 4·2 | — | — | — |
| COTAC 744 | $\gamma$Ni + $\gamma'Ni_3Al$ | NbC | bal. | 2·0 | 10·0 | 4·0 | 4·9 | — | 0·6 | 10 | — | — | — | — | — |
| $\gamma$–$\gamma'$–$Cr_3C_2$ | $\gamma$Ni + $\gamma'Ni_3Al$ | $Cr_3C_2$ | bal. | — | 12·3 | 6·9 | — | — | 1·8 | — | — | — | — | 0·5 | — |
| (Co,Cr)–$Cr_7C_3$ | Co,Cr solid soln | $Cr_7C_3$ | — | bal. | 41·0 | — | — | — | 2·4 | — | — | — | — | — | — |
| $\gamma$–$\gamma'$–$\delta$ | $\gamma$Ni + $\gamma'Ni_3Al$ | $\delta$ $Ni_3Nb$ | bal. | — | 6·0 | 2·5 | 20·1 | — | 0·06 | — | — | — | — | — | — |
| $\gamma$–$\gamma'$–$\alpha$ | $\gamma$Ni + $\gamma'Ni_3Al$ | $\alpha$ Mo | bal. | — | — | 6·0 | — | — | — | — | — | — | 32 | — | — |
| W/FeCrAlY | FeCrAlY | W-1% $ThO_2$ * | — | — | 20–24 | 5 | — | — | — | — | — | — | — | 1 | bal. |

* Composition depends on volume fraction of reinforcement, matrix quoted

# Author index

# Subject index